EARTH WORKS

*Recommended
Fiction and Nonfiction
about Nature and
the Environment
for Adults and
Young Adults.*

Jim Dwyer

Neal-Schuman Publishers, Inc.
New York   London

Published by Neal-Schuman Publishers, Inc.
100 Varick Street
New York, NY 10013

Printed and bound in the United States of America

Library of Congress Cataloging-in-Publication Data

Dwyer, Jim 1949–
    Earth works : recommended fiction and non-fiction about nature and the environment for adults and young adults / by Jim Dwyer.
        p.   cm.
    Includes bibliographical references and index.
    ISBN 1-55570-194-9
    1. Environmental sciences—Bibliography.   I. Title.
Z5863.P6D88   1996
[GE 105]
016.3637—dc20   94-36284

Earth Works is respectfully dedicated to my mother, Ellen Dwyer, and the memory of my father, Bill Dwyer, for instilling a love of reading and the outdoors and for teaching me to write on both sides of the paper. To the late Chuck Furstenburg, my Scoutmaster, for emphasizing camping, nature lore, and conservation over close-order drill and starched uniforms. To one of America's great conservationists, David Brower, for challenging his audiences to dedicate a year of their lives to improving the environment. (I dedicated two, Dave, but I wish you'd warned me about the sixty hour work weeks!) And to everyone willing to take action for the sake of future generations of all species.

Contents

Fiction

Acknowledgements

The author would like to sincerely thank the following people for their assistance and recommendations.

California State University, Chico: Lorraine Mosley, Jo Ann Bradley, and Karen Seaman of Interlibrary Loan for their friendly faces and indefatigable work tracking down copies of books. Director of Collections Division Bill Post for administrative support and flexibility. Supervising Library Assistant Stan Griffith and the entire Bibliographic Services staff for keeping productive. Head of Acquisitions and Collection Development Meta Nissley for providing administrative coverage during my sabbatical. Prof. Andrea Lerner for assistance with the Native American chapters. Librarian Colleen Power for assistance with the Science Fiction and Fantasy chapter. Profs. Sarah Emily Newton and Susan Place for their interest and suggestions. Prof. Claire Farrer for her example of intellectual rigor, joy in learning, and deep compassion. Librarian Kent Stephens and Curriculum Assistant Marci McAndress for recommending many good books.

Friends and associates: Allison Travis Bee for critiquing the chapter introductions and providing valuable writing and editing suggestions. Bobby Plapinger and Annie Hoy, Rod and Becky Slade, Barbara O'Neill and Chris Adams, Roy and Rebecca Conant, Brian Aveney, my brother Bill, and my mother Ellen for putting me up and putting up with me during my travels. Lorraine Anderson, compiler of *Sisters of the Earth* , and Jean Troy Smith, author of *Called to Healing* (SUNY Press, 1995) for encouragement and assistance with the Ecofeminism chapters. Lynnette Hutting of the Butte Environmental Council for assistance with the Environmental Education section. Julie Cunningham for introducing me to Robert Meredith. Dave Foreman, author of *Books of the Big Outside*, for his encouragement and advice.

Library staffs: Univ. of Oregon, Univ. of Washington, Chico branch of Butte County Library, Eugene Public Library (PL), Seattle PL, Sacramento PL, and especially Salem PL, one of Oregon's little-known treasures. Pleasant Valley HS, Chico, CA Librarian Peter Milbury for electronically querying other school librarians for their suggestions, and the many librarians who responded.

Bookstores: Conant and Conant Books (my favorite) and Powell's, Portland; The Blue Dragon, Ashland, OR; Smith Family Books and The Hungry Head, Eugene; Left Bank Books and Shorey's, Seattle; The Bookstore, Phoenix Books, and Tower Books, Chico; Ned Ludd Books, Tucson.

Professors and librarians at other universities: Lawrence Clark Powell, former library director at UCLA and the Univ. of Arizona, for providing a role model through excellence in librarianship and literary scholarship and his love of the West. Cheryl Burgess Glotfelty, Univ. of Nevada and Pres. of the Assn. for the Study of Literature and the Environment; Diane Quantic, Wichita State Univ. and Past Pres. of the Western Literature Assn.; Steve Tatum, Univ. of Utah and Pres. of the Western Literature Assn.; William Meredith, Penn State Univ. and author of *The Environmentalist's Bookshelf*; Tom Lynch, Holy Names College; Harold P. Simonson, Univ. of Washington; Glen Love, Louise Westling, and Tom Stave, Univ. of Oregon; Patricia Vanderbergh, Univ. of California; William Howarth, Princeton Univ.; Dorys C. Grover, East Texas State Univ.; Paul T. Bryant, Radford Univ.; Betsy Braun, Univ. of Chicago; Michelle Muraine, Hampshire College; Susanna Van Essen, Univ. of Tasmania.

Authors: Linda Hasselstrom, David Brin, Joseph Bruchac, Barbara Kingsolver, Vonda S. McIntyre, Ann McCaffrey, and Marge Piercy for their suggestions and encouragement.

Publishers: Fulcrum Press, North American Press, and Pfeifer-Hamilton for providing review copies of books. Thanks also to many other publishers for their catalogs.

The Professional Bibliographic Software User Support staff, particularly Mary LaFountaine, for their assistance.

The staff of Neal-Schuman Publishers, particularly Margo Hart and Bettina Versaci-Young.

Introduction

The Earth is currently facing an environmental crisis unmatched since the last major extinction. Stabilization or potential improvements to the natural environment require a fundamental change in human values and behavior. The "environment vs. economy" argument is outmoded and counterproductive. Nature is bountiful and natural resources are sustainable, but sustainability requires a shift from short-range to long-range thinking.

The increasing level of public awareness about environmental problems since the first Earth Day 25 years ago is reflected in several writing and publishing trends. This awareness combined with the emergence of deep ecology, the green movement, ecofeminism, and "the new nature writing," has generated thousands of new books. There has also been a dramatic increase in reprints or new editions of hundreds of classic works and of environmental themes in almost all genres of fiction. Many of the books which chronicle these fascinating and at times controversial trends fill the pages of *Earth Works*.

Earth Works is a guide to the best new and classic books about nature and the environment for adults and young adults. It describes popular trade and scholarly books that are appropriate for the general public as well as specialists. The listings include a significant number of fiction titles, with each section containing a selection of books for adults and one for teens. In short, *Earth Works'* comprehensive listings are geared for the general audience

Why a Book about Earth Works

Individuals *can* make a difference. Taking some form of action, however limited, can have a positive effect on the local, and by extension, the global environment and also help alleviate feelings of depression or hopelessness. Effective action is based upon a symbiosis of knowledge and commitment. Combining a personal experience of the environment with the knowledge and experience of others increases understanding. Commitment has intellectual and emotional aspects. The arts, including fiction, can enrich people's understanding and engage their emotions. Learning can be entertaining and entertainment enlightening.

Unfortunately, the amount of information about the environment can seem almost as complex and daunting as the environmental crisis itself. Therefore, librarians, educators, and activists have a responsibility to apply their knowledge and skills to direct people to the most relevant and helpful information. Therefore *Earth Works* is structured in a way to help people find specific information, relate it to other information, and use their knowledge to develop lifestyle changes or action plans

Intended Audiences and Uses

Earth Works is intended for use by librarians, bookstore proprietors, environmental educators, middle school teachers, scholars, environmental activists, and other professionals. General readers can use this book to select the most appropriate books from a sometimes bewildering array of choices.

Librarians and bookstore operators are encouraged to use this book as a reader's advisory tool, and as a tool to assess and build their collections. As of March, 1995, all the books listed here were in print. Two-thirds were published in the nineties, including 20 percent with a 1994 imprint. Prices are not included because they may vary or change. However, International Standard Book Numbers (ISBN'S) are listed. This makes it easy to verify editions and determine prices from sources such as *Books in Print.*

Environmental educators and other teachers can find nearly 70 books about their specialty in the Environmental Education section. It includes catalogs of curriculum materials and directories to relevant professional groups. Educators can use the other chapters to identify books on a wide variety of topics. Scientific books, literary or philosophical nature writing, and selected works of fiction can be combined to engage students intellectually and emotionally. The overall outline for the book can be modified to form a syllabus for introductory environmental studies or environmental literature classes.

Earth Works can help bridge gaps and update knowledge for scholars who aren't always aware of relevant works outside of their immediate specialty or of emerging ideas and trends, due to the highly interdisciplinary nature of environmental studies. It can also be used to help define the canon of essential works in a given subfield, or to develop specialized reading lists for students or associates. This is particularly true in the emerging field of environmentally oriented literary criticism known as ecocriticism. *Earth Works'* inclusion of nearly one thousand works of fiction and a list of criteria for defining and selecting ecofictional works—*works depicting environmental issues, nature as a main theme, earth centered themes, or the effects of nature on human consciousness or behavior*—is intended to assist in the creation of an ecofictional canon and body of criticism. A small but select body of relevant literary criticism is included as the final chapter, Literary Studies.

Ultimately, though, the purpose of this book is to assist environmental organizations and individual activists. My motivation for compiling *Earth Works* was to provide knowledge to encourage environmentally responsible lifestyles and to stimulate effective, widespread activism. Note that the largest chapter in this book is Environmental Action, which includes over 400 books on an incredibly broad range of activities and tactics.

Sources

Most books were selected from existing bibliographies, library and environmental journals, standard review sources, the collections of academic and public libraries and bookstores, and conference exhibits. The rest were recommended by environmental activists, teachers, librarians, and others who were queried via electronic discussion groups and professional organizations. Dave Foreman's *Books of the Big Outside* (entry 1701), Lorraine Anderson's *Sisters of the Earth* (1713) and Donald Edward Davis' *Ecophilosophy* (1699) were particularly fruitful sources.

I have read current or earlier editions of approximately 20 percent of the adult books over the last quarter century and read many of the classic young adult works in my youth. I examined most of the other books in libraries, bookstores, or via interlibrary loan. The remainder are included on the basis of their inclusion in several bibliographies or "best" lists or for garnering positive reviews in several reliable sources..

General Organization

Ecology, once a somewhat obscure subfield of life sciences, is now a household word, but large segments of the general public have little understanding of ecological processes. Books listed in the first two chapters of *Earth Works*, General Works and Natural History, are intended to provide basic knowledge. This knowledge empowers readers to critically evaluate news stories or political commentaries about the environment and enhances appreciation of literary and philosophical nature writing. Books about negative human impacts on the environment that are not limited to a specific type of environment or region are in the Environmental Crisis section.

The Specific Environments chapter covers books about deserts, forests, prairies, and other biomes. The Activities and Issues section is about human activities such as agriculture or energy use which effect all types of environments and broad issues such as endangered species or waste disposal. The Cultural Factors section includes political, social, and economic considerations, relations between industrialized nations and the Third World, and Native American environmental concepts and issues. The Philosophical and Literary Nature Writing includes aspects of the humanities relevant to understanding, appreciating, and finding one's proper place in relation to the natural world. Books on ecofeminism, which often include philosophical and social commentary, are also listed there. The Environmental Action chapter offers many options for putting knowledge to use.

Fiction Chapters

The organization of the fiction chapters is somewhat similar to that of the nonfiction, beginning with Anthologies and General Works followed by disaster books about the Environmental Crisis. The novels about animals reflect natural history, although sometimes in a somewhat distorted manner. Books with cultural or historic perspectives include those on Native Americans and the indigenous peoples of Australia and Oceania, the Frontier, the New West, and Country Life. Ecofeminism has influenced modern fiction and ecofeminist authors have made important contributions to contemporary mysteries, science fiction, and books on life in prehistoric times. Novels about Environmental Action provide entertainment and inspiration.

Subject Location

There is considerable subject scattering because many books are really about several related subjects. Although there is an entire section of 47 books primarily about Endangered Species, for example, other books with this topic as a secondary subject can be found in the Natural History, Whales and Dolphins, Human-Animal Relations, Animal Fiction, Mystery, and Environmental Action Fiction chapters. The broader or more interdisciplinary a subject, is the more likely it will be found in several different places.

Terms in the Subject Index are based upon the *Library of Congress Subject Headings*, but were sometimes edited for the sake of brevity, to correct misleading headings (e.g. "Man—Influence on nature" became "Human influences on nature,") to add new headings (e.g. "Ecotage,") or to clarify existing ones (e.g. books about overpopulation are entered under "Overpopulation," not the broader LC heading "Population.")

Also for the sake of brevity, there are no cross references in the already lengthy Subject Index. Users are advised to look under related headings—for all books on bears see Bears, Black bear, Brown bear, Grizzly bear, and Polar bear. There is no specific heading for Bear conservation, but the broader heading Wildlife conservation may refer to books which include information on the preservation of bear populations.

Certain types of reference books (atlases, dictionaries, directories, encyclopedias, handbooks, quotations, and yearbooks) and literary anthologies are indexed under both subject and form—Allaby's *Dictionary of the Environment* is indexed as Environmental Science—Dictionaries; Ecology—Dictionaries; Natural History—Dictionaries; and Dictionaries.

Inclusions and Exclusions

Due to space limitations only books are included.

Periodicals, videos, sound recordings, and other media resources have also proliferated in the last decade. Information on these resources can be found in some of the bibliographies listed in the Bibliographies section and in some directories. Online resources are continually growing and being more heavily used. As of this time the best way to access them is through the WorldWideWeb.

Books requiring significant previous knowledge of the subject or those for specialists that may also be suitable for general readers are indicated in the annotations. Titles requiring substantial knowledge of higher mathematics, engineering, physics, or law were excluded. Over 2,000 other books I examined were excluded because they were obsolete, out of print, inaccurate, misleading, or of poor quality. One should not assume that any title excluded is of low quality, because it may have never come to my attention.

Local guidebooks were excluded, but some regional and national guides can be found in the Ecotourism and Recreation or Natural History sections. Lavish photoessays were only included in cases where there was also significant text.

Except for a few bilingual English-Spanish titles, all of the books are English translations. The primary emphasis is the global environment and North America, with secondary emphasis on environmental relations between industrialized nations and the Third World. There are relatively few books here exclusively on the industrialized nations of Europe and Asia.

Request For Information

The database developed to create *Earth Works* is being continuously updated with future editions in mind. I'm particularly interested in finding more bilingual editions and more books on environmental equity. If you would like to recommend any new books, if any of your favorites were not included in this edition, or if you have any other comments, please contact me at one of the following addresses:

Internet: jim_dwyer@macgate.csuchico.edu (make sure you use an underline mark, not a dash, between jim and dwyer)
Postal: Jim Dwyer
    c/o Bibliographic Services Dept.
    Meriam Library
    CSU-Chico
    Chico, CA, 95929-0295

Many thanks in advance for your interest and contributions.

How to Use This Book

One of the key principles of ecology, in Muir's words, is that "everything is hitched to everything else." Environmental and nature books are often highly interdisciplinary. Matthiessen's eclectic classic *The Snow Leopard*, for example, could be considered natural history, travel, adventure, philosophy, religion, or literary nature writing. Given the presence of multiple subjects in many books, there are literally thousands of ways *Earth Works* might have been organized.

A hypermedia approach to organizing this material would be far preferable to a hierarchical one. The closest method I could think of to mimic hypermedia in a printed volume was to provide broad, general chapters with many relatively short sections accompanied by detailed indexes. The indexes perform a collocation function and are the best way to find a title, all the works of an author, or all the books on a specific subject.

The Table of Contents serves as a general guide and is a good starting point for browsers. Chapter or section introductions describe the scope of each section, identify a few key books or authors, and refer to important related works in other chapters.

Again using Matthiessen as an example, the author index reveals that his books can be found under Natural History, Freshwater, Mountain, Wilderness, Endangered Species, Social Factors, Native American, Religion and Spirituality, General Fiction, Native American Fiction, and Environmental Action Fiction. Although this is a fairly extreme example it should be noted that the works of many prolific authors such as Muir, Abbey, Stegner, Zwinger, George, Paulsen, or Pringle are scattered throughout this volume. Over 60 authors have fiction and nonfiction books listed here. There is a much larger overlap between poetry and literary nature writing, but books of environmental poetry are excluded from *Earth Works*.

Recommendations

Although all of the books listed in *Earth Works* are recommended, particularly strong titles are marked HR for "Highly Recommended." Some annotations specify select audiences—for example, "Essential for environmental educators." Others may qualify the best uses for a book, such as "Somewhat dated, but useful for the study of wildlife conservation history."

Nearly a quarter of the books are appropriate for both young adults and adults. These can be easily identified by the presence of a # sign at the end of the entry. Age or grade level is indicated for many of the young adult titles. OT stands for "Older Teens" meaning a high school reading level and MT means "Mature Teens" indicating a relatively high reading level and the emotional depth to deal with controversial, challenging, or sexual issues. AMT means a book is recommended for Adults and Mature teens.

Sample Entry

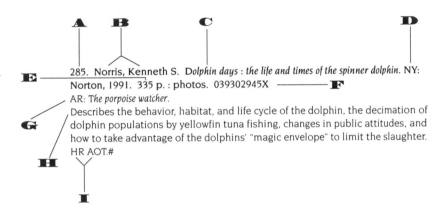

A   Entry number
B   Author(s)
C   Title
D   Place, publisher, date, variant copyright date
E   Pagination and illustrations
F   International Standard Book Number(s)
G   Notes. AR=Also recommended. Most of these are out-of-print books by the same author. Some are books by the same author that were published in late 1994 or 1995 and well reviewed. These will receive complete entries in supplements or later editions of *Earth Works*.
H   Annotation
I   Reading level and recommendation. HR=Highly Recommended. OT=Older Teens, high school level. MT=Mature Teens. #=Appropriate for a broad range of adults and young adults. AMT=Adults and Mature Teens. AOT=Adults and Older Teens. *An asterisk after a name indicates an enrolled member of a Native American tribe. Many citations for young adult books have inclusive grade levels, e.g GR 6-9 means sixth through ninth grade.

Earth Works Outline

NONFICTION

FICTION

Nonfiction

1. General Works

This section contains current general reference works such as dictionaries, encyclopedias, almanacs, and directories. Reference books specific to a certain subject appear in the the the relevant section—for example, The *Oxford Dictionary of Natural History* (entry 99) is in Natural History, and *Environmental Literacy: Everything You Need to Know About Saving Our Planet* (1420) in Environmental Action. Many other directories are listed in the Activism and Direct Action section. Bibliographies appear in the Bibliographies section of Environmental Action.

General anthologies are also listed here. Please note that although *Green Perspectives*, *The Endangered Earth*, and *Being in the World* were designed for college classroom use, they are also highly recommended for all adults. Many other anthologies are listed in the Nature Writing chapter, in Literary Anthologies or in other relevant sections, e.g. *Words from the Land* (157) in Natural History, or *Sisters of the Earth* (1713) in Ecofeminist Fiction.

Books for adults:

1. Allaby, Michael. *Dictionary of the environment*. 3d ed. NY: New York Univ., 1989. 423 p. 081470591X 0814705979 (pbk)
A very thorough dictionary with over 4,000 clear definitions, good cross-referencing, and a table of environmental disasters.

2. ___. ed. *Thinking green : an anthology of essential ecological writing*. London: Barrie & Jenkins, 1989. 260 p. 0712634894
A diverse anthology with 46 articles or extracts on a wide variety of topics by Rachel Carson, Lewis Mumford, Barry Commoner, George Perkins Marsh, Paul Ehrlich, Kenneth Boulding, Amory Lovins, Barbara Ward, Rene Dubos, E. F. Schumacher, Malthus, Dickens, Morris, Hobbes, and others. HR.

3. Anzovin, Steven. *Preserving the world ecology*. NY: Wilson, 1990. 236 p. 0824207904
A compilation of essays dealing with world ecological and environmental issues along with a consideration of their political and possible preservational aspects.

4. Ashworth, William. *The encyclopedia of environmental studies*. NY: Facts on File, 1991. 470 p. : ill. 0816015317
Includes over 3,000 brief entries written in non-technical language. Very comprehensive for a relatively short reference book with such a wide scope. For all libraries or home use. #

5. Axelrod, Alan, and Phillips, Charles. *The environmentalists : a biographical dictionary from the 17th century to the present*. NY: Facts on File, 1993. 258 p. : ill. 0816027153
Provides over 600 profiles of people, organizations, and agencies from the 17th century on. Primarily American, but worldwide in scope. Most entries are less than one page and include references to further reading. For all libraries and environmental educators. #

6. Brown, Lester Russell. ed. *The world watch reader on global environmental issues*. NY: Norton, 1991. 336 p. : ill. 0393030075
Includes chapters on restoring degraded land, mending the Earth's shield, all types of pollution, environmental problems and prospects in eastern Europe and China, water scarcity, sustainable agriculture and forestry, nuclear power, wind power, automobiles, bicycles, and related topics. HR.

7. Cartledge, Bryan. ed. *Monitoring the environment*. NY: Oxford, 1992. 216 p. 0198584083 0198584121 (pbk)
Contains essays by Richard Southwood, Michael Heseltine, John Mason, Crispin Tickell, James Lovelock, John Woods, Ghillean T. Prance, and John Phillipson about environmental monitoring, developing an accurate assessment of the world's environmental quality, and related subjects.

8. Cunningham, William P. ed. *Environmental encyclopedia*. Detroit: Gale, 1994. 981 p. : photos. 0810349868
Contains over 3,000 entries on both the scientific and social aspects of environmental science, ecology, the environmental movement, legislation, and related subjects. A.k.a. the *Gale environmental encyclopedia*. For all libraries and environmental organizations. #

9. Durrell, Lee. *State of the ark*. Garden City, NY: Doubleday, 1986. 224 p. : ill. 0385236670
Maps, text, tables, and illustrations covering the impact of human activity on the environment, the state of the wild, the state of various species, biotic regions, and the conservation movement. #

10. *The environment encyclopedia and directory*. London: Europa, 1994. 381 p. : ill., maps. 0946653941
Provides the broadest worldwide scope of all environmental encyclopedias

or directories. Includes a glossary, international directory of organizations and individuals, and a bibliography of 1,000 periodicals. A.k.a. the *Europa environment encyclopedia and directory*. #

11. Franck, Irene M., and Brownstone, David M. eds. *The green encyclopedia*. NY: Prentice Hall, 1992. 485 p. : ill. 0133656853 0133656772 (pbk)
Contains over 1,000 short entries on ecological processes, conservation and preservation, endangered species, environmental problems, major disasters, treaties, legislation, animal rights, ecotourism, etc. Many entries have resource and action guides.

12. Goldsmith, Edward, and Hildyard, Nicholas. eds. *The Earth report : the essential guide to global ecological issues*. LA: Price, Stern, Sloan, 1988. 240 p. 0895866781
Includes entries ranging from Acceptable Daily Intake (of chemicals) to Zero Population Growth. Contains very useful graphs and charts. Global in scope.

13. Hammond, Allen. ed. *The information please environmental almanac*. Washington: World Resources Inst., annual. 656 p. : ill. 0395677424 (1994 ed.)
A well organized and comprehensive almanac covering the state of the planet, water, wastes, energy transportation, air pollution, grassroots activism, cold war cleanup, ecotourism, forests and wetlands, industry, green city rankings, ozone depletion, greenhouse warming, biodiversity, and area profiles. HR.

14. Hoyle, Russ. ed. *Gale environmental almanac*. Detroit: Gale, annual. 684 p. : ill., maps. 0810388774 (1993)
Includes encyclopedia entries, biographies, essays, congressional votingrecords, glossary, and directories of environmentally responsible companies and college environmental programs. HR

15. Katz, Linda Sobel; Orrick, Sarah; and Honig, Robert. eds. *Environmental profiles : a global guide to projects and people*. NY: Garland, 1993. 1083 p. : ill. 0815300638
Organized on a country-by-country basis, this book covers 7,000 projects, provides biographies of 3,000 people, and contains extensive factual information on 14 key issues. Possibly the best current biographical guide and directory in the field. HR.

16. Lean, Geoffrey; Markham, Adam; and Hinrichsen, Don. *Atlas of the environment*. NY: Prentice Hall, 1990. 192 p. : col. maps. 0874367689
Includes over 200 maps and diagrams on major biomes, population, urbanization, food, water, health, the education gap, industrialization, indige-

nous peoples, desertification, pollution, tropical forest destruction, toxic wastes, ozone depletion, the greenhouse effect, energy resources, biological diversity, wildlife, conservation, fisheries, etc. #

17. Levy, Walter, and Hallowell, Christopher. eds. *Green perspectives : thinking and writing about the environment.* NY: HarperCollins, 1994. 490 p. : ill. 0065015002
Contains nearly 60 selections by Thoreau, Dickinson, Marsh, Muir, Austin, Burroughs, Frost, Hemingway, Steinbeck, Leopold, Eiseley, Carson, Dubos, Abbey, Silko, McPhee, Dillard, Le Guin, and Bass. Includes "Questions for writing and thinking" with each selection. Designed for college use, but also for high schools, libraries, and individuals. HR. #

18. Mason, Robert J., and Mattson, Mark T. *Atlas of United States environmental issues.* NY: Macmillan, 1990. 252 p. : ill., maps. 0028972619
Contains 150 maps illustrating issues related to agriculture, coastal zone management, air quality, noise and light pollution, the economics and politics of the environment, etc. HR. #

19. Merchant, Carolyn. ed. *Major problems in American environmental history.* Lexington, MS: Heath, 1993. 568 p. : ill., maps. 0669249939
Contains nearly 200 historical and modern documents on diverse but related issues such as Native American ecology, soil depletion, farm ecology, mining, grassland exploitation, wilderness preservation, and urban pollution.

20. Morgan, Sarah, and Okerstrom, Dennis. eds. *The endangered Earth : readings for writers.* Boston: Allyn & Bacon, 1992. 480 p. 0205132189
Intended to represent the entire spectrum of the environmental movement and demonstrate the interrelatedness of the issues. Each selection has an introduction, the text, study questions, writer's strategies, and further writing possibilities. Authors include Eiseley, Thoreau, McPhee, Dillard, Austin, Carson, Lopez, Abbey, and others. Designed for college classroom use, but also useful for all adults, particularly writers. HR.

21. Nat'l Geographic Society. *Living on the Earth.* Washington: The Society, 1988. 320 p. : maps, photos. 087044736X
Focuses on the human factor in several major habitiats: deserts, arctic, islands, forests, grasslands, rivers, highlands, and coasts. Covers changes effecting each habitat and the relations betweeen ecology and technology, politics, and climate change. #

22. Oxford Univ. Press. *The illustrated encyclopedia of world geography.* NY: Oxford, 1990-1993. 11 v. : col. photos, maps. 019521059X
The volumes on Planet management, Animal life, Plant life, and especially

Nature's last strongholds are particularly relevant to the study of ecology, natural history, and environmental issues. HR. #

23. Rodes, Barbara K., and Odell, Rice. A *dictionary of environmental quotations*. NY: Simon & Schuster, 1992. 335 p. 0132105764
Contains approximately 3,700 quotations in 143 categories arranged alphabetically from Acid Rain to Zoos.

24. Slovic, Scott, and Dixon, Terrell. eds. *Being in the world : an environmental reader for writers*. NY: Macmillan, 1993. 726 p. 0024117617
Contains selections from Edward Abbey, Peter Matthiessen, Barry Lopez, Alice Walker, Eudora Welty, Maxine Kumin, Lewis Thomas, Jack Kerouac, Joan Didion, and others. Provides background material and discussion questions for each selection. Designed for college environmental studies and creative non-fiction writing, but also useful for all adults. HR.

25. Trzyna, Thaddeus C., and Gotelli, Ilze M. *World directory of environmental organizations : a handbook of national and international organizations and programs, governmental and non-governmental, concerned with protecting the Earth's resources*. 3d ed. Claremont, CA: Cal. Inst. of Public Affairs, 1989. 176 p. 091210287X
International in scope, covering organizations in over 200 countries. Useful for libraries, organizations, and adults seeking organizations to join or support. HR.

26. Wade, Nicholas, and Dean, Cory. eds. *The New York Times book of ecoliteracy : everything you need to know from your backyard to the ocean floor*. NY: Times, 1994. 388 p. : ill. 0812922158
Contains three to seven page articles from the *Times*, 1988-1993, on a broad range of environmental topics. Repitition of information and some inconsistencies between the pieces weaken an otherwise excellent collection. This is v. 2 of the *New York Times book of science literacy*.

Books for young adults:

27. Bramwell, Martyn. *The Simon & Schuster young readers' book of planet Earth*. NY: Simon & Schuster, 1992. 192 p. : col. ill. 0671778315 (pbk) 0671778307
Discusses Earth's place in the solar system, its atmosphere, weather, geology, oceans, natural history, and environmental issues. GR 4-8.

28. Holland, Barbara, and Lucas, Hazel. *Caring for planet Earth : the world around us*. Batavia, IL: Lion, 1990. 40 p. : ill. 0745913504
Describes different environments around the world, the interdependence of plants, animals, and people, the depletion of natural resources and wildlife

species, and efforts to protect the Earth's natural environment. GR 5-8.

29. Gamlin, Linda. *Life on Earth*. NY: Gloucester, 1988. 36 p. : ill. 0531171205
Considers the question, "What is life?" A series of two page sections on
such specifics as food chains, growth, populations, endangered species,
conservation, and wildlife preservation are presented in a context of identifying and solving environmental problems. GR 5-9.

30. Leinwand, Gerald. *The environment*. NY: Facts on File, 1990. 122 p.
081602099X
Surveys environmental problems in the United States, examines legal and
social aspects of environmentalism, and discusses political plans in place
to deal with the environmental decline.

31. Mattson, Mark T. *The Scholastic environmental atlas of the United States*. NY:
Scholastic, 1993. 80 p. : col. ill., maps. 059049354X 0590493558 (pbk)
Text plus maps, photographs, charts, and graphs describe environmental
conditions in the United States, both in the past and today. GR 4-8.

32. Middleton, Nick. *Atlas of environmental issues*. NY: Facts on File, 1989. 63
p. : col. maps, photos. 081602023X
Describes and explains major environmental issues of the world today
including soil erosion, deforestation, mechanized agriculture, oil pollution
of the oceans, acid rain, overfishing, and nuclear power. GR 6-9.

33. Coote, Roger. ed. *Atlas of the environment*. Austin: Raintree Steck-
Vaughn, 1993. 96 p. : ill., maps. 0811472507
Maps, photographs, and text examine the planet's endangered environ-
ment and possible future. GR 6-8.

34. Thaler, Mike. *Earth mirth : the ecology riddle book*. Ill. by Rick Brown. NY:
Freeman, 1994. 64 p. : cartoons. 0716765217
This book contains approximately 60 riddles, for example, "What do you
call a one eyed monster that recycles newspaper? A recyclops." Each is
illustrated with a cartoon. Recommended for punsters GR 3-8.

Ecology and Environmental Science

This section contains textbooks and general works about natural processes
and the scientific study of the environment. Some offer secondary coverage

of environmental problems, conservation, or activism, but their emphasis is clearly scientific rather than social or exhortatory. Most are readily accessible to adults with a minimal background in general science or environmental issues. Books which are limited to the ecology of specific types of environments can be found in Chapter 3. Those which emphasize corrective actions are in Chapter 7.

Books for adults:

35. Allen, T. F. H., and Hoekstra, T. W. *Toward a unified ecology.* NY: Columbia Univ., 1993. 384 p. 0231069197
Focuses on the need for ecology to become a more predictive discipline and the need to consider large-scale systems.

36. Art, Henry Warren. ed. *The dictionary of ecology and environmental science.* NY: Holt, 1993. 632 p. : over 100 ill. 0805020799
Includes 8,000 fairly brief definitions of words from ecology and the related fields of geology, biology, genetics, nuclear science, chemistry, forestry, and engineering. #

37. Arthur, Wallace. *The green machine : ecology and the balance of nature.* Cambridge: Blackwell, 1990. 257 p. : ill. 0631178538 (pbk) 0631169865
A "bottom-up" introduction to ecology which begins four billion years ago by showing how an undisturbed, pre-human world ecosystem maintained balance despite enormous changes. Arthur then demonstrates how plants, animals, and humans interact in the modern ecosphere.

38. Begon, Michael; Harper, John L.; and Townsend, Colin R. *Ecology : individuals, populations, and communities.* 2d ed. Boston: Blackwell, 1990. 945 p. : ill. 0865421110
This college level text takes an integrated approach, using complex situations to bring natural history, environmental physiology, behavior, field experimentation, and mathematical modelling together to solve ecological problems. For adults with some background in biology and math.

39. Brewer, Richard. *The science of ecology.* Philadelphia: Saunders, 1988. 922 p. : ill., maps. 0030099447
An excellent textbook which explains and interrelates ecological concepts and practices. For intermediate to advanced general adult readers, teachers, and specialists. HR.

40. Colinvaux, Paul A. *Introduction to ecology.* NY: Wiley, 1973. 621 p. : ill. 0471164984
Features the ecological aspects of geology, animals, and plants as parts of

natural systems, the natural checks on population, and an evolutionary view of the pattern of life.

41. ___. *Why big fierce animals are rare : an ecologist's perspective*. Princeton, NJ: Princeton Univ., 1988, c1978. 256 p. 0691023646
Provides a fresh and upbeat approach to a variety of ecological issues such as niches, the social life of plants, succession, the efficiency of living things, why the sea is blue, biodiversity and stability, and the place of people within nature. HR AOT. #

42. Ehrlich, Paul R., and Roughgarden, Jonathan. *The science of ecology*. NY: Macmillan, 1987. 710 p. : ill., maps. 0023317000
This intermediate to advanced text interrelates the principles and methodologies of ecology. Covers the individual and the environment, populations, social interactions, population interactions, organization of communities, distribution of communities, and ecosystems. HR

43. Haila, Yrjo, and Levins, Richard. *Humanity and nature : ecology, science, and society*. Concord, MS: Pluto, 1992. 270 p. 1853050415 0745306691 (pbk)
Makes distinctions between the nature, the science, the idea, and the movement known as ecology. Examines the interfaces of these aspects, ecological patterns, health as part of the ecosystem, agricultural ecology, the social history of nature, and political ecology.

44. Jackson, Jerome A. *Nature's habitats*. Washington: Starwood, 1991. 117 p. : col. photos. 0912347783
Illustrates the web of interdependence between plants, wildlife, and landscape. Emphasizes the subtle but strong connections between all living things.

45. Krebs, Charles J. *Ecology : the experimental analysis of distribution and abundance*. 4th ed. NY: HarperCollins, 1994. 801 p. : ill. 0065004108
This college level text provides an introduction to ecological concepts and processes. Discusses distribution and abundance at the community level. Emphasizes population and biogeography.

46. McIntosh, Robert P. *The background of ecology : concept and theory*. NY: Cambridge Univ., 1986, c1985. 383 p. 0521270871
Combines historical, scientific, and social perspectives on the antecedents of ecology, its development as a field of study, dynamic ecology, quantitative community ecology, populations, ecosystems ecology and big biology, and ecology and the environmental movement. This multifaceted book is HR.

47. Miller, G. Tyler. *Living in the environment : an introduction to environmental*

science. 7th ed. Belmont, CA: Wadsworth, 1992. 705 p. : ill., maps.
0534165605
AR: *Resource conservation and management, Replenish the Earth*.
A popular, comprehensive text covering ecology, demography, food, pollution,
natural resources, environmental problems, and environmental economics
and politics. For college classroom use and environmental educators. HR.

48. ___. *Sustaining the Earth : an integrated approach*. Belmont, CA:
Wadsworth, 1994. 360 p. : ill. 0534214320
Treats environmental science as an interdisciplinary study, combining
ideas from the natural sciences and social sciences. Emphasizes interconnectedness. This college level text doesn't require much math and is recommended for classroom use, other environmental educators, and
individuals.

49. Odum, Eugene Pleasants. *Fundamentals of ecology*. 3d ed Philadelphia:
Saunders, 1971. 574 p. : ill 0721669417
AR: *Ecology, the link between the natural and the social sciences*.
Covers basic ecological principles and concepts, terrestrial and marine
habitats, and ecology as applied to natural resources, public health, radiation, and human society. Somewhat dated, this classic text is most useful
to those interested in the history of the study of ecology. HR.

50. Owen, Denis Frank. *What is ecology?* 2d ed. NY: Oxford, 1980. 234 p. : ill.
0192191551 0192891405 (pbk)
This textbook covers a middle ground between those providing a background to ecology as part of a course in biology and those oriented
towards presenting environmental dilemmas and catastrophes.

51. Parker, Sybil P., and Corbitt, Robert A. eds. *McGraw-Hill encyclopedia of
environmental science & engineering*. 3d ed. NY: McGraw-Hill, 1993. 749 p. : ill.
0385236689
An excellent, comprehensive work with thorough explanations. Includes
many tables, graphs, and charts. For medium to large libraries and professional environmentalists and engineers. HR.

52. Ricklefs, Robert E. *Ecology*. 3d ed. NY: Freeman, 1990. 896 p. : ill.
0716720779
This classic college-level ecology text covers the physical environment,
adaptation, organisms in their physical and biological environments, ecological genetics and evolution, population genetics and ecology, and the
ecology of communities and ecosystems. HR.

53. Smith, Robert Leo. *Ecology and field biology*. 4th ed. NY: Harper & Row,

1990. 922 p. : ill., maps. 0060463317 (student ed.) 0060463325 (teacher ed.)
This college level text covers the ecosystem and the community, aquatic
and terrestrial habitats, population ecology, natural selection and speciation, and animal behavior. Emphasizes the biological aspect of ecology and
the ecological aspects of field biology.

54. Southwick, Charles H. ed. *Global ecology*. Sunderland, MS: Sinauer,
1985. 323 p. 0878938109
AR: *Ecology and the quality of our environment*.
A collection of papers containing opposing views on pollution, acid rain,
groundwater contamination, erosion, desertification, deforestation, overpopulation, the role of technology in the environment, and related subjects.

55. Stevenson, L. Harold, and Wyman, Bruce C. *The Facts On File dictionary
of environmental science*. NY: Facts On File, 1991. 294 p. : ill. 0816023174
RASD/ALA Outstanding Reference Source, 1993.
Includes succinct, clear definitions of over 3,000 terms in non-technical language. HR for libraries or home use. # (RASD=*Reference and Adult Services Division*)

56. Tudge, Colin. *Global ecology*. NY: Oxford, 1991. 173 p. : col. photos.
0195209044
Considers conditions for life, the waters of the world, the land, energy and
the atmosphere, food webs, populations and communities, and the impact
of people on the environment. Provides both an introduction to ecological
processes and inspiration for environmental action.

Books for young adults:

57. Anderson, Margaret Jean. *Food chains : the unending cycle*. Hillside, NJ:
Enslow, 1991. 64 p. : ill. 0894902903
Explores the concept of food chains and discusses their importance in
nutrient cycles and in the maintenance of ecological balance. GR 6-10.

58. Dethier, V. G. *The ecology of a summer house*. Amherst: Univ. of Massachusetts, 1984. 133 p. : ill. 0870234218 0870234226 (pbk)
Describes the many animals living in Dethier's summer house on the
Maine Coast, such as bugs, squirrels, wasps, crickets, bats, and mice. Tells
how they establish interdependent ecological communities indoors as well
as outdoors. #

59. Hughey, Pat. *Scavengers and decomposers : the cleanup crew*. NY: Atheneum,
1984. 56 p. : ill. 0689310323
Outstanding Science Trade Book for Children, 1984. Describes the charac-

teristics and habits of various insects, birds, and other animals that clean up waste materials in the environment and thus ensure that life as we know it continues on the Earth. HR GR 4-8.

60. McLaughlin, Molly. *Earthworms, dirt, and rotten leaves : an exploration in ecology.* NY: Atheneum, 1986. 86 p. : ill. 0689312156
Outstanding Science Trade Book for Children, ALA Notable Children's Book, 1986.
Examines the earthworm and its environment, suggesting experiments to introduce basic ecological concepts as demonstrated by the earthworm's survival in its habitat. HR GR 4-8.

61. Reed, Willow. *Succession : from field to forest.* Hillside, NJ: Enslow, 1991. 64 p. : ill. 0894902717
Discusses the different kinds of plant succession, the relationship between this process and people, and the effect on plant communities. GR 7-10.

62. Silverstein, Alvin, and Silverstein, Virginia B. *Life in a tidal pool.* Boston: Little, Brown, 1990. 60 p. : ill. 0316791202
Describes the varied forms of shore life found in and around tidal pools and discusses their struggle for survival. Explains threats to tidal pools from pollution and overcollecting of specimens. GR 5-9.

Environmental Crisis

Books about the current Environmental Crisis tend to combine scientific and social aspects. They often inspire conservation and environmental action such as Rachel Carson's classic *Silent Spring* (653) Carson specifically examined the effects of DDT on birds, but used that as a springboard to explore the general environmental crisis. These books reflect a wide range of beliefs and attitudes from environmental activists like Gordon and Suzuki to industrial apologists Langone and Ray. A debate on whether there is a crisis, entitled *Scarcity or Abundance?* (1049) is listed in the Philosophy section. Books describing specific aspects of the crisis can be located in the Specific Environments or the Activities and Issues chapters.

Books for adults:

63. Bernards, Neal. ed. *The Environmental crisis : opposing viewpoints.* San Diego: Greenhaven, 1991, c1986. 288 p. : ill. 0899081754 0899081509 (pbk)
Presents opposing views on questions of environmental protection and of

damage resulting from pollution, toxic wastes, pesticides, and refuse. AOT. #

64. Brower, Kenneth. ed. *One Earth : photographed by more than 80 of the world's best photojournalists.* SF: Collins, 1990. 192 p. : col. photos 0002157306
A superb photoessay emphasizing the interconnectedness of nature and the many consequences of environmental degradation. HR. #

65. Carson, Rachel. *Silent spring.* 25th anniversary ed. Boston: Houghton Mifflin, 1987, c1962. 368 p. 0395453909
Carson documents the ecological consequences of pollution, pesticides, and over-population. This environmental classic is HR AOT. #

66. Collins, Carol C. ed. *Our food, air, and water : how safe are they?* NY: Facts on File, 1984. 231 p. : ill. 087196967X
A collection of editorials and cartoons from American newspapers, 1980-1984, on a variety of related topics such as pesticides, PCBs, food irradiation, asbestos, lead, auto emissions, passive smoking, greenhouse effect, acid rain, groundwater contamination, sewage, hazardous wastes, and dioxins.

67. Feshbach, Murray, and Friendly, Alfred. *Ecocide in the* USSR : *health and nature under siege.* NY: Basic Books, 1992. 376 p. 0465016642
Describes massive environmental problems in the former USSR: air pollution, groundwater contamination, resource depletion, soil contamination, and toxic and nuclear wastes. Analyzes their effects on current economic and political conditions, and potential costs of the cleanup.

68. Foster, John Bellamy. *The vulnerable planet.* NY: Monthly Review, 1994. 144 p. : photos, maps. 0853458758 085345874X (pbk)
Outlines the extent of the environmental crisis, examines its social and economic causes and consequences, and argues for an environmental revolution that would reorganize production processes and incorporate ecological laws into daily life through social control of the economy.

69. Freedman, Bill. *Environmental ecology : the impacts of pollution and other stresses on ecosystem structure and function.* San Diego: Academic, 1989. 424 p. : ill. 0122665406
Very specifically describes how population and developmental stresses on the environment have created air pollution, toxic waste, acidification, forest decline, eutrophication of fresh water, loss of species diversity, and ecological dysfunctionality. For adults with some scientific background.

70. Gordon, Anita, and Suzuki, David T. *It's a matter of survival.* Cambridge: Harvard, 1991. 278 p. 0674469704
Forecasts worsening environmental conditions over the next decade that

cannot be corrected via minor technofixes, but must be addressed through major changes in lifestyles, industrial processes, and government policies, and through greater utilization of conservation, recycling, and solar energy.

71. Goudie, Andrew. *The human impact on the natural environment.* 4th ed. Cambridge: MIT, 1994. 388 p. : ill. 0262071568 0262571013 (pbk)
These impacts include deforestation, desertification, erosion, flooding, species loss, habitat destruction, ozone loss, greenhouse effect, toxic waste, etc. An excellent textbook for advanced senior high students and all adults, especially conservationists. HR. #

72. Langone, John. *Our endangered Earth : what we can do to save it.* Boston: Little, Brown, 1992. 197 p. : ill. 0316514152
Discusses the environmental crisis, focusing on overpopulation, pollution, ozone depletion, global warming, and disappearing wildlife. The book suffers from Langone's role as an industrial apologist. Recommended only as an example of applying bandaids when major surgery is required.

73. McMichael, Anthony J. *Planetary overload : global environmental change and the health of the human species.* NY: Cambridge Univ., 1993. 352 p. 0521441382 0521457599 (pbk)
McMichael provides a broad biological, historical, and social analysis of environmental degradation and its relation to overpopulation, ozone destruction, the greenhouse effect, food production, infection, soil and freshwater depletion, etc.

74. Mitchell, John G. *The man who would dam the Amazon, and other accounts from afield.* Lincoln: Univ. of Nebraska, 1990. 368 p. 0803231474
Contains twelve lively accounts of environmental problems around the world which first appeared in *Audubon* or *Wilderness.*

75. Mungall, Constance, and McLaren, D. J. eds. *Planet under stress : the challenge of global change.* NY: Oxford, 1991. 344 p. : ill. 0195407318
Results of a Royal Society of Canada survey asking experts from the natural sciences, the social sciences, and the humanities to provide perspectives on the evolutionary history of this planet, its changing dynamics, and possible ecological changes as a consequence of industrial and societal waste.

76. Naar, Jon, and Naar, Alex J. *This land is your land : A guide to North America's endangered ecosystems.* NY: HarperPerennial, 1993. 388 p. : maps. 0060552980
Covers rivers, wetlands, oceans, grasslands, forests, deserts, public lands, endangered species, conservation legislation, biodiversity, and restoration. An ecosystem approach is taken throughout, key issues are highlighted, and environmental activism is encouraged.

77. Palmer, Tim. ed. *California's threatened environment : restoring the dream.* Washington: Island, 1993. 305 p. : maps. 1559631732 1559631724 (pbk) Covers what must be done to stop the state's continuous environmental degradation and the ongoing erosion of quality of life from pollution, over-crowding, and overdevelopment. Offers remedies to a variety of problems. Particularly recommended for concerned Californians.

78. Ray, Dixy Lee, and Guzzo, Louis R. *Trashing the planet : how science can help us deal with acid rain, depletion of the ozone, and nuclear waste.* NY: Harper-Perennial, 1992. 206 p. 0060974907 Spans from yellow journalistic attacks on environmentalist enemies to textbook discussions of environmental mechanisms. Included here only as an example of a cleverly disguised anti-environmental book.

79. Repetto, Robert C. ed. *The Global possible : resources, development, and the new century Global Possible Conference.* New Haven: Yale, 1985. 538 p. 0300633826 AR sequel: *World enough and time.* What is considered "possible" in these revised conference papers is that environmental degradation can be stopped through rapid cooperative planning and action by business, government, and the public. Specific recomendations are offered.

80. Trager, Oliver. *Our poisoned planet : can we save it?* NY: Facts on File, 1989. 216 p. : ill. 0816022496 Examines environmental issues through the words and images of the nation's leading editorial writers and cartoonists. Good for term papers. #

81. Wagner, Travis. *In our backyard : a guide to understanding pollution and its effects.* NY: Van Nostrand Reinhold, 1994. 320 p. : ill. 0442014996 Using everyday language and a question and answer format, this book describes the extent of global pollution, how the components of the environment interrelate, major sources of pollution, and remediation efforts.

Books for young adults:

82. Bright, Michael. *Pollution and wildlife.* NY: Gloucester, 1992. 32 p. : ill. 0531173844 Discusses the causes and various types of pollution and their impact on insects, animals, fish, and birds. GR 5-9.

83. Herda, D. J. *Environmental America series.* Brookfield, CT: Milbrook, 1991. 5 v., each 64 p. : ill. Contents: *Northwestern states* (1878841106) — *South cen-*

tral states (1878841092) — *Southeastern states* (1878841076) — *Southwestern states* (1878841114) — *Northeastern states* (1878841068).
Focuses on environmental problems and steps being taken to counteract the damage. Each book includes suggestions for action, toll-free hotlines, and a directory of organizations. HR GR 5-9.

84. Markham, Adam. *The environment*. Vero Beach, FL: Rourke, 1988. 48 p. : col. photos 0865922861
Examines how various ecosystems are being destroyed by human activities, efforts to stop the destruction, and the cost of implementing environmental policies. GR 5-8.

85. Peckham, Alexander. *Changing landscapes*. NY: Gloucester, 1991. 36 p. : col. ill. 0531172899
Examines how the Earth's resources have been misused and wasted through short-sightedness, technological progress, overpopulation, and overuse of resources and farmland. The rapid rate of change in technology and civilization doesn't take into account the very slow changes naturally occuring to the Earth. GR 5-9.

86. Pedersen, Anne. *The kids' environment book : what's awry and why*. Santa Fe: John Muir, 1991. 181 p. : ill. 0945465742
Examines environmental problems, humankind's historic relationship with the Earth and its living species, how industrialization has dramatically changed our planet, and what must be done to repair the damage. GR 5-10.

87. Pringle, Laurence P. *Living in a risky world*. NY: Morrow, 1989. 105 p. : ill. 0688043267
Discusses the risks our society faces every day as we deal with frightening uncertainties about food, transportation, clothing, air, disease, natural disasters, and environmental degradation in almost every aspect of our lives. HR GR 5-9.

Gaia Hypothesis

The Gaia Hypothesis proposes that the entire Earth is a self-regulating biological entity, thus providing a comprehensive new way of understanding our home planet. It has widespread philosophical and practical implications. The revised, 1987 edition of Lovelock's seminal *Gaia: A New Look at Life on Earth* is still an excellent introduction.

Books for adults:

88. Allaby, Michael. A *guide to Gaia : a survey of the new science of our living Earth*. NY: Dutton, 1990, c1989. 181 p. 0525248226
A good nontechnical review of Gaia theory. Seeing Earth as a single super-organism allows dispassionate analysis about a variety of environmental problems and separates the crucial ones from those unreal or less significant.

89. Barnaby, Frank. *The Gaia peace atlas*. NY: Doubleday, 1988. 271 p. : maps 0385241909
Environmentalists, peace activists, politicians, scientists, and academics with a Gaian orientation offer their views on the relationships between peace and environmental issues, with the goal of a peaceful and environmentally sustainable world in mind. HR.

90. Chapple, Christopher. ed. *Ecological prospects : scientific, religious, and aesthetic perspectives*. Albany: SUNY, 1994. 236 p. 0791417395 0791417409 (pbk)
This book addresses the idea that all living forms on Gaia merit protection. Contains 13 revised conference papers by various authors on "The Definition and Preservation of Living Systems" and "Religion, Aesthetics, and Ecology."

91. Gribbin, John R. *Hothouse Earth : the greenhouse effect and Gaia*. NY: Grove Weidenfeld, 1990. 272 p. : ill. 0802113745
Provides a comprehensive explanation of the mechanics and implications of global warming within a fascinating summary of the workings of the Earth, including the carbon cycle, weather patterns, and the roles of oceans and forests.

92. Joseph, Lawrence E. *Gaia : the growth of an idea*. NY: St. Martin's, 1990. 276 p. 031204318X
A good introduction to the Gaia hypothesis, tracing the scientific and popular debate about it throughout the 1980s.

93. Lovelock, J. E. *Gaia : a new look at life on Earth*. NY: Oxford, 1987, c1979. 157 p. 0192860305
AR: *The ages of Gaia*.
The seminal presentation of the Gaia hypothesis: Earth behaves like a single, self-regulating biological organism. This book is not as technical as many more recent volumes on Gaia and has retained great popularity. HR.

94. Myers, Norman. ed. *Gaia, an atlas of planet management*. Rev., updated ed. NY: Anchor, 1993. 272 p. : col. photos, maps. 0385426267

AHR:*The Gaia atlas of future worlds, The sinking ark.*
Divided into the "themes" of land, ocean, elements, evolution, humankind, and civilization, each considered in terms of potential, crisis, and management. Contains contributions by over 70 environmentalists. HR.

95. Schneider, Stephen Henry, and Boston, Penelope J. eds. *Scientists on Gaia.* Cambridge: MIT, 1993. 433 p. : ill 0262193108
Contains 44 conference papers on the philosophical, empirical, and theoretical foundations of Gaia, mechanisms through which planetary homeostasis could occur, applicability of the hypothesis to other planets, posssible destabilization by outside forces, and public policy implications. Some scientific background required.

96. Sheldrake, Rupert. *The rebirth of nature : the greening of science and God.* Rochester, VT: Destiny, 1994, c1991. 260 p. 0892815108
A biochemist's version of the Gaia hypothesis, contending that evolving nature has an inherent memory and that people are rejecting secular humanism and the mechanistic worldview that brought about the environmental crisis in favor of a "new animism." This controversial blend of biology, mythology, history, and religion is HR.

97. Thompson, William Irwin. ed. *Gaia, a way of knowing : political implications of the new biology.* Hudson, NY: Lindisfarne, 1988. 244 p. 0940262231
Conference papers considering specific aspects of the Gaia hypothesis and its political implications. Political and social systems should reflect a knowledge of interrelatedness in the natural world.

2. Natural History

Natural history is something of a catchall phrase refering to the scientific study of nature. It includes elements of biology, zoology, botany, ecology, evolution, and related fields. While in the past it was considered a branch of scientific inquiry, today the phrase often refers to scientific nature books written for general readers rather than scientists.

The work of ancient philosophers and scientists can be found here under Pliny the Elder and in *Ancient Natural History*. Non-scientific Medieval and Enlightenment views can be found in the Philosophy and Religion sections.

The widespread popularity of natural history seems to have increased in the past quarter-century as evidenced by television programs such as *Nature*, *National Geographic*, and Jacques Cousteau specials, the reprinting of many classic natural history books, and the emergence of superb new writers such as Ackerman, Wallace, Matthiessen, Schaller, Nabhan, Williams, Lopez, and Patterson. See the New Nature Writing section for books by these authors. The works of classic literary naturalists such as Thoreau, Burroughs, Beston, Hay, Finch, and Hubbell can be found in the Nature Writing section.

Because naturalists repond to nature emotionally as well as intellectually, it is difficult if not impossible to draw a firm line between "pure" natural history and the more "personal" literary nature writing. It may be theoretically correct to think of natural history as writing about nature, while nature writing is writing about nature's effect on the author, but this distinction can't be practically applied since so many books include both.

General Natural History

The first section contains general natural books. Nearly half of those written for adults are also suitable for older teens. Pringle, Ryden, and the Facklams are among the leading contemporary natural history authors for young adults. Maxwell's delightful *Ring of Bright Water* and the Kohl's fascinating *The View from the Oak* were writtten primarily for teenagers, but are very highly recommended for all ages.

Books for adults:

98. Ackerman, Diane. *The moon by whale light and other adventures among bats, penguins, crocodilians, and whales.* NY: Vintage, 1992, c1991. 249 p. 0394585747
Personal views on four animals in particular, animals in general, how animals and people interrelate, and nature writing. Shorter versions of these essays appeared in the *New Yorker.* HR AOT. #

99. Allaby, Michael. ed. *The Oxford dictionary of natural history.* NY: Oxford, 1985. 688 p. : ill. 0192177206
A very thorough and well written dictionary. HR.

100. Attenborough, David. *Life on Earth : a natural history.* New ed. NY: Readers House, 1993, c1979. 319 p. : col. photos. 1568530021
Demonstrates how bats, moles, whales, dolphins, hedgehogs and anteaters may be related to the same primitive animal. This serves as a microcosm to understand the evolutionary process and biodiversity. HR. #

101. Audubon, John James. *Audubon reader : the best writings of John James Audubon.* Ed. by Scott Sanders. Bloomington: Indiana Univ., 1986. 245 p. 0253310814 0253203848 (pbk)
Includes writings on natural history in general, birds in particular, and Audubon's life. HR for birders.

102. Bedichek, Roy. *Adventures with a Texas naturalist.* Rev. ed. Austin: Univ. of Texas, 1994, c1961. 330 p. : ill. 0292703111
AR: *Karankaway country.*
Reminiscent of Thoreau in its combination of natural history and meditative philosophy. Considers the importance of diversity and stability in nature and demonstrates how civilization typically both destabilizes nature and limits diversity.

103. Bruant, Nicolas. *Wild beasts.* SF: Chronicle, 1993. 144 p. : photos. 0811803384
These black and white photos of zebras disappearing in fog, rhinos wallowing in the dust, lions sleeping, gorillas staring into the distance, etc. are dramatic and evocative, much like Ansel Adams' landscape photographs. HR. #

104. Burton, Robert. *Nature's night life.* London: Blandford, 1989, c1982. 160 p. : ill. 0713721294
AR: *The mating game.*
Describes the habits and natural histories of insects, fishes, owls, possums, bats, and a wide variety of other nocturnal animals. Explains how

some previously diurnal animals have adapted to nocturnal conditions and how to observe animals at night. AOT. #

105. Colesberry, Adrian, and McLean, Brass. *Costa Rica : the last country the gods made.* Helena, MT: SkyHouse, 1993. 152 p. : photos, map. 1560441917
This photoessay explores the people and environment of Costa Rica: its beauty, ecology, history, and efforts to create natural preserves.

106. Darwin, Charles. *The origin of species by means of natural selection, or, The preservation of favored races in the struggle for life.* NY: Modern Library, 1993, c1859. 1000 p. 067960071
Darwin's theory of evolution and related theories have been very influential in the development of conservation biology, biodiversity, and other scientific and environmental thinking. For adults with some science background. *The illustrated Origin of species* abridged by Richard Leakey (240 p.) is better for teens and for most adults. HR.

107. ___. *The voyage of the Beagle.* NY: New American Library, 1988, c1839. 456 p. 0451626206
AR: Alan Moorehead's *Darwin and the Beagle.*
This journal of the three year scientific expedition, mostly to South America and the Galapagos Islands, foreshadows Darwin's later theories of natural selection and evolution.

108. Dolan, Edward F. *Animal folklore : from black cats to white horses.* NY: Ivy, 1992. 201 p. 0804105529
AR: *Animal rights.*
Dolan considers the origins, meanings, applications, and veracity of hundreds of quotes and anecdotes. #

109. Durrell, Gerald Malcolm. *Marrying off mother and other stories.* NY: Arcade, 1992. 197 p. 1559701803
When two legged, four legged, and winged animals get together, comedy ensues. These stories are embellishments of factual incidents from Durrell's life

110. Durrell, Gerald Malcolm, and Durrell, Lee. *Gerald & Lee Durrell in Russia.* NY: Simon & Schuster, 1986. 191 p. : col. photos, map. 0671612980
How a father-daughter naturalist team and a film crew toured the former Soviet Union to film a BBC TV series and what they discovered about that vast region's natural history and animals. #

111. Eckstrom, Christine K. *Forgotten edens : exploring the world's wild places.* Photos by Frans Lanting. Washington: Nat'l Geographic Soc., 1993. 202 p. :

col. photos. 0870448668
Lush photography and essays depicting some of the Earth's most spectacular areas: Borneo, South Georgia (Antarctica), Hawaii, Madagascar, and Okavango (Kalahari). #

112. Evans, Howard Ensign. *Pioneer naturalists : the discovery and naming of North American plants and animals*. NY: Holt, 1993. 294 p. : ill. 0805023372
AR: *Life on a little-known planet*; *Wasp farm*.
Provides brief biographical sketches of over 70 18th and 19th century naturalists, their studies and adventures, and the species they discovered and named. Includes Agassiz, Comstock, Douglas, Franklin, Fremont, LeConte, Jefferson, Merriam, and Xantus.

113. Ferrari, Marco. *Colors for survival : mimicry and camouflage in nature*. Charlottesville, VA: Thomasson-Grant, 1993. 144 p. : col. photos. 156566048X
Depicts how a wide range of insects, birds, reptiles, and mammals use or even change their coloration to hide from predators. Also covers how they can mimic other animals or emit smells or tastes that repel their adversaries.

114. Fletcher, Colin. *The man who walked through time*. NY: Vintage, 1989, c1968. 247 p. : map, photos. 0679723064
This account of a two month backpacking trip through the entire length of the Grand Canyon includes vivid descriptions of the Canyon's natural history and geography, Fletcher's interactions with animals, and his philosophical odyssey back into geologic time. This contemporary classic is HR AOT. #

115. Forsyth, Adrian. *Mammals of the American North*. Camden East, Ont.: Camden House, 1985. 351 p. : col. ill., maps. 0920656420
Packed with color photographs and range maps for each species. A standard reference to the animals of the Nearctic. HR.

116. French, Roger Kenneth. *Ancient natural history : histories of nature*. NY: Routledge, 1994. 368 p. 0415115450 (pbk) 0415088801
Examines the relationship between the physical world, the gods, philosophy, and ancient society. Considers how natural history was treated differently by the Greeks, Romans, Jews, and Christians. Examines the work of Aristotle, Theophrastus, Strabo, Pliny, and others.

117. Gehlbach, Frederick R. *Mountain islands and desert seas : a natural history of the U.S.-Mexican borderlands*. College Station: Texas A&M Univ., 1993. 298 p. : photos. 0890961182 0890965668 (pbk)
Describes the natural history, habitats, and ecological communities along

the border from the lower Rio Grande to the Sonoran Desert. Stresses the importance of conservation to maintain the biodiversity of the area.

118. Gibbons, Whit. *Keeping all the pieces : perspectives on natural history and the environment.* Washington: Smithsonian, 1993. 182 p. 1560982241
A collection of columns stressing the interconnectedness of all living things, the importance of a biologically complex environment, and the value of endangered species.

119. Gould, Stephen Jay. *Eight little piggies : reflections in natural history.* NY: Norton, 1993. 479 p. : ill. 039303416X
Library Journal Best Sci-Tech Book, 1994.
Considers characters from the history of science and some evolutionary oddities, but the main focus is the theme of environmental destruction and consequent extinctions. HR.

120. Hanson, Jeanne K. *The beastly book : 100 of the world's most dangerous creatures.* NY: Prentice Hall, 1993. 232 p. : ill. 0671850229
Attempts to delineate fact from folklore in presenting information on 100 mammal, reptile, fish, and insect species which humans consider dangerous or frightening. #

121. Henson, Paul, and Usner, Donald J. *The natural history of Big Sur.* Berkeley: Univ. of Cal., 1993. 416 p. : ill. 0520074661
Part 1 covers the area's geology, climate, shoreline, flora, fauna, and human settlements and their effect on the environment. Part 2 is a detailed guide to the 13 parks and wilderness areas in this rugged, coastal region.

122. Hitching, Francis. *The neck of the giraffe : Darwin, evolution, and the new biology.* NY: New American Library, 1983, 1987 printing. 258 p. : ill. 0451622324
A century after Darwin proposed the theory of evolution it is coming under fire again, not just from creationists, but from various members of the scientific community. This book attempts to present all sides of the controversy.

123. Johnson, Cathy. *The nocturnal naturalist : exploring the outdoors at night.* Chester, CT: Globe Pequot, 1989. 226 p. : ill. 087106524X
The artist-naturalist makes the most of insomnia in nocturnal excursions and demonstrates how to study nature after dark. Good complement to Diana Kappel-Smith's *Night life* and David Rains Wallace's *The dark range.* AOT.*

124. Kappel-Smith, Diana. *Night life : nature from dusk to dawn.* Boston: Little, Brown, 1990. 312 p. 0316483001
Another naturalist explores nature after dark. A good complement to Cathy

Johnson's *The nocturnal naturalist* and David Rains Wallace's *The dark range.* #

125. Keeney, Elizabeth. *The botanizers : amateur scientists in nineteenth-century America.* Chapel Hill: Univ. of North Carolina, 1992. 206 p. 0807820466
Throughout the 19th century, thousands of Americans took to the field to collect, identify, and preserve plant specimens. Keeney considers the intellectual, social, and religious motivations of the botanizers, their cooperation with professional botanists, and the eventual divergence of the two groups.

126. Lawrence, R. D. *The zoo that never was.* Toronto: HarperCollins, 1991. 308 p. 000637705X
AR *The green trees beyond, a memoir.*
Describes how the Lawrence farm in Ontario became a wild menagerie with racoons, skunks, lynx, otters, porcupines, geese, ducks, and other species. An account of the friendship-rivalry between their labrador retriever and a bear is particularly fascinating.

127. Lorenz, Konrad. *On life and living : Konrad Lorenz in conversation with Kurt Mundl.* NY: St. Martin's, 1991. 166 p. 031205937X
AR: *King Solomon's ring : new light on animal ways.*
Considers the effects of Chernobyl, threats to forests and oceans, the perils of overpopulation, and the future of humanity. Lorenz consistently shows a visionary insight into how nature and civilization interact and develop. HR.

128. Lutts, Ralph H. *The nature fakers : wildlife, science & sentiment.* Golden, CO: Fulcrum, 1990. 255 p. : ill. 1555910548
Although most of the natural histories and literary nature essays around the turn of the century may appear credible, many were exaggerated or based on hearsay, rumor, or pure imagination. This is a study of fraud in natural history and the efforts of Burroughs, Roosevelt and others to expose it.

129. Matthiessen, Peter. *Shadows of Africa.* NY: Abrams, 1992. 120 p. : ill. 0810938286
AR: *Sand rivers.*
Combines essays from three of his books with over 70 striking illustrations by Mary Frank to create a stunning evocation of African wildlife. Encourages readers to support wildlife conservation. HR AOT. #

130. ___. *The tree where man was born.* NY: Viking, 1994, c1972. 448 p. : photos 0140239340
Portrays the natural history of animals in the wild, their interactions with humans, and a variety of people including safari hunters, herders, and scientists.

131. ___. *Wildlife in America*. NY: Viking, 1987. 332 p. : photos. 0670819069
Matthiessen contrasts the historic abundance of American wildlife with the present extinction of species and habitat loss. He also traces the rise of the American conservation movement. HR.

132. Morris, Desmond. *Animalwatching : a field guide to animal behaviour*. London: Cape, 1990. 256 p. : col. photos. 0224027948
Explores the endlessly varied and ingenious strategies animals have devised to get along with each other and their world. Covers such basic activities as feeding, mating, and grooming, and illuminates fascinating, obscure, and exotic behaviors found in every niche of the animal kingdom.

133. Nabhan, Gary Paul. *Saguaro : a view of Saguaro National Monument & the Tucson Basin*. Tucson: SW Parks & Monuments Assn., 1986. 74 p. : col. photos. 091140869X
Nabhan's descriptions of the natural history of this spectacular desert region are accentuated by George H. Huey's lush color photos.

134. ___. *Songbirds, truffles, and wolves : an American naturalist in Italy*. NY: Pantheon, 1993. 227 p. 0679415858
Nabhan reveals what is useful in the old ways of Italian country life, what remains wild in a longtime inhabited and civilized country, and what aspects of ancient science have found their way into contemporary culture.

135. Nat'l Geographic Society. *The curious naturalist*. Washington: The Society, 1991. 288 p. : 300 col. photos. 0870448617 (reg. ed.) 0870448625 (deluxe ed.)
Seven great naturalists (John Hay, Ted Levin, Michael Parfit, David Rains Wallace, Ann Zwinger, Douglas Chadwick, Diane Ackerman) reveal how to view, experience, and understand nature. HR AOT. #

136. Neill, William; Murphy, Pat; and Ackerman, Diane. *By nature's design : photography*. SF: Chronicle, 1993. 119 p. : photos. 0811803295
This photoessay presents the order and patterns in nature, including similar patterns used by very different organisms and the existence of patterns where none is readily apparent. #

137. Nyhuis, Allen W. *The zoo book : a guide to America's best*. Albany, CA: Carousel, 1994. 288 p. : photos. 0917120132
Includes advice on where to go, what to bring, how to maximize entertainment and fun, and reviews of 53 zoos visited in the last three years. Also covers wildlife parks, aquariums, and aquatic parks. HR.

138. Olson, Sigurd F. *Listening point*. NY: Knopf, 1990, c1958. 242 p. : ill. 0394433580
AR: *Open horizons, Wilderness days*.
The things which inspire Olson's sense of wonder include rain, wind, plants, animals, rocks, woods, beaches, and other natural pheneomena. This lyrical account is for AOT. #

139. Patent, Dorothy Hinshaw. *How smart are animals?* San Diego: Harcourt Brace Jovanovich, 1990. 189 p. : ill. 0152367705
Discusses recent research on levels of intelligence in both wild and domestic animals, including learning and instinct, thinking and consciousness, language, and learning to understand animals. Younger readers should see Margery Faclam's *What does the crow know?* (172) HR. #

140. Peattie, Donald Culross. *Flowering Earth*. Bloomington: Indiana Univ., 1991, c1939. 260 p. : ill. 0253206626
AR: *Green laurels: lives and achievements of great naturalists*.
While tracing the evolutionary path of plant life, Peattie provides observations on natural history and philosophy along with autobiographical musings. Recommended primarily for AOT with an interest in historical botany. #

141. ___. *The road of a naturalist*. Boston: G.K. Hall, 1986, c1941. 315 p. 083982890X
Covers his travels, adventures, and nature studies in various parts of America intermingled with observations on American Indian environmental thought and practice, humanity's place in nature, and how to be an effective naturalist.

142. Pliny the Elder. *Natural history, a selection*. Ed. by John F. Healy. NY: Penguin, 1991. 399 p. 0140444130
AR: Aristotles's *Historia animalium*.
Translation of selections from *Naturalis historia*. Observations on natural history and philosophy from one of the greatest writers of Roman antiquity.

143. Quammen, David. *The flight of the iguana : a sidelong view of science and nature*. NY: Anchor, 1989, c1988. 302 p. 0385263279
A sometimes humorous and always fascinating acount of plants, animals, and processes which may seem odd to many people, but are a "perfectly natural" part of nature. Includes stories about giant earthworms, barkless dogs, fatal attraction bedbugs, and vegetarian pirhanas. #

144. ___. *Natural acts : a sidelong view of science and nature*. NY: Schocken, 1985. 221 p. 0805239677
Sequel: *The flight of the iguana*.
A unique vision of nature and science. #

145. Seton, Ernest Thompson. *Adventures in the wild : the worlds of Ernest Thompson Seton*. NY: Arrowood, 1991, c1976. 204 p. : ill. 0884860345
A collection as remarkable for Seton's little known but brilliant paintings as for his lively prose. Covers his entire 70 years work in the Arctic, the Rockies, the deserts, and the eastern woodlands. #

146. ___. *Lives of the hunted, containing a true account of the doings of five quadrupeds & three birds, and, in elucidation of the same, over 200 drawings*. Berkeley: Creative Arts, 1987, c1905. 360 p. : ill. 0887390544
AR: *Biography of a grizzly*.
Contains "biographical" stories about a ram, sparrow, bear, teal, wolf, rat, coyote, and chickadee.

147. ___. *Wild animals I have known*. Toronto: McClelland & Stewart, 1991, c1898. 254 p. : ill. 0771098731
Contains stories about wolves, crows, rabbits, dogs, foxes, mustangs, and partridges. Readers are warned that Seton's anthropomorphized animals caused John Burroughs to call this book *Wild animals I ALONE have known*. Entertaining, but not always accurate. #

148. Shreeve, James. *Nature : the other Earthlings*. NY: Macmillan, 1987. 288 p. : col. photos, maps. 0625349201
Companion to the *Nature* television program.
Describes how a wide variety of species have each developed, adapted to the environments, and interacted with other species in their own unique way. Each chapter focuses on a single animal group and also serves as a guide to the flora and fauna of its environment. #

149. Stanwell-Fletcher, Theodora C. *Driftwood Valley*. Intro. by Wendell Berry. NY: Penguin, 1989, c1946. 384 p. : ill. 0140170103
John Burroughs Medal, 1947.
A journal of a three year specimen collecting expedition to the Driftwood Valley of British Columbia. HR.

150. Streshinsky, Shirley. *Audubon : life and art in the American wilderness*. NY: Villard, 1993. 407 p. : ill 0679408592
Describes Audubon's early life and travels, his "great idea" of painting every indigenous bird and beast; selling paintings door to door until he was better known and could publish his work; pioneering work in bird conservation, and his boundless devotion to nature.

151. Terres, John K. *From Laurel Hill to Siler's Bog : the walking adventures of a naturalist*. New ed. Chapel Hill: Univ. of North Carolina, 1993, c1969. 232 p. : ill. 0807844268

John Burroughs Medal, 1971.
Follows the natural cycles of the year in this engaging study of the varied wildlife in North Carolina's Mason Farm reserve. HR. #

152. ___. *Things precious & wild : a book of nature quotations.* Golden, CO: Fulcrum, 1991. 297 p. 1555910726
Includes nearly 700 quotations from 140 naturalists, conservationists, philosophers, and poets.

153. Thomas, Lewis. *The lives of a cell; notes of a biology watcher.* NY: Bantam, 1984, c1974. 192 p. 0553275801
AR: *The fragile species.*
Philosophical, personal, and scientific observations on a variety of biological phenomenon, including Earth itself. Foreshadows the Gaia hypothesis. HR.

154. ___. *The medusa and the snail : more notes of a biology watcher.* NY: Viking, 1994, c1979. 192 p. 0140243194
Addresses issues surrounding recent biological research and applies them to the unique complexity of the human condition. Includes observations on disease, life, and death in nature. HR

155. Tinbergen, Niko. *Curious naturalists.* Rev. ed. Amherst: Univ. of Massachusetts, 1984, c1974. 268 p. : ill. 0870234560
AR: *The animal in its world, Social behavior in animals.*
Tinbergen presents very interesting accounts of his experiences observing birds, bees, flowers, insects, and other naturalists. Combines natural history lore with human interest stories. HR AOT. #

156. Trimble, Stephen. *The sagebrush ocean : a natural history of the Great Basin.* Reno: Univ. of Nevada, 1989. 280 p. : ill., maps. 0874171288
In this eloquent synthesis of natural history and personal experience Trimble focuses on biogeography: what lives where and why. An excellent introduction to the ecology and spirit of the Great Basin. HR.

157. ___. ed. *Words from the land : encounters with natural history writing.* Salt Lake City: Peregrine Smith, 1989. 303 p. : photos. 0879052422
Essays by Ed Abbey, Wendell Berry, Annie Dillard, Gretel Ehrlich, Robert Finch, John Hay, Edward Hoagland, Sue Hubbell, Barry Lopez, John Madson, Peter Matthiessen, John McPhee, Gary Nabhan, David Quammen and Ann Zwinger. Trimble's introduction is a fascinating meditation on nature writing. HR.

158. Tweit, Susan J. *Pieces of light : a year on Colorado's front range.* Niwot, CO: R. Rinehart, 1990. 248 p. : ill. 0911797726

AR:*The great southwest nature fact book.*
A vegetation ecologist uses a daily journal account of a year spent in Colorado's Front Range to present the changing faces of and relationships between flora and fauna, including people.

159. Wallace, David Rains. *Bulow Hammock : mind in a forest.* SF: Sierra Club, 1988. 170 p. : ill. 0871566761
"Amidst alligators, armadillos, white ibis, golden silk spiders, and an immense variety of flora...Wallace experiences parallels between the natural history of a woodland ecosystem and the evolution of human consciousness."-Intro. AOT. #

160. ___. *The untamed garden and other personal essays.* NY: Collier, 1988, c1986. 203 p. 0020298919
AR: *The dark range, The wilder shore.*
Essays on "Growing things, Field notes, Natural history and conservation, and Wildlife." Wallace finds his place in nature, ranging from his backyard garden to wilderness areas in California, Florida, Connecticut, and Japan. For a broad range of adults including gardeners, hikers, naturalists, and conservationists. HR.

161. *The way nature works.* NY: Macmillan, 1992. 359 p. : col. ill. 0025081101
Covers geologic history, the atmosphere, evolution, reproduction, food seeking, movement and shelter, attack and defense, animal communication, and biomes. Each section presents a series of double page spreads dealing with a key subject in depth. AOT. #

162. White, T. C. R. *The inadequate environment : nitrogen and the abundance of animals.* NY: Springer-Verlag, 1993. 425 p. : ill. 038756828X
White challenges the theory that the competition for energy is the most important factor in ecological interactions, arguing that "the universal hunger for nitrogen is the misery that drives the ecology of all organisms."

163. Wilson, Edward Osborne. *Biophilia.* Cambridge: Harvard, 1984. 157 p. 0674074416
Biophilia means literally "love of nature" and includes the innate tendency to focus on life and living processes. Wilson believes we have an inherent need for nature, that our separation from it is both mentally and spiritually harmful, and that separation is what makes the environmental crisis possible.

Books for young adults:

164. Batten, Mary. *Nature's tricksters : animals and plants that aren't what they*

seem. SF: Sierra Club, 1992. 54 p. : ill. 0316083712
Describes the use of trickery and camouflage by many animal species to protect themselves, to obtain food, and to attract mates.

165. Brooks, Bruce. *Making sense : animal perception and communication.* NY: Farrar Straus Giroux, 1993. 73 p. : ill. 0374347425
Discusses animals' six senses and how they use them to perceive and react to the world around them.

166. ___. *Nature by design.* NY: Farrar Straus Giroux, 1991. 74 p. : col. photos 0374303347
Describes functional structures built by such animals as the beaver, termite, and tailorbird.

167. ___. *Predator!* NY: Farrar, Straus, Giroux, 1991. 74 p. : col. photos. 0374361118
Survival in the wild creates a hierarchy of predators and their prey. This interaction forms the basis for a complex ecological process known as the food chain.

168. Carwardine, Mark. *The illustrated world of wild animals.* NY: Simon & Schuster, 1988. 64 p. : col. ill. 0671665634
Illustrations, maps, and text introduce both common and unusual animals from different regions of the world. HR.

169. Cranfield, Ingrid. *Animal world.* NY: Dorset, 1991. 64 p. : col. ill. 0880296933
A comprehensive survey of animals and their habits around the world. GR 4-8.

170. Facklam, Margery. *Do not disturb : the mysteries of animal hibernation and sleep.* Ill. by Pamela Johnson. SF: Sierra Club, 1989. 47 p. : ill. 0316273791
Why do different animals require vastly different amounts of sleep? Why does sleep vary with age? What are the purposes of hibernation, and how does it work? These and related questions are considered. GR 4-8.

171. ___. *Partners for life : the mysteries of animal symbiosis.* SF: Sierra Club, 1989. 48 p. : ill. 0316259837
Examines partnerships between two different species of animals that provide one or both of the partners with food, protection, transportation, or a way to keep clean. GR 4-8.

172. ___. *What does the crow know? : the mysteries of animal intelligence.* SF: Sierra Club, 1994. 48 p. : ill. 0871565447

Raises the issue of whether or not animals are capable of thought, learning, remembering, and creativity, with examples of animal behavior that appears to be truly intelligent. This good introduction to the subject is recommended for GR 5-9. Older readers should see Patent's *How smart are animals?* (139)

173. Feltwell, John. *Animals and where they live.* NY: Grosset & Dunlap, 1988. 64 p. : ill. 0448192187
Describes animal life and behavior in relation to habitat, showing how they adapt to natural conditions. Shows which kinds of animals live in which regions of the world. GR 5-9.

174. George, Jean Craighead. *Animals who have won our hearts.* NY: Harper-Collins, 1994. 56 p. : ill. 0060215437
A collection of stories about animals who became beloved and famous, including Balto the sled dog, who found his way through a blinding snowstorm, and Koko the gorilla, who learned sign language.

175. Kerrod, Robin. *Animals around the world.* NY: Prentice Hall, 1992. 96 p. : col. ill., maps. 0130333824
Demonstrates how animals are distributed around the world, how their behavior is effected by their habitat, and how human activity helps some species and threatens others. GR 4-8.

176. Kohl, Judith, and Kohl, Herbert R. *The view from the oak : the private worlds of other creatures.* SF: Sierra Club, 1988, c1977. 110 p. : ill. 0316501379
AR: *Pack, band, and colony.*
Attempts to enable us to view the world of ticks, flies, birds, jelly fish, and other animals through their senses, not our own. This book was written for young adults, but is HR for all ages. #

177. *The living world.* NY: Oxford, 1993. 160 p. : col. ill. 0199101426
This reference source covers the biology, behavior, and evolution of plants, animals, and humans. GR 4-8.

178. Mason, George Frederick. *Animal tracks.* Hamden, CT: Linnet, 1988, c1943. 95 p. : ill. 0208022139
A nature guide presenting pictures of 44 North American animals and their tracks. Also discusses their habitat and behavior. #

179. Morris, Desmond. *The world of animals.* NY: Viking, 1993. 128 p. : col. ill. 0670851841
Morris describes the natural history and behavior of 30 different wild animals. Strives to increase our understanding of and empathy with other animals. #

180. Nat'l Wildlife Federation. *Incredible animals* A *to* Z. Washington: The Federation, 1985. 95 p. : col. photos. 0912186666
A photographic survey, done in alphabetical order, of some of the world's most unusual animals, inhabitants of the land, sea, and air, with descriptions of their physical characteristics and behavior.

181. Parker, Steve. *Animal babies*. Emmaus, PA: Rodale, 1993. 196 p. : col. photos, maps. 0875965954
Provides two-page descriptions of 55 different species, mostly mammals, living in nine separate types of habitats. The lengthy introduction covers animal courtship, mating, birth, and care of newborns. #

182. ___. *The Random House book of how nature works*. NY: Random House, 1993. 124 p. : ill. 0679837000 0679937005
A guide to the way plants and animals function and struggle to survive on Earth. #

183. Patent, Dorothy Hinshaw. *Feathers*. NY: Cobblehill/Dutton, 1992. 64 p. : col. photos. 0525650814
Describes and depicts birds' feathers from structure, type, and color to various uses. GR 4-8.

184. Pellowski, Anne. *Hidden stories in plants : unusual and easy-to-tell stories from around the world, together with creative things to do while telling them*. NY: Macmillan, 1990. 93 p. : ill. 0027706117
Outstanding Science Trade Book for Children, 1990.
Presents myths, legends, tales, and folklore about plants. Describes how to use plants to make ornaments, toys, disguises, dolls, and musical instruments. The stories and crafts are well matched. HR GR 5-10.

185. Pringle, Laurence P. *Home : how animals find comfort and safety*. NY: Scribner, 1987. 71 p. : ill. 0684185261
AR: *The controversial coyote*, *Estuaries*, *The gentle desert*, *Throwing things away*, *Water: the next resource war*.
Examines the characteristics of the many different types of places in which animals make their home. GR 4-9.

186. Seddon, Tony, and Bailey, Jill. *The living world*. Garden City, NY: Doubleday, 1987, c1986. 160 p. : ill. 0385237545
Presents snippets of lesser-known information about plants and animals such as adaptation, interdependence, food chains, behavior, symbiosis, endangered species, preservation, and various habitats. GR 4-8.

187. Sibbald, Jean H. *Homes in the sea : from the shore to the deep*. Minneapo-

lis: Dillon, 1986. 95 p. : ill. 0875183042
An Introduction to the various animals that inhabit the sea.

188. Sierra Club. *The Sierra Club book of small mammals*. SF: Sierra Club, 1993.
68 p. : col. photos. 0871565250
An introduction to such small mammals as lemurs, gibbons, tapirs, and
aardvarks. Explains what all mammals have in common, how they evolved,
and what threats they face.

189. Stephen, David. *Living with wildlife*. Edinburgh: Canongate, 1989. 189
p. : photos. 0862412153
AR: *Alba the last wolf, Birds of the world, Bodach the badger, Nature's way, The six
pointer buck.*
Covers the wildlife of Scotland, conservation efforts, and Stephen's career
as a naturalist and author of both nonfiction and fiction. #

190. Tomb, Howard, and Kunkel, Dennis. *Microaliens : dazzling journeys with
an electron microscope*. NY: Farrar, Straus & Giroux, 1993. 79 p. : photos.
0374349606
Text and photographs taken with an electron microscope examine such
items as bird feathers, fleas, skin, mold, and blood. Contains chapters on
what you will find in the air, the water, the yard, the home, on you, and
inside you. #

191. Walters, Martin. *The Simon & Schuster young readers' book of animals*. NY:
Simon & Schuster, 1991, c1990. 191 p. : col. ill. 0671731297 (pbk)
0671731289
A guide to the world of animals, beginning with the simplest forms of life
and working through the classes in order of increasing complexity. GR 4-8.

Animal Species and Families

This section lists books about animal species and families such as wolves,
bears, and apes. Naturalists such as Goodall, Adamson, and Lawrence are
associated with certain species or families (primates, felines, and wolves,
respectively) and have done a great deal to increase public knowledge of
and concern for these animals. Unfortunately, many other books about animals can be found in the Endangered Species section. Also see the
Human-Animal Relations section for coverage of related social and political issues.

Books for adults:

192. Adamson, Joy. *Born free, a lioness of two worlds.* NY: Pantheon, 1987, c1960. 220 p. : col. photos, map. 0394561414
AR: *Pippa's challenge, Queen of Shaba, The spotted sphinx.*
The story of the lioness Elsa who was raised in captivity and then taught to hunt and kill so that she could return to live freely in the wild. *Living free* and *Forever free* continue Elsa's story, cover the birth and raising of her kittens, and include a plea for wildlife preservation. HR. #

193. Bass, Rick. *The ninemile wolves : an essay.* NY: Ballantine, 1993. 162 p. : ill., map. 034538251X
A study of the biological, economic, and political aspects of the reintroduction of wolves into the northern Rocky Mountains and an emotional inquiry into the proper relationship between people and nature. A good companion to Lopez's *Of wolves and men*, this may be Bass' best work to date. HR AOT. #

194. Brooks, Bruce. *On the wing : the life of birds from feathers to flight.* NY: Scribner's, 1989. 192 p. : col. photos. 0684191199
A companion to the *Nature* TV program. ALA Best Book for Young Adults. Stories about the bird physiology and behavior of birds which attempt to show that birds follow certain behavioral patterns for reasons we can easily understand. HR AOT. #

195. Brown, David E., and Murray, John A. compilers. *The last grizzly and other Southwestern bear stories.* Tucson: Univ. of Arizona, 1988. 184 p. 0816510679
A collection of essays from 1826-1926 illustrating changes in how we think of and interact with this species, now extinct in the Southwest. The later authors argue for the reintroduction of the grizzly. AOT. #

196. Craighead, Frank C. *Track of the grizzly.* SF: Sierra Club, 1982, c1979. 261 p. : ill. 0871563223
The story of the Craighead brothers' scientific study of grizzly bears. Covers natural history and behavior and critiques the National Park Service's methods of wildlife management. HR.

197. Dary, David. *The buffalo book : the full saga of the American animal.* Athens: Swallow, 1989. 434 p. : ill. 0804009317
Covers the buffalo's origins and habits, its importance to both Indians and whites, and the efforts of determined people to bring it back from the edge of extinction.

198. Dobie, J. Frank. *The mustangs.* Austin: Univ. of Texas, 1984, c1952. 376

p. : ill. 0292750811
AR: *Rattlesnakes, The voice of the coyote.*
Covers how the mustang was introduced to North America, legends and
folklore about it, its natural history, how large herds developed and effected the environment, and related topics. May be Dobie's most enduring natural history book. HR.

199. **Domico, Terry, and Newman, Mark.** *Kangaroos : the marvelous mob.* NY:
Facts on File, 1993. 202 p. : col. photos. 0816023603
A journalist and a photographer follow a pack of kangaroos to observe their
behavior, incredible physical capabilities and feats, and the cruelty inflicted
upon them by hunters and others. #

200. **Erickson, Laura.** *For the birds : an uncommon guide.* Duluth, MN: Pfeifer-
Hamilton, 1994. 1 v. (unpaged) : ill. 0938586912
Includes 365 day-by-day sketches introducing over 250 bird species and
providing advice on bird watching. Provides space for the reader's field
notes. Especially for birders. AOT. #

201. **Feazel, Charles T.** *White bear : encounters with the master of the Arctic ice.*
NY: Ballantine, 1992, c1990. 223 p. : photos. 0345373979
Explores the history, myths, and science of the polar bear from the ice age
through the stone age and into the modern world.

202. **Fossey, Dian.** *Gorillas in the mist.* Boston: Houghton Mifflin, 1983. 444
p. : photos. 0395282179
Covers her 15 year study of three generations of gorillas in east Africa.
While her descriptions of the natural history of gorillas, their social organization, and parallels between human and ape behavior are highly informative, her description of friendships with individual apes is what makes this
a classic. HR.

203. **Goodall, Jane.** *In the shadow of man.* Photos by Hugo van Lawick.
Boston: Houghton Mifflin, 1988, c1971. 361 p. : photos 0395331455
AR: *Innocent killers, Solo.*
After months of frustration, Goodall gained the trust of the chimpanzees of
Gombe and was able to observe their behavior, hierarchy, social interac-
tions, communications, and use and making of tools. Many similarities
between human and chimp society are analyzed. AOT. #

204. ___. *Through a window : my thirty years with the chimpanzees of Gombe.*
Boston: Houghton Mifflin, 1990. 416 p. : photos. 0395500818
Describes the interactions of chimpanzees with each other and with peo-
ple. Includes fascinating observations on human behavior. Updates *In the*

shadow of man and *The chimpanzees of Gombe*. Younger readers should see her *My life with the chimpanzees*. AOT#

205. Gordon, Nicholas. *Murders in the mist : who killed Dian Fossey?* London: Hodder & Stoughton, 1993. 254 p. : photos. 0340598808
Originally Fossey's graduate assistant and chief tracker were charged with her murder, but this investigation links her death to the highest levels of the corrupt and brutal Rwandan government.

206. Hansen, Kevin. *Cougar : the American lion*. Flagstaff, AZ: Northland, 1992. 129 p. : col. photos, maps. 0873585445
A forest ranger combines the latest research on cougars with his personal experiences with them to present the cougar's way of life, dismiss myths about it, demonstrate how its existence is inextricably connected with our own, and argue for its preservation. HR for all ages. #

207. Herrero, Stephen. *Bear attacks : their causes and avoidance*. NY: Lyons, 1985. 287 p. : ill., maps. 0941130827
A thorough study of attacks on humans by grizzly and black bears. An essential guide to traveling safely in bear country.

208. House, Adrian. *The great safari : the lives of George and Joy Adamson*. NY: Morrow, 1993. 465 p. 0688101410
This critique of the Adamsons refutes their popular image as dedicated conservationists, presenting them as autocratic opportunists whose work has negative impacts on African wildlife and people.

209. Hyde, Dayton O. *Don Coyote : the good times and the bad times of a maligned American original*. NY: Ballantine, 1989, c1986. 228 p. : photos. 0345347078
A rancher's positive experiences with coyotes lead him to reconsider his negative attitude of the coyote as a pest, and also his role as being responsible to the land, rather than dominating it. "His" coyotes keep down his rodent population, interact well with humans and dogs, and do not molest the cattle. #

210. Kofalk, Harriet. *No woman tenderfoot : Florence Merriam Bailey, pioneer naturalist*. College Station: Texas A&M, 1989. 225 p. 0890963789
Kofalk uses 60 years of family correspondence and other source material in this biography of an important naturalist who was a leader in the movement to study birds in their natural environments. Includes information on ornithology and wildlife conservation in the late 19th and early 20th centuries.

211. Laidler, Keith, and Laidler, Liz. *Pandas : giants of the bamboo forest*. NY:

Parkwest, 1994. 208 p. : photos. 0563363614
Accompanies a BBC TV program. Covers the natural history, biology, and habitat of the giant panda and the red panda, and efforts to preserve them. Not as detailed as Schaller's *The last panda*, but a good, nontechnical overview. AOT. #

212. Lawrence, R. D. *The ghost walker.* Toronto: HarperCollins, 1991, c1983. 242 p. 0006377041
Lawrence tracks the puma in British Columbia's rugged Selkirk Mountains and returns with a book on the natural history and behavior of the puma, how to observe wildlife in its natural habitat, and winter survival techniques. AOT.

213. ___. *In praise of wolves.* Toronto: HarperCollins, 1991, c1986. 245 p. : photos. 0006377033
Lawrence describes their shrinking population and habitat, individual and group behavior, their important niche in the ecological web, and their partial domestication. An appreciation of the wolf and a call to end wolf hunting and preserve this legendary beast. AOT. #

214. ___. *Trail of the wolf.* Emmaus, PA: Rodale, 1993. 160 p. : col. photos 0875965946
Lawrence details the natural history and behavior of the wolf, pack formation and social structure, differences between wolves and other canines, wolf-human relationships, folklore and legends, and the need for preservation. Combines natural science and personal narrative very effectively. AOT. #

215. Leydet, Francois. *The coyote : defiant songdog of the West.* Rev. ed. Norman: Univ. of Oklahoma, 1988, c1977. 221 p. 0806121688
AR: *The last redwoods.*
Leydet observes one coyote throughout its life as a case study of coyote development, behavior, interaction, and natural history. His descriptions of the coyote's place in nature and human attempts at extermination may cause one to reconsider which species is the actual "predator."

216. Liotta, P. H. *Learning to fly : a season with the peregrine falcon.* NY: Ballantine, 1990, c1989. 241 p. 0449219186
An Air Force officer spends a season helping young falcons reintroduced to the Adirondacks learn how to fly. Drawing from literature, philosophy, myth, legend, falcon lore and aerodynamic theory, Liotta offers a remarkable investigation into the mysteries and meaning of flight.

217. Lopez, Barry Holstun. *Of wolves and men.* NY: Scribner, 1978. 309 p. : ill.

0684156245
Lopez reveals not only the wolf of the scientist, but that of the Eskimo and Indian, the wolf killer, werewolves and feral children, the wolf of folklore and fable. He creates a compelling picture of both the wolf as animal and the image that society has created. Notable for both its scientific accuracy and skillful blend of fact and fable. HR AOT. #

218. McCall, Karen. *Cougar : ghost of the Rockies.* SF: Sierra Club, 1992. 146 p. : col. photos. 0871565641
Text and photos follow a pregnant cougar released into the Idaho wilderness over the course of a year. Describes the cougar's habitat and life cycle "from the cougar's point of view," recounts myths and legends about the cougar, and makes a plea for its preservation. #

219. McNamee, Thomas *The grizzly bear.* NY: Penguin, 1990, c1984. 314 p. 0140128123
Follows a mother bear and her cubs near Yellowstone from birth in April to hibernation in late October, studying physiology, mating, cub rearing, social relations, feeding, predation, aggression, and habitat requirements. Analyzes controversies about grizzlies and their struggle to survive in the modern world.

220. Millar, Margaret. *The birds and the beasts were there.* Santa Barbara: Capra, 1991. 241 p. 0884963241
A personalized account of the wildlife at her feeding stations, the great destruction of the Coyote fire, and the process of the canyon healing itself from the blaze.

221. Montgomery, Sy. *Walking with the great apes : Jane Goodall, Dian Fossey, Birute Galdikas.* Boston: Houghton Mifflin, 1991. 280 p. : photos. 0395515971
This biography of the "primate trimates" studies their lives, work, ideas, and values. Describes how they have introduced themselves to animals in the wild, rather than work in zoos and laboratories. They serve as exemplars of ecofeminism in the field. #

222. Moss, Cynthia, and Colbeck, Martyn. *Echo of the elephants : the story of an elephant family.* NY: Morrow, 1992. 192 p. : col. photos. 0688121039
This intimate photoessay account of Echo, a 40 year old elephant matriarch, includes herd behavior, journeys to escape drought and ecological devastation, and the birth of a baby elephant. HR. #

223. Moss, Cynthia. *Elephant memories : thirteen years in the life of an elephant family.* NY: Ivy, 1992, c1988. 299 p. : photos, maps. 0804110891
Based on observations of an extended elephant family for 13 years, Moss

describes their natural history and behavior, focusing primarily on their matriarchal social organization. HR.

224. Mowat, Farley. *Never cry wolf.* NY: Bantam, 1993, c1963. 164 p. 0553273965
AR: *Born naked, My father's son.*
By mimicking their behavior, a field biologist gains the wolves' trust and learns their ways. His discovery that huge caribou kills are caused by hunting, not wolves, puts him in harm's way from hunters and the Canadian government. This exciting and inspirational book is HR AOT. #

225. ___. *Woman in the mists : the story of Dian Fossey and the mountain gorillas of Africa.* NY: Warner, 1988, c1987. 409 p. : photos, maps. 0446387207
Mowat explores Fossey's diaries and interviews colleagues to present her inner life and its relation to her work. Fossey's reported eccentricities are minimized and the charge that she may have been to blame for her own murder is disputed.

226. Murray, John A. ed. *The great bear : contemporary writings on the grizzly.* Anchorage: Alaska Northwest, 1992. 245 p. 0882403923
Contains 16 essays, articles, or stories about grizzly bears written by Adolph Murie, John McPhee, Richard Nelson, Ed Abbey, William Kittredge, A. B. Guthrie, Aldo Leopold, Rick Bass and others.

227. Nabhan, Gary Paul. ed. *Counting sheep : twenty ways of seeing desert bighorn.* Tucson: Univ. of Arizona, 1993. 261 p. 0816513856
Presents the diverse views of zoologists, anthropologists, historians, geographers, hunters and journalists on the elusive bighorns. This is as much about how the bighorns roam the hearts and minds of the observers as it is about the sheep being counted. A good book for people considering careers as naturalists. HR AOT. #

228. Nichols, Michael. *The great apes : between two worlds.* Washington: Nat'l Geographic Soc., 1993. 200 p. : photos 0870449478
Nichols updates studies of the status of gorillas, orangutans, and chimpanzees. Some progress has been made in their treatment in zoos and labs, but conditions in their native habitats have deteriorated due to deforestation, poaching, and warfare. HR AOT. #

229. O'Brien, Dan. *The rites of autumn : a falconer's journey across the American West.* NY: Anchor, 1989, c1988. 192 p. : map. 0385265599X
To teach an injured falcon chick to survive in the wild, O'Brien taught her to hunt and took her along the migration route from Montana to San Padre Island, Texas. Ultimately, it is the story of his search to find the meaning of

wildness, domesticity, and the balance between them. AOT. #

230. Paine, Stefani. *The world of the sea otter.* SF: Sierra Club, 1993. 132 p. : 67 col. photos 0871565463
Both a scientific account and a celebration of the sea otter. Covers its natural history and ecology, near extermination from the fur trade, unsuccessful and succesful attempts to reintroduce it, and its natural return to areas where it had been eliminated off the California coast and Alaska. #

231. Peacock, Doug. *Grizzly years : in search of the American wilderness.* NY: Holt, 1990. 288 p. : col. photos. 0805004483
Describes 20 years of watching and interacting with grizzlies in Montana, detailing their habitats, natural history, behavior, mating, social hierarchy and methods of communication. Provides a personal narative about how the wilderness has transformed his life. Includes powerful arguments for wildlife preservation. HR.

232. Peterson, Dale, and Goodall, Jane. *Visions of Caliban : on chimpanzees and people.* Boston: Houghton Mifflin, 1993. 367 p. : photos. 0395537606
Library Journal Best Sci-Tech Book, 1994.
Traces the relations between chimpanzees and humans and the changes in human attitudes toward chimps from 1600 to the present. The humanity of chimps is captured in accounts of their use of tools and medicinal plants, their sense of self, and their devotion to one another. HR.

233. Peterson, Roger Tory. *Birds over America.* Rev. ed. NY: Dodd, Mead, 1983, c1964. 342 p. : col. ill. 0396082696
John Burroughs Award, 1964.
Primarily an entertaining account of the notable people and birds he has encountered over a distinguished career. Secondarily a natural history of American birds and a plea for habitat preservation. HR, as are his field guides.

234. Rabinowitz, Alan. *Jaguar : one man's battle to establish the world's first jaguar preserve.* NY: Anchor, 1991, c1986. 370 p. : photos. 0385415192
Rabinowitz combines hard science, adventure, humor, and socio-political criticism in his study of the jaguars of Belize. Considers the implications of research on wildlife populations. HR.

235. Richards, Alan J. *Birds of prey : hunters of the sky.* Philadelphia: Courage, 1992. 144 p. : 162 col. ill. 1561381764
Divides raptors into ten groups, describing their common characteristics and differences, the natural history of various species, the ecology of predation, migration, and social behavior. AOT. #

236. Ryden, Hope. *America's last wild horses*. Rev. ed. NY: Lyons & Burford, 1990, c1970. 333 p. 1558210814
AR for younger readers: *Mustangs, a return to the wild*.
Presents the history of wild horses in America, covers current controveries, and provides a plea for treating them in an ethical manner.

237. ___. *God's dog : a celebration of the North American coyote*. NY: Lyons & Burford, 1989, c1979. 321 p. : photos. 15558210466
AR for younger readers: *The wild pups : the true story of a coyote family*.
Covers habitats, behavior, breeding, feeding, predation, and their importance to other animals and to society. AOT. #

238. ___. *Lily Pond : four years with a family of beavers*. NY: HarperPerennial, 1990, c1989. 256 p. : photos, maps. 0060973447
Describes how family members interact, how disputes are settled by shoving matches, and how the entire colony can mobilize to avoid calamity. Ryden muses on the proper relationship between people and wild animals, realizing she must leave before they become too trusting of human intruders. #

239. Savage, Candace Sherk. *Peregrine falcons*. SF: Sierra Club, 1993, c1992. 145 p. : 100 col. photos. 0871564610
Briefly recounts the history of the falcon in art, literature, religion, and recreation. Chronicles its near extinction, its resurgence since the ban on DDT, and continuing environmental threats to its existence. By extension, this is a plea to preserve all wild species.

240. ___. *Wild cats : lynx, bobcats, mountain lions*. SF: Sierra Club, 1993. 136 p. : col. ill. 0871564548
Includes incredibly vivid and sometimes graphically violent photos of three species of cat in America's wilds. HR for the non-squeemish. #

241. Schaller, George B. *Golden shadows, flying hooves : with a new afterword*. Chicago: Univ. of Chicago, 1989, c1973. 340 p. : photos. 0226736504
A study of the wildlife of the Serengeti plain, focusing primarily on lions and their prey. Considers various aspects of animal and human behavior including the origin of hunting by humans.

242. ___. *The year of the gorilla*. Chicago: Univ. of Chicago, 1988. 300 p. : photos. 0226736504
AR: *The Serengeti lion, Mountain monarchs*.
Describes the natural history and behavior of the mountain gorilla, efforts to preserve gorilla populations, and the natural history and landscape of Central Africa.

243. Schaller, George B., and Richardson, Nan. *Gorilla : struggle for survival in the Virungas*. NY: Aperture, 1989. 112 p. : col. photos. 0893813109
Depicts the tale of the last few hundred mountain gorillas. The story of a people, the land, and the animals that struggle to maintain a delicate ecological balance.

244. Scott, Jonathan. *Kingdom of lions*. Emmaus, PA: Rodale, 1992. 192 p. : col. photos. 0875965504
Scott brings us face to face with Kenya's hyenas, buffalo, impala, leopards, warthogs, and especially its lions. This vivid photoessay is HR. #

245. Shaw, Harley G. *Soul among lions : the cougar as peaceful adversary*. Boulder, CO: Johnson, 1989. 140 p. 1555660533
A Fish & Game biologist considers the natural history and behavior of cougars, the politics of wildlife preservation, and the public debate about cougars. HR.

246. Tomkies, Mike. *Wildcat haven : the complete story of my wilderness wildcats and Liane, a cat from the wild*. London: Cape, 1987. 224 p. : ill. 0224025023
AR: *Between Earth and paradise*, *A world of my own*, *Alone in the wilderness*.
A Scottish naturalist describes how he became the first known person to rescue, raise, and breed wildcats, and then re-release them into the wild.

247. Vermeij, Geerat J. *A natural history of shells*. Princeton, NJ: Princeton, 1993. 207 p. : ill. 069108596X
Describes the geometry of shells, the "economics" of their construction and maintenance, how they work, methods of shell predators, how shells are used for protection, their historical geography, and their evolutionary adaptive qualities.

Books for young adults:

248. Calabro, Marian. *Operation grizzly bear*. NY: Four Winds, 1989. 118 p. : ill. 0027162419
Describes a 13 year study by John and Frank Craighead in Yellowstone Park in which their use of the radio-tracking collar and other innovations added to the scope of human knowledge about the grizzly. GR 5-9.

249. Cerullo, Mary M. *Sharks : challengers of the deep*. NY: Cobblehill, 1993. 57 p. : col. photos. 0525651004
Describes the physical characteristics, behavior, and varieties of sharks and dispels common myths about these unusual fish.

250. Clark, Margaret Goff. *The vanishing manatee*. NY: Cobblehill, 1990. 64 p. : col. photos. 0525650245
Outstanding Trade Book for Children, 1990.
Introduces the playful marine mammal and discusses its future and its relationship with humans. Threats to it include pollution, habitat loss due to development, and motorboat propellers. HR GR 4-8.

251. Goodall, Jane. *My life with the chimpanzees*. Boston: Houghton Mifflin, 1993, c1988. 124 p. : ill. 0395618495
Goodall relates her exciting life among the chimpanzees while briefly describing the natural history of the apes and chimpanzee-human relations in general. GR 5-8. Older readers should see *Through a window*.

252. Grace, Eric S. *Elephants*. SF: Sierra Club, 1993. 62 p. : col. photos. 0871565382
Describes the physical characteristics of the elephant, how it searches for food and behaves in its natural environment, and how it interacts with people.

253. Harvey, Diane Kelsay, and Harvey, Bob. *Fishing with Peter = Pescando con Pedro*. Ill. by Carol Johnson. Wilsonville, OR: Beautiful America, 1993. 1 v. (unpaged) : col. ill. 0898025923 089802630X (pbk)Text in English and Spanish.
Discusses the way of life of brown pelicans and other fishing birds in the unique salt water environment of a mangrove cay along the coast of Belize, a habitat which is now threatened. GR 5-8.

254. ___. *A journey of hope = una jornada de esperanza*. Ill. by Carol Johnson. Wilsonville, OR: Beautiful America, 1991. 1 v (unpaged) : col. ill. 0898026032
Text in English and Spanish. Chronicles the perilous journey of a hatchling sea turtle across the beach to the sea. GR 5-8.

255. Johnson, Sylvia A., and Aamodt, Alice. *Wolf pack : tracking wolves in the wild*. Minneapolis: Lerner, 1985. 96 p. : col. photos. 0822515776
ALA Notable Children's Book, Outstanding Science Trade Book for Children, 1985.
Describes the social interaction of wolves in a pack as they share the work of hunting, maintaining territory, and raising young. Provides a history of the wolf in folklore and explains why the wolf is endangered. HR GR 5-10.

256. Johnston, Ginny, and Cutchins, Judy. *Scaly babies : reptiles growing up*. NY: Morrow, 1988. 40 p. : col. photos. 068809998X (pbk) 0688073069
Parenting Reading-Magic Award Best Book of the Year, Outstanding Science

Trade Book for Children, Texas Bluebonnet Award Master List, 1988.
Describes the physical characteristics and behavior of a variety of baby reptiles as they struggle to survive and grow to adulthood. HR GR 5-10.

257. ___. *Slippery babies : young frogs, toads, and salamanders*. NY: Morrow,
1991. 40 p. : col. photos. 0688096069
Describes the physical characteristics and behavior of a variety of baby
amphibians as they struggle to survive and grow to maturity. HR GR 5-10.

258. Lawrence, R. D. *Wolves*. SF: Sierra Club, 1990. 62 p. : col. photos,
maps. 0316516767
Follows the life cycle of wolves, from their birth, through their early months
of growth and development, to their daily lives as adults. Discusses stereotypes and misperceptions about wolves. HR.

259. Love, John A. *Sea otters*. Golden, CO: Fulcrum, 1992. 148 p. : ill.
1555911234
A detailed introduction to the behavior, natural history, efforts to preserve,
and environmental threats to this popular sea mammal. #

260. Maxwell, Gavin. *Ring of bright water*. NY: Viking Penguin, 1987, c1960.
236 p. 1850895910
An account of the author's early years in the wilds of the western Scottish
highlands and of two otters transported from Iraq. The company of the
otters gave him insights into nature and his place in it. HR. #

261. McClung, Robert M. *Whitetail*. NY: Morrow, 1987. 82 p. : ill. 0688061265
0688061273
A fawn is born and grows up, dealing with the dangers present in nature,
learning to avoid cars, hunters, snowmobiles, and dogs, and occasionally
being aided by a friendly young man. GR 7-9..

262. Miller, Joaquin. *True bear stories*. Intro. by William Everson. NY: Capra,
1987, c1900. 80 p. : ill. 0884962598
The Bard of the Sierras relates factual but exciting stories about bears in
the Cascades, Sierras, and Coast Range. For all ages. #

263. Patent, Dorothy Hinshaw. *Buffalo : the American bison today*. NY: Clarion,
1986. 73 p. : col. photos. 0899193455
Outstanding Science Trade Book for Children, 1986.
Describes the life of American bison today on the National Bison Range in
Montana and in other parks and preserves, with an emphasis on how
humans carefully manage each herd. HR GR 4-7.

264. ___. *Prairie dogs*. NY: Clarion, 1993. 63 p. : ill. 0395565723
Discusses the habits and life cycle of prairie dogs and examines their place in the ecology of their grassland environment. GR 4-8.

265. Pringle, Laurence P. *Batman : exploring the world of bats*. NY: Scribner's, 1991. 42 p. : photos 0684192322
Describes Merlin Tuttle's interest in bats, his study of them in their natural habitat, and his work to protect them through such efforts as the organization he founded, Bat Conservation International. The important ecological role that bats play and their endangered status is emphasized. GR 5-8.

266. ___. *Bearman : exploring the world of black bears*. NY: Scribners, 1989. 42 p. : col. photos. 068419094X
Outstanding Science Trade Book for Children, 1989.
Examines the physical characteristics, habits, and natural environment of the American black bear. Particularly focuses on the work of Lynn Rogers as he studies and works with bears in the field. HR GR 5-8.

267. ___. *Wolfman : exploring the world of wolves*. NY: Scribner, 1983. 71 p. : photos 068417832X
A biography of wildlife biologist David Mech who has spent 25 years studying the wolf. Includes information on behavior, natural history, endangered status, and efforts to preserve the wolf. GR 5-8.

268. Reed, Don C. *Wild lion of the sea : the Steller sea lion that refused to be tamed*. SF: Sierra Club, 1992. 115 p. : ill. 0316736619
Describes the training of a Steller sea lion who comes to California's Marine World as a pup and eventually grows to be a 2,000-pound giant that becomes too difficult for his trainers to handle.

269. Ryden, Hope. *America's bald eagle*. NY: Lyons & Burford, 1992, c1985. 63 p. : ill. 1558211411
AR: *The little deer of the Florida Keys*.
Presents information about the physical characteristics, habitat, parental behavior, predatory activities, and flight of the bald eagle. GR 4-8.

270. ___. *The beaver*. NY: Lyons & Burford, 1992, c1986. 62 p. : col. photos. 155821142X
AR for older readers: *Lily pond*.
Describes the physical characteristics and habits of the beaver and illustrates the beneficial effects of its work on the environment that it inhabits. GR 4-8.

271. ___. *Bobcat*. NY: Lyons & Burford, 1992, c1983. 62 p. : photos. 1558211438

Describes the habits and characteristics of the stubby-tailed spotted wild-cat that is native to North America. Companion to her novel, *Bobcat year.* #

272. ___. *Mustangs : a return to the wild.* Missoula: Mountain, 1984. 111 p. : col. photos. 0878421769
AR for adults: *America's last wild horses.*
Ryden follows wild horses across the Great Basin on foot, learns details of their social structure and behavior, and describes efforts to preserve existing herds. GR 4-8.

273. ___. *Your cat's wild cousins.* NY: Lodestar, 1991. 48 p. : col. photos. 0525673547
AR: *Your dog's wild cousins.*
Common links are explored between domestic cats and various types of wild cats that inhabit the Earth. GR 4-8.

274. Sattler, Helen Roney. *Giraffes, the sentinels of the Savannas.* NY: Lothrop, Lee & Shepard, 1990. 80 p. : ill. 068808284X 0688082858
Discusses the physical characteristics, habits, natural environment, and relationship to human beings of giraffes. Includes a glossary giving the popular and scientific name of each species and details of their size and appearance.

Whales and Dolphins

There is widespread public interest in books about our cetacean cousins. Most contain information on preserving whales and dolphins in addition to their natural history. Some compare human and cetacean consciousness and communications. Varawa's *The Delicate Art of Whale Watching* also contains both philosophical and practical advice on nature study in general.

Books for adults:

275. Cafiero, Gaetano, Jahoda, Maddalena, Manferto, Valeria, and Falcone, Monica. *Giants of the sea : whales, dolphins, and their habits.* Charlottesville, VA: Thomasson-Grant, 1993. 143 p. : col. photos. 1565660463
Depicts the lives, behavior, and interrelationships of a wide variety of whales and dolphins. #

276. Carrighar, Sally. *The twilight seas.* NY: Truman Talley, 1989, c1975. 179 p. 0525484922

Covers the natural history and life cycle of the blue whale and efforts to stop its extermination. #

277. Connor, Richard C., and Petersen, Dawn Micklethwaite. *The lives of whales and dolphins*. NY: Holt, 1994. 288 p. : ill., photos. 0805019529
Depicts the birth of a sperm whale and the lives of several species of whales and dolphins. Covers communications, play, fighting, migration, care of the young, feeding, relations with other species, and relations with humans.

278. Day, David. *The whale wars*. SF: Sierra Club, 1987. 168 p. : photos, maps. 0871567784
Includes background information on whaling, a history of the anti-whaling movement, a survey of current activity, and a consideration of anti-whaling methods and tactics.

279. Doak, Wade. *Encounters with whales & dolphins*. Dobbs Ferry, NY: Sheridan House, 1989, c1988. 250 p. : col. photos. 0911378863
Covers communication between whales, dolphins, and people. Includes the use of dolphin suits, solitary dolphins who seek human companions, dolphin tribes in close rapport with humans, sonic communication with whales, mother whales presenting their babies to be stroked by humans, and orca establishing trust by gently mouthing human limbs. #

280. Gormley, Gerard. *A dolphin summer*. NY: Taplinger, 1985. 196 p. 0800822641
Describes the first eight months of a dolphin's life from the dolphin's perspective: early experiences, sights and sounds, how its mother cares for it, and how it learns to relate to other animals and to its overall environment. This pleasing blend of scientific fact and personal experience is HR AOT. #

281. Hand, Douglas. *Gone whaling : a search for orcas in the Northwest waters*. NY: Simon & Schuster, 1994. 256 p. : ill. 0671768409
Hand journeys to the Vancouver Aquarium, the Center for Marine Research in the San Juan Islands, a hydrophonic laboratory on Vancouver Island, and the Queen Charlotte Islands in search of "killer whales," the people who study them, and native craftspeople. Should appeal to readers interested in whales and/or northwest folk art.

282. May, John. *The Greenpeace book of dolphins*. NY: Sterling, 1990. 159 p. : col. photos. 0806974842
Describes several species of dolphins, their behavior and habitat, and the threats to their existence. A strong case is made for for pollution abatement since toxic threats to dolphins also effect other organisms, including humans. HR.

283. Mowat, Farley. A *whale for the killing*. NY: Bantam, 1986. 240 p. 0553267523
Mowat relates the painful, tragic story of his attempt to save a trapped whale from the sport of those who poured hundreds of rounds of ammunition into the animal. It symbolizes the plight not only of whales, but of all species hunted to the point of extinction.

284. Nollman, Jim. *Dolphin dreamtime*. NY: Bantam, 1990, c1987. 219 p. 0553344277
Original title: *Animal dreaming*.
This account of the art and science of interspecies communication focuses on dolphins, emphasizing the importance of not imposing human assumptions or values on other species. HR.

285. Norris, Kenneth S. *Dolphin days : the life and times of the spinner dolphin*. NY: Norton, 1991. 335 p. : photos. 039302945X
AR: *The porpoise watcher*.
Describes the behavior, habitat, and life cycle of the dolphin; the decimation of dolphin populations by yellowfin tuna fishing; changes in public attitudes; and how to take advantage of the dolphins' "magic envelope" to limit the slaughter. AOT. #

286. Obee, Bruce. *Guardians of the whales : the quest to study whales in the wild*. Photos by Graeme Ellis. Anchorage: Alaska Northwest, 1992. 169 p. : col. photos, map. 0882404288 0882403982 (pbk)
This photoessay depicts the "comeback of the whale" due to preservation efforts and cetacean research being conducted on the Pacific Coast of North America. Focuses primarily on orcas, Pacific grays, humpbacks, and minkes.

287. Payne, Roger. *Among whales*. NY: Macmillan, c1993. 288 p. : photos 0025952455
Covers cetacean biology, migration, courtship, social life, cetacean-human interactions and play, convergent human-whale evolution, and related topics. Demonstrates that whale-watching and related tourism are far more profitable than whaling. HR.

288. Pryor, Karen, and Norris, Kenneth S. *Dolphin societies : discoveries and puzzles*. Berkeley: Univ. of Cal., 1991. 397 p. 0520067177
Contains 18 articles by various authors on all aspects of dolphin life, relations with humans and other animals, political aspects of dolphin conservation, and the technological and methodological aspects of studying dolphins. Places particular emphasis on interactions of dolphins in large groups.

289. Sylvestre, Jean-Pierre. *Dolphins & porpoises : a worldwide guide*. NY: Sterling, 1993. 160 p. : col. photos, maps. 080698791X
An excellent reference source with information on the natural history, habitats, ecology, location of, and threats to 43 different species. HR AOT. #

290. Varawa, Joan McIntyre. *The delicate art of whale watching*. Rev. ed. SF: Sierra Club, 1991, c1982. 144 p. 0871565501
AR *Mind in the waters*.
Provides both philosophical and practical advice on how to watch whales and, by extension, all of nature. Media-created expectations and "the impediments of observation" such as cameras and binoculars can be a hindrance. HR AOT. #

Books for young adults:

291. Mallory, Kenneth, and Conley, Andrea. *Rescue of the stranded whales*. NY: Simon & Schuster, 1989. 63 p. : col. photos 0671671227
Describes the rescue, rehabilitation, and successful release of three young pilot whales that were stranded on a Cape Cod beach during the winter of 1986. GR 5-9.

292. McClung, Robert M. *Thor, last of the sperm whales*. Hamden, CT: Linnet, 1988, c1971. 62 p. : ill. 0208021868
AR: *Gorilla*; *Lily, a giant panda of Sichuan*; and *Rajpur, last of the bengal tigers*.
Traces the life cycle of a sperm whale from his birth in tropical waters to his near-death by an explosive harpoon. GR7-9.

293. McCoy, J. J. *The plight of the whales*. NY: Watts, 1989. 144 p. : photos. 0531107787
Discusses how whales and other marine mammals have become endangered due to whaling, pollution, and other causes. Highlights the activities of conservation groups.

294. Morton, Alexandra. *In the company of whales, from the diary of a whale watcher*. Victoria: Orca, 1993. 63 p. : col. photos. 1551430002
Morton provides clues to whale behavior through journal entries, notes, and photos. Also depicts the life and work of a scientist in the field.

295. Reed, Don C. *The dolphins and me*. SF: Sierra Club, 1989. 135 p. : ill. 0316736597
The author, a former diver, describes his experiences with and observations of the dolphins whose underwater world he shared for more than 13 years at Marine World.

296. Robson, Frank D. *Pictures in the dolphin mind.* Dobbs Ferry, NY: Sheridan House, 1988. 135 p. : photos. 0911378782
AR: *Thinking dolphins, talking whales; Stranded.*
A naturalist's personal account of human association with dolphins from the Dolphin People of Aboriginal dreamtime to the horrors of modern Japan. Provides new ideas on the causes of mass strandings, practical advice on assisting beached cetaceans, and inquiries into interspecies communication. #

297. Whitfield, Philip. *Can the whales be saved? : questions about the natural world and the threats to its survival answered by the Natural History Museum.* NY: Viking Kestrel, 1989. 96 p. : ill. 0670827533
Questions and answers about the balance of nature, and the ways that human behavior affects this delicate system. The title is somewhat misleading since the book covers a broad range of ecological and environmental questions. GR 4-8.

3. Specific Environments

This chapter considers the ecology and relevant environmental issues of specific types of environments. Certain problems created by human over-population and industry are common to all environments: pollution, erosion, loss of biological diversity, toxic and radioactive waste disposal, etc. Some of these books also include information on what can be done to help save specific environments, but books emphasizing activism are in the Environmental Action chapter. Some works of noted naturalists and nature writers about specific types of environments can also be found in this chapter.

Atmosphere and Climate

Problems specific to the atmosphere and world climate include ozone depletion, the greenhouse effect, acid rain, global warming, and world climate change. According to high school librarians I spoke with, air pollution and toxic waste disposal are the most common environmental concerns of teenagers. This is reflected in the fact that nearly as many books are published for young adults as for adults. The best known naturalists in this section are C.L. Rawlins and Vicki McVey.

Books for adults:

298. Bates, Albert K. *Climate in crisis : the greenhouse effect and what we can do.* Summertown, TN: Book Co., 1990. 228 p. : ill. 091399067
Describes what the greenhouse effect is, it's causes, and what can be done about it. Contains a foreword by VP Al Gore.

299. Bridgman, Howard A. *Global air pollution : problems for the 1990s.* London: Bellhaven, 1990. 261 p. : ill. 1852930993
Considers acid rain, threats to stratospheric ozone, increasing oxidants, aerosols, carbon dioxide and trace gas warming, nuclear winter, long range transport of air pollutants, and urban air quality. Designed as upper level college text, but accessible to adults with background in meteorology or climatology.

300. Brown, Michael Harold. *The toxic cloud : the poisoning of America's air.* NY: Perennial, 1988, c1987. 307 p. 0060915064 0060915099
Brown demonstrates the many calamitous health effects of windborne pollution. He discusses the creation of "cancer alleys," and the connivance of industry and government in the proliferation of these hazards.

301. Bryner, Gary C. *Blue skies, green politics : the Clean Air Act of 1990.* Washington: CQ, 1993. 203 p. 087187668X
The author looks at motives of congress and the executive branch in attempting to revise the Clean Air Act.

302. Cagin, Seth, and Dray, Philip. *Between Earth and sky : how CFCs changed our world and endangered the ozone layer.* NY: Pantheon, 1993. 430 p. : ill. 0679420525 0374154910
CFCs were invented in 1928, used widely in refrigeration, aerosols, and styrofoam production, then discovered to be eroding the ozone layer. Despite industrial and government opposition, the Montreal Protocol limiting CFC production was signed in 1987, but CFCs still pose a significant threat.

303. Cline, William R. *The economics of global warming.* Washington: Inst. for Int'l Economics, 1992. 399 p. 088132132X
Covers the scientific basis for the greenhouse effect, global warming in the very long term, economic benefits of limiting global warming, economic models of carbon reduction costs, carbon abatement costs, time discounting, cost-benefit synthesis, and international policy strategy. Somewhat technical, but accessible for adults with a background in economics.

304. Cohen, Richard E. *Washington at work : back rooms and clear air.* NY: Macmillan, 1992. 191 p. : photos. 0023231904
Cohen goes behind the scenes to reveal how the 1990 Clear Air Act were enacted into law. Provides an interesting case study of how Washington works.

305. Dornbusch, Rudiger, and Poterba, James M. eds. *Global warming : economic policy responses.* Cambridge: MIT, 1991. 389 p. 026204126X
Contains conference papers and related comments on global warming, economic approaches to reducing it, tax policies, prospects for international cooperation, Pacific Rim global warming initiatives, and options for slowing Amazon jungle clearing.

306. Elsom, Derek M. *Atmospheric pollution : a global problem.* 2d ed. Cambridge: Blackwell, 1992. 422 p. : ill., maps. 0631185399
Covers the nature, sources and effects of atmospheric pollution and national and international approaches to atmospheric pollution control.

This college level text is HR.

307. Gribbin, John R. *The hole in the sky : man's threat to the ozone layer.* Rev. ed. NY: Bantam, 1993. 216 p. 0553275372
Describes what the ozone layer is and its importance to all life on Earth, the extent and speed of ozone depletion, past and present experiments, what chemicals are destroying it and where they come from, and what can be done to stop the damage.

308. Kahan, Archie M. *Acid rain : reign of controversy.* Golden, CO: Fulcrum, 1986. 238 p. 1555910033
Explores the basics of acid rain including its causes, suspected and verified effects, possible industrial alternatives to reduce it, and the implications of those changes.

309. Kemp, David D. *Global environmental issues : a climatological approach.* 2d ed. NY: Routledge, 1994. 304 p. : ill. 0415103096
An interdisciplinary approach to issues such as global warming, ozone depletion, drought, and acid rain, emphasizing both societal and environmental components. Jargon free. HR.

310. Leggett, Jeremy. ed. *Global warming : the Greenpeace report.* NY: Oxford, 1990. 554 p. : ill. 0192861190
Outlines urgent government policies and measures to halt global warming. Contains 19 essays by Anne Ehrlich, Amory Lovins, Norman Myers, Stephen Schneider, and others.

311. Lyman, Francesca. *The greenhouse trap : what we're doing to the atmosphere and how we can slow global warming.* Boston: Beacon, c1990. 190 p. 0807085022 0807085030 (pbk)
Lyman analyzes how the greenhouse effect is being created and its negative effects. She concludes that in order to reduce or contain global warming it is essential that industrial nations substantially reduce their use of fossil fuels.

312. Maunder, W. J. *Dictionary of global climate change.* NY: Chapman & Hall, 1992. 240 p. 041203901X
This dictionary sponsored by the Stockholm Environment Institute is designed to be useful for both general readers and specialists.

313. McKibben, Bill. *The end of nature.* NY: Anchor, 1990. 226 p. 0385416040
Contends that global warming, rainforest destruction, and pollution are merely symptomatic of our disregard for nature. We must renew widespread respect for nature to avoid total catastrophe. HR.

314. Mello, Robert A. *Last stand of the red spruce.* Washington: Island, 1990. 217 p. : ill. 0933280378
Examines the causes and consequences of acid rain on forests, focusing on the red spruce population of the Appalachian Mountains. Includes an examination of legal process and suggestions for possible action.

315. Meszaros, Erno. *Global and regional changes in atmospheric composition.* Boca Raton: Lewis, 1993. 175 p. : maps 0873716620
Summarizes changes in atmospheric composition caused by human activities. Tells consequences of such modifications.

316. Miller, E. Willard, and Miller, Ruby M. *Environmental hazards : air pollution : a reference handbook.* Santa Barbara: ABC-Clio, 1989. 250 p. 0874365287
Library Journal Best Reference Book, 1989.
Covers a wide spectrum of topics related to air pollution such as acid rain, the greenhouse effect, and ozone layer depletion. HR AOT. #

317. Minger, Terrell J. ed. *Greenhouse glasnost : the crisis of global warming.* NY: Ecco, 1990. 292 p. : photos. 0880012587
Contains 24 papers by American and Soviet participants in the Sundance Symposium on Global Climate Change. Authors include Carl Sagan, Stephen Schneider, James Lovelock, Paul and Anne Ehrlich, and Robert Ornstein.

318. Mitchell, George J. *World on fire : saving an endangered Earth.* NY: Scribner, 1991. 247 p. 0684192314
The former Senator's descriptions of the greenhouse effect, rain forest destruction, ozone layer loss, and acid rain contain some political hyperbole. However, his proposals for a world atmosphere fund financed by fossil fuel consumers, and forgiveness of debt to ecologically responsible Third World countries are of considerable interest.

319. Newton, David E. *Global warming : a reference handbook.* Santa Barbara: ABC-CLIO, 1993. 183 p. : ill. 0874367115
Explores this controversial topic, presenting scientific evidence that both supports and undermines the claim that human activity is the cause of atmospheric heating. Alternative causes are considered.

320. Park, Chris C. *Acid rain : rhetoric and reality.* NY: Routledge, 1990, c1987. 272 p. : ill. 0685263134
Examines the implications of recent scientific studies and sets the political debate in the main polluting countries, Britain and America, into an international context. Evidence is drawn from around the world, with particular emphasis on damage in Scandinavia and West Germany. Somewhat techni-

cal, but accessible to adults with some science background.

321. Rawlins, C. L. *Sky's witness : a year in the Wind River Range.* NY: Holt, 1993. 326 p. : ill. 0805015973
Rawlins actually becomes the "self-appointed inspector of snowstorms and rainstorms" that Thoreau joked about being, but he is serious in his quest to document the effects of pollution and acid precipitation in the Rockies. HR.

322. Schmandt, Jurgen; Clarkson, Judith; and Roderick, Hilliard. eds. *Acid rain and friendly neighbors : the policy dispute between Canada and the United States.* Rev. ed. Durham, NC: Duke Univ., 1988. 0822308703
A variety of papers on the scientific, social, economic, and political factors which have prevented a successful international agreement on acid rain.

323. Schneider, Stephen Henry. *Global warming : are we entering the green-house century?* NY: Random House, 1990. 317 p. 0679730516
AR: *The coevolution of climate and life, The Genesis strategy.*
A climatologist examines the consequences of failing to halt the green-house effect: increasing temperatures, rising sea levels, agricultural losses, smog, forest fires, and dramatic weather changes. Provides excellent advice on citizen activism and lobbying. HR.Books for young adults:

324. Baines, John D. *Acid rain.* Austin: Steck-Vaughn, 1990. 48 p. : col. pho-tos. 0811423859
AR: *Conserving the atmosphere.*
Discusses how pollution creates acid rain, the vast contamination of forests and lakes throughout the world by it, and efforts to control it. Con-cludes with an effective plea for energy conservation in order to decrease the pollution that creates acid rain. GR 4-9.

325. Dolan, Edward F. *Drought : the past, present, and future enemy.* NY: Watts, 1990. 144 p. : photos. 0531109003
Discusses the causes and effects of droughts and examines the possibility of future droughts due to such phenomena as the greenhouse effect. Also discusses weather modification as a solution to this problem.

326. ___. *Our poisoned sky.* NY: Cobblehill, 1991. 121 p. : photos. 0525650563
Explains how pollutants are ruining our atmosphere and what is being done about them. Suggest ten things readers can do to help solve the problem.

327. Facklam, Margery, and Facklam, Howard. *Changes in the wind : Earth's*

shifting climate. San Diego: Harcourt Brace Jovanovich, 1986. 128 p. : ill. 0152161155
Examines the factors causing changes in the Earth's climate, including ocean currents, the destruction of the rain forests, and the greenhouse effect. Discusses predictions for the future.

328. Gay, Kathlyn. *Air pollution*. NY: Watts, 1991. 144 p. : ill. 0531130029
AR: *The greenhouse effect.*
Examines the growing problem of air pollution, its effect on the ozone and stratosphere and what we can do to improve air quality.

329. ___. *Ozone*. NY: Watts, 1989. 128 p. : ill. 0531107779
Discusses the dual problem of too much ozone in the troposphere, creating smog, and the depletion of the ozone layer in the stratosphere, which shields harmful ultraviolet rays.

330. Gutnik, Martin J. *The challenge of clean air*. Hillside, NJ: Enslow, 1990. 64 p. : ill. 0894902725
Explores the scientific, technological, and social background of the many environmental problems plaguing the atmosphere. GR 7-10.

331. Harris, Jack C. *The greenhouse effect*. NY: Crestwood, 1990. 48 p. : col. photos. 0896865436
Discusses the causes of the greenhouse effect, its effects on the planet, and how it can be stopped. GR 4-7.

332. Johnson, Rebecca L. *The greenhouse effect : life on a warmer planet*. Minneapolis: Lerner, 1990. 112 p. : col. photos. 0822515911
Discusses what the greenhouse effect is, research into its causes, and its possible impact on our planet. Possible remedies include energy conservation, reforestation, and alternate energy. GR 7-9.

333. Knapp, Brian J. *Drought*. Austin: Steck-Vaughn, 1990, c1989. 48 p. : ill. 081142376X
Examines the causes and effects of droughts and ways to prevent future disasters involving drought, particularly in the developing countries of the world. GR 5-9.

334. Koral, April. *Our global greenhouse*. NY: Watts, 1991, c1989. 64 p. : col. ill. 053115601X
Discusses the origins, possible results, and prevention of the environmental problem known as the greenhouse effect. As with Peckham's *Global warming*, emphasis is on international aspects of the problem. GR 4-8.

335. McVey, Vicki. *The Sierra Club book of weatherwisdom*. Boston: Little Brown, 1991. 104 p. : ill., maps. 0316563412
Discusses climates and seasons, wind and rain, warm and cold fronts, atmospheric pressure, and weather prediction. Features activities, games, and experiments. HR GR 4-8.

336. Peckham, Alexander. *Global warming*. NY: Gloucester, 1991. 32 p. : col. ill. 0531172740
Discusses the issue of the greenhouse effect and possible solutions to the global warming trend, stressing the international aspects of the problem and the need for international cooperation toward solutions. GR 5-8.

337. Pringle, Laurence P. *Global warming*. NY: Arcade, 1990. 46 p. : col. ill. 1559700122
Outstanding Science Trade Book for Children, 1990.
Examines the greenhouse effect, focusing on its complex causes and potential impact on global weather, ecology, and economy. This is the best young adult book on the subject and is HR GR 4-8.

338. ___. *Rain of troubles : the science and politics of acid rain*. NY: Macmillan, 1988. 121 p. : photos. 0027753700
Outstanding Science Trade Book for Children, 1988.
Discusses the discovery, formation, transportation, and effects on plant and animal life of acid rain and how economic and political forces have delayed action needed to reduce this slow poison from the sky. HR GR 5-10.

339. Sandak, Cass R. *A reference guide to clean air*. Hillside, NJ: Enslow, 1990. 128 p. 089490261X
Discusses the many environmental problems, in alphabetical order, that are presently plaguing the air and atmosphere. GR 7-10.

340. Stubbs, Harriet S.; Klinkhammer, Mary Lou; Knittig, Marsha; and Eclov, Homer. *The Acid rain reader*. Raleigh, NC: Acid Rain Foundation, 1989. 18 p. : ill. 0935577122
A brief consideration of the causes and effects of acid rain for GR 3-8.

341. Tesar, Jenny E. *Global warming*. NY: Facts on File, 1991. 111 p. 0816024901
Discusses the gradual warming of the planet, its possible causes and effects, and some solutions. Environmental action tips included. HR.

342. Wheeler, Jill C. *For the birds : a book about air*. Edina, MN: Abdo & Daughters, 1993. 31 p. : ill. 1562391968
Discusses air pollution and suggests things we can do to help clean the air. GR 3-7.

Arctic and Antarctic Regions

The polar regions are beset by radioactive fallout, spills from oil drilling and shipping, the effects of mining, toxic waste, poaching and overhunting, and the rape-ruin-and-run attitude of Alaska governor Wally Hickel, industry, and some settlers. Naturalists associated with the Arctic or Alaska include Jean Aspen, Sally Carrighar, Barry Lopez, Robert Marshall, John McPhee, John Muir, Adolph Murie, and Susan Zwinger. Other northern naturalists can be found in the anthologies *The Great Land* (350), *A Republic of Rivers* (360), and *In the Dreamlight* (1725)

Books for adults:

343. Aspen, Jean. *Arctic daughter : a wilderness journey.* Birmingham, AL: Menasha Ridge, 1993, c1988. 205 p. : ill., maps. 0897321219
A second generation explorer chronicles building a cabin and trying to live off the land in the rugged Brooks Range. Includes observations on land and nature in Alaska and memories of growing up in a household of women adventurers.

344. Brown, Dave, and Crane, Paula. *Who killed Alaska?* Far Hills, NJ: New Horizon, 1991. 336 p. 0882820699
Analyzes the massive degradation of Alaska and Brown's struggles when he blew the whistle on the fraud, waste, and collusion between big oil and the state government. HR.

345. Campbell, David G. *The crystal desert : summers in Antarctica.* Boston: Houghton Mifflin, 1992. 297 p. 039558969X
The "banana belt" of the Antarctic Peninsula is home to a wide variety of land and sea life. Covers the evolution and natural history of this area and considers the effects of human activities ranging from exploration and conservation to oil, whaling, and sealing.

346. Carrighar, Sally. *Icebound summer.* Lincoln: Univ. of Nebraska, 1991, c1953. 262 p. 0803263473
Chronicles nine years of observing the courtship and mating patterns of a variety of arctic animals.

347. ___. *Wild voice of the North.* Lincoln: Univ. of Nebraska, 1991, c1959. 191 p. : photos. 0803263473
AR: *Home to the wilderness, One day on Beetle Rock, One day at Teton Marsh. Wild heritage.*
The story of an Alaskan sled dog's life in the Arctic. #

348. Fradkin, Philip L. *Wanderings of an environmental journalist in Alaska and the American West*. Albuquerque: Univ. of New Mexico, 1993. 273 p. 0826314163
AR *California, the golden coast; Sagebrush country*.
Demonstrates the impact humans have on the environment and explores how natural landscapes determine human history. This collection of articles by a Pulitzer Prize winning journalist is HR.

349. Hatfield, Fred. *North of the sun : a memoir of the Alaskan wilderness*. NY: Carol, 1992, c1990. 184 p. 0806513719
Covers his experiences studying, living in, and attempting to preserve the Alaskan environment. #

350. Hedin, Robert, and Holthaus, Gary H. eds. *The Great land : reflections on Alaska*. Tucson: Univ. of Arizona, 1994. 300 p. 0816514178 0816514372 (pbk)
An anthology of nature and historical writings by Muir, Hoagland, Dos Passos, and others. Features dazzling accounts of encounters with whales, sea otters, grizzly bears, spawning salmon and other natural wonders. HR.

351. Lopez, Barry Holstun. *Arctic dreams : imagination and desire in a northern landscape*. NY: Bantam, 1989. 496 p. 0553346694
American Book Award. Lopez discovers both incredible natural wonders and the devastation of the environment and native cultures. He contends that civilized values have estranged modern people from nature and trivialized the very cosmology which would allow us to live harmoniously. HR AOT. #

352. Marshall, Robert. *Alaska wilderness; exploring the Central Brooks Range*. 2d ed., edited by George Marshall. Berkeley: Univ. of Cal., 1970, c1956. 173 p. : photos. 0520017102
This account of explorations in the Brooks Range in 1939 takes the reader deep into an unspoiled wilderness and into Marshall's soul as he expresses his joy and wonder. HR.

353. May, John. *The Greenpeace book of Antarctica : a new view of the seventh continent*. NY: Doubleday, 1989. 192 p. : col. ill. 0385262809
Prose and photos depicting the natural wonders and surprising aridity of Antarctica. May observes many environmental problems, offering alternatives and a call to action. HR.

354. McPhee, John. *Alaska : images of the country*. NY: Promontory, 1992. 145 p. : col. photos. 0883940604
Text from McPhee's *Coming into the country* with lavish color photos by Galen Rowell. HR AOT. #

355. ___. *Coming into the country.* NY: Noonday, 1991, c1977. 438 p. 0374522871
McPhee takes us across the Alaskan wilderness by foot, canoe, and helicopter, in an exploration of wilderness and the human spirit. HR AOT. #

356. Miller, Debbie S. *Midnight wilderness : journeys in Alaska's Arctic National Wildlife Refuge.* SF: Sierra Club, 1990. 238 p. : photos, maps. 0871567156
Miller describes her travels in America's largest refuge, examining various wildlife species, their behavior and interactions, the history of the area, its use by Inuit and Indian people, and the threat posed to it by the Prudhoe Bay oil fields. #

357. Muir, John. *The cruise of the Corwin.* SF: Sierra Club, 1993, c1917. 226 p. 0871565234
A posthumous collection of newspaper articles and journal entries from Muir's expedition to Alaska in 1881.

358. ___. *Travels in Alaska.* SF: Sierra Club, 1988, c1915. 248 p. 0871567830
Muir worked sporadically between 1896 and his death in 1914 on this account of his early travels. His vivid accounts of nature and the effect of the massive landscape on his consciousness are powerful. Mature, reflective tone. HR.

359. Murie, Adolph. *A naturalist in Alaska.* Ill. by Olaus Murie. Tucson: Univ. of Arizona, 1990, c1961. 302 p. : ill. 0816511683
John Burroughs Medal. Describes the natural history and ecological balance of the bear, wolf, lynx, wolverine, Dall sheep, and arctic fox. Murie employs a personalized approach combining science, philosophy, esthetics, and sociology. HR.

360. Murray, John A. ed. *A Republic of rivers : three centuries of nature writing from Alaska and the Yukon.* NY: Oxford, 1990. 325 p. : photos, maps. 0195061020
Contains 48 selections written from 1741 to 1990. Authors include James Cook, George Vancouver, Otto Von Kotzebue, John Muir, John Burroughs, Jack London, Bob Marshall, Margaret and Adolph Murie, Kenneth Brower, Annie Dillard, Edward Abbey, Barry Lopez, and David Rains Wallace. HR.

361. Rice, Larry. *Gathering paradise : Alaska wilderness journeys.* Golden, CO: Fulcrum, 1990. 303 p. : photos, maps. 1555910572
Describes adventures and insights gained from crossing ten of Alaska's most remote wilderness areas via foot, raft, and kayak. Conveys a strong sense of excitement and respect for the wilderness. #

362. Upton, Joe. *Journeys through the Inside Passage : seafaring adventures along the coast of British Columbia and Alaska.* Anchorage: Alaska Northwest, 1992. 189 p. 0882403664
Combines 15 years of experience traveling this still largely unspoiled area with stories of other travelers in a brilliant depiction of history, natural history, and travel.

363. Walker, Tom. *Denali journal : a thoughtful look at wildlife in Alaska's majestic national park.* Harrisburg, PA: Stackpole, 1992. 190 p. : photos. 0811724379
Combines journal entries, government reports, field diaries, and interviews to provide a year-long view of the area. Unlike most books about Denali, it does not focus on mountain climbing, but on wildlife and natural history.

364. ___. *River of bears.* Stillwater, MN: Voyageur, 1993. 155 p. : photos. 0896581780
Describes a 30 year experiment in the citing, care, and preservation of bears in the McNeil River State Game Sanctuary. Analyzes bear behavior, concluding that humans have little to fear from bears if they do not intentionally enrage them.

365. ___. *Shadows on the tundra : Alaskan tales of predator, prey and man.* Harrisburg, PA: Stackpole, 1990. 177 p. : photos. 0811717240
The author draws upon his twenty-five years experience in Alaska to show how important wildlife and the outdoors are to individuals.

366. Weeden, Robert B. *Messages from Earth : nature and the human prospect in Alaska.* Fairbanks: Univ. of Alaska, 1992. 189 p. : maps. 0912006560
Describes the exploitative thinking and behavior that has rendered environmental devastation in Alaska. Offers an alternative ethical and behavioral path for living gently, comfortably, and substainably in the north.

367. Wright, Charles S. *Silas : the Antarctic diaries and memoir of Charles S. Wright.* Ed. by Colin Bull. Columbus: OSU, 1993. 418 p. : ill., maps. 0814205488
This collection of journals, letters, and memoirs details Scott's last voyage to Antarctica and the discovery of more animal and plant life than expected in this frigid area.

368. Zwinger, Susan. *Stalking the ice dragon : an Alaskan journey.* Tucson: Univ. of Arizona, 1991. 219 p. 0816512027
"We witness the author coming into her own as a person working toward the recovery of a world in which all things, animate and inanimate, are linked in natural unity."-Intro. AOT. #

Books for young adults:

369. Foster, Janet. *Journey to the top of the world*. NY: Prentice Hall, 1987. 96 p. : col. photos. 0135114454
Text and photographs present the plant and animal life encountered on a journey across the Canadian Arctic and introduce some of the permanent inhabitants and scientists who work there. GR 3-7.

370. Hiscock, Bruce. *Tundra, the Arctic land*. NY: Atheneum, 1986. 135 p. : ill. 0689312199
Outstanding Science Trade Book for Children, 1986.
Describes the geography of the tundra and the animals, plants, birds, and people who have adapted to life on these arctic plains. HR GR 4-8.

371. Lambert, David. *Polar regions*. Morristown, NJ: Silver Burdett, 1988. 48 p. : col. ill. 0382095022
Describes the origins, geographic features, seasons, plant and animal life, exploration, and natural resources of the Arctic and Antarctic. Also covers environmental problems such as overhunting, pollution, and radioactive fallout. GR 5-8.

372. Stone, Lynn M. *Arctic tundra*. Vero Beach, FL: Rourke, 1989. 48 p. : col. ill. 0865924368
Examines the arctic tundra as an ecological niche and describes the plant and animal life supported there. Some coverage of conservation issues. GR 4-8.

373. Williams, Terry Tempest, and Major, Ted. *The secret language of snow*. SF: Sierra Club, 1984. 129 p. : ill. 039486574X 0394965744 (lib. bdg.)
Examines over a dozen different types of snow and snowy conditions through the vocabulary of Inuit people. Discusses how snow affects the plants, animals, and people of the Arctic. The inherent ecological and multicultural awareness and superior writing and illustration make this book HR GR 3-8.

Deserts

Deserts figure in two worldwide environmental crises. The first is the loss of natural desert landscapes which support diverse and complex ecosystems to urban and agricultural growth, and the concurrent diversion of desert water to development. The second is the rapid desertification of prairies, grasslands, and other landscapes, particularly in Africa and the Third World, leading to drought, famine, and further urbanization. Spare but nurturing types of desert are being replaced by less fruitful ones. Naturalists associated with the desert include Edward Abbey, Bruce Berger, Charles Bowden, Joseph Wood Krutch, John Charles Van Dyke, and Ann Zwinger. Recommended works in other chapters pertaining to deserts include those by Nabhan (227, 660-662), Abbey (577-581), and Reisner (836)

Books for adults:

374. Abbey, Edward. *Beyond the wall : essays from the outside*. NY: Holt, Rinehart & Winston, 1984. 203 p. 0030693012 (pbk) 0030692997
Contains essays written for ten photo books including *Appalachian wilderness*, *The Hidden Canyon*, and *Slickrock*. The locales may vary, but Abbey is always a provocative defender of wilderness. HR.

375. ___. *Desert solitaire : a season in the wilderness*. NY: Simon & Schuster, 1990, c1968. 269 p. 0671695886
AR: *Cactus country, Slickrock*.
Combines a detailed description of Arches National Park and the insights gained on solo journeys there, with a ringing plea to protect the wilderness. Probably Abbey's most important and influential work of nonfiction. HR AMT. #

376. Alcock, John. *The masked bobwhite rides again*. Tucson: Univ. of Arizona, 1993. 186 p. : ill. 0816513872 0816514054 (pbk)
Demonstrates the complex ecological web of the Sonoran Desert and the effects of human habitation and use. Offers hope for desert preservation, despite the economic forces arrayed against it, in the story of the reintroduction of the endangered masked bobwhite. HR.

377. ___. *Sonoran Desert summer*. Tucson: Univ. of Arizona, 1994, c1990. 187 p. : ill. 0816514380
An introduction to the hardy animals and plants of this region and how they behave and interact to survive extreme conditions of heat and aridity.

378. ___. *Sonoran Desert spring*. Tucson: Univ. of Arizona, 1994, c1985. 180 p. : ill. 0816513996
Vivid depiction of the regeneration of desert life.

379. Allan, Tony, and Warren, Andrew. eds. *Deserts : the encroaching wilderness : a world conservation atlas*. NY: Oxford, 1993. 176 p. : maps, photos. 0195209419
Library Journal Best Sci-Tech Book Award, 1994.
Examines the full range of human and ecological factors affecting world desert climates. Stresses their variety and biodiversity. HR.

380. Arritt, Susan. *The living Earth book of deserts*. Pleasantville, NY: Reader's Digest, 1993. 224 p. : col. photos, maps. 0895775190
Provides background information on the natural history and ecology of deserts, and considers four diverse types of deserts: (hot) subtropical, polar and high-altitude, cold-winter, and cool coastal. An appendix lists the topography, geology, climate and life-forms of 20 major deserts.

381. Berger, Bruce. *The telling distance : conversations with the American desert*. NY: Anchor, 1991. 243 p. : ill. 0385413947
Western States Book Award, 1990.
Fifty highly personal essays, narratives, and meditations considering the people, myths, legends, ecology, and biogeography of the Southwestern deserts. HR.

382. Bowden, Charles. *Blue desert*. Tucson: Univ. of Arizona, 1986. 175 p. 0816510059
The most interesting and paradoxical creature in the desert walks on two legs, and in so doing, inevitably changes it.

383. ___. *Desierto : memories of the future*. NY: Norton, 1993, c1991. 225 p. 0393310094
A highly personalized account of environmental and cultural changes in the Southwest resulting from unwise land development, the drug trade, and corruption. Bowden confesses his conflict over loving both nature and the Anglo and Hispanic cultures that are destroying it. HR.

384. ___. *Mezcal*. Tucson: Univ. of Arizona, 1988. 152 p. 0916510733
Bowden searches for home through the desert and the seamy underbelly of American culture: drugs, sex, alcohol, and politics. HR for adventurous adults.

385. Bowden, Charles, and Dykinga, Jack W. *The Sonoran Desert*. NY: Abrams, 1992. 167 p. : col. photos. 0810938243
Primarily lavish color photos by Dykinga of the desert landscape, plants, and animals in all seasons, with text and ecological observations by Bowden. HR for adults, and probably the best Bowden book for teens. #

386. Flegg, Jim. *Deserts : miracle of life*. NY: Facts on File, 1993. 160 p. : photos. 0816029024
Covers desert ecology, biology, plant and animal adaptation to harsh conditions, and human settlement. International in scope.

387. Hyde, Philip. *Drylands : the deserts of North America*. NY: Park Lane, 1990, c1987. 173 p. : col. photos. 0517032899

Depicts various types of deserts, changes to them over 25 years, and the complexity and diversity of desert flora and fauna. Combines an artistic and factual appreciation of deserts with a plea for their preservation.

388. Kappel-Smith, Diana. *Desert time : a journey through the American Southwest*. Boston: Little, Brown, 1992. 262 p. : maps. 0316482986
The book takes you on a journey through the American Southwest—along the way the environment, wildlife, and habitat of this region are explored.

389. Krutch, Joseph Wood. *The desert year*. Tucson: Univ. of Arizona, 1985, c1952. 270 p. : ill. 0816509239
This classic work uses an almost lyrical style to describe the animals, plants, natural cycles, climate, and seasons of the SW American desert. Despite being over 40 years old, it still seems fresh and is HR AOT. #

390. ___. *The forgotten peninsula : a naturalist in Baja California*. Tucson: Univ. of Arizona, 1986, c1961. 277 p. : photos, map. 0816509875
Describes this region's unique history and natural history, desert ecology, whale breeding and migration, and threats posed by overhunting or commercial development. Originally written for travelers, today it is primarily of historical interest.

391. ___. *The voice of the desert : a natural interpretation*. NY: Morrow, 1980, c1955. 223 p. 0688059341
AR: *Grand Canyon : today and all its yesterdays*, *The best nature writing of Joseph Wood Krutch*, *The great chain of life*.
Krutch describes his transformation from urban pessimist to crusading conservationist as he explores the biology, ecology, and wonders of the desert. He rejects mechanistic consciousness for a more intuitive, natural one. HR AOT. #

392. Sears, Paul Bigelow. *Deserts on the march*. Washington: Island, 1988, c1935. 241 p. 0933280467AR: *The living landscape*.
Discusses the destruction of prairies, the despoiling of natural waterways, massive forest clearing, wildlife slaughter, and their inevitable consequence: desertification. HR.

393. Sheldon, Charles. *The wilderness of the Southwest : Charles Sheldon's quest for desert bighorn sheep and adventures with the Havasupai and Seri Indians*. Ed. by Neil B. Carmony and David E. Brown. Salt Lake City: Univ. of Utah, 1993. 263 p. : ill., maps. 0874804175
Reports on five expeditions, 1912-1922, mostly in the Sonoran desert, but also in the Grand Canyon. While focusing on Bighorn sheep, intersperses comments on natural history, the importance of conservation, and the native populations.

394. Van Dyke, John Charles. *The autobiography of John C. Van Dyke : a personal narrative of American life*, 1861-1931. Ed. by Peter Wild. Salt Lake City: Univ. of Utah, 1993. 274 p. 0874803926
Covers the many facets of his life as art historian, lawyer, librarian, champion of the desert, and associate of Twain, Muir, Carnegie, and other luminaries. Wild's commentaries are particularly interesting. HR.

395. ___. *The desert.* Salt Lake City: Peregrine Smith, 1991, c1901. 271 p. 087905395X
This account of his three year solo journey through the Colorado and Sonoran deserts for three years includes observations on natural history, desert preservation, philosophy, and the potential costs of industrial progress to the environment. This classic is HR.

396. ___. *The Grand Canyon of the Colorado : recurrent studies in impressions and appearances.* Salt Lake City: Univ. of Utah, 1992, c1920. 218 p. 0874803888
Confronted by the size of the Grand Canyon, Van Dyke rejects its use for hydro power or mining. The true power of the Canyon is to put humanity in its proper place of being in awe of a greater power.

397. Wild, Peter. ed. *The desert reader : descriptions of America's arid regions.* Salt Lake City: Univ. of Utah, 1991. 236 p. : ill. 0874803667
AR: *Pioneer conservationists of Eastern America, Pioneer conservationists of Western America.*
Contains short selections from Papago and Pima legends and the work of Cabeza de Vaca, Horace Greeley, John Wesley Powell, Clarence Dutton, John C. Van Dyke, Mary Austin, D.H. Lawrence, H. Frank Dobie, Aldo Leopold, Joseph Wood Krutch, Wallace Stegner, Edward Abbey, Ann Zwinger, and others. HR AOT. #

398. Zwinger, Ann. A *desert country near the sea : a natural history of the Cape region of Baja California.* NY: Harper, 1983. 399 p. : ill. 0060152087
This account of her adventures in Baja offers acute observations on its mountains, reefs, geology, botany, marine biology, and villages. #

399. ___. *The mysterious lands : an award-winning naturalist explores the four great deserts of the Southwest.* NY: Dutton, 1989. 388 p. 0452265134
An exploration of the Chihuahuan, Sonoran, Mojave, and Great Basin deserts. Explains the uniqueness of each desert and how plants and animals survive extreme conditions of heat, cold, and aridity.

400. ___. *Wind in the rock : the canyonlands of southeastern Utah.* Tucson: Univ. of Arizona, 1986, c1978. 258 p. 0816509859
Zwinger explores the five canyons of the Grand Gulch Plateau, recounts the

area's natural and cultural history, evokes the dramatic landscape, and captures the moments of terror and euphoria.

Books for young adults:

401. George, Jean Craighead. *One day in the desert.* NY: HarperCollins, 1983. 48 p. : ill. 0690043414
Explains how the animals and people of the Sonoran Desert, including a mountain lion, a roadrunner, a coyote, a tortoise, and members of the Papago Indian tribe, adapt to and survive the desert's merciless heat. GR 4-8.

402. Lerner, Carol. A *desert year.* NY: Morrow, 1991. 48 p. : col. ill. 0688093833
AR: *Seasons of the tallgrass prairie.*
A seasonal study of the animals and plants found in the southwestern deserts. Emphasizes ecological communities, interdependence, and adaptation. GR 4-8.

403. Lye, Keith. *Deserts.* Morristown, NJ: Silver Burdett, 1987. 48 p. : col. ill. 0382095014
Describes different kinds of deserts and the plants, animals, and people that make their home in them. Less information on ecology than in Lerner or George, but more on human use of the desert and its impact. GR 5-9.

404. Moore, Randy, and Vodopich, Darrell S. T*he living desert.* Hillside, NJ: Enslow, 1991. 64 p. : ill. 0894901826
Describes the different types of deserts found throughout the world, discussing climate, plant and animal life, and the influence of humans. GR 7-10.

405. Williams, Terry Tempest. *Coyote's canyon.* Salt Lake City: Peregrine Smith, 1989. 96 p. : col. photos. 0879052457 (pbk) 0879051280
Evokes the beauty and mystery of southern Utah's desert canyons, describes their natural history, and provides some information on National Parks of the region. #

Forests

Temperate and boreal forests are threatened by deforestation, soil erosion and subsequent riparian damage, huge economic booms and busts in logging dependent areas, and the rapid destruction of plant and animal diversity as complex forest systems are replaced with monocrop tree farms. Books by Bowden, McPhee, and Olson deal with these issues. Peattie's classic books on American forests can also be found here.

Books for adults:

406. Agee, James K. *Fire ecology of Pacific Northwest forests*. Washington: Island, 1993. 493 p. 1559632291
A source book for natural area managers on restoring or maintaining fire in the natural areas of the Pacific Northwest. Provides a natural baseline for understanding the effects of natural or altered fire regimes.

407. Anderson, Chris. *Edge effects : notes from an Oregon forest*. Iowa City: Univ. of Iowa, 1993. 185 p. : ill. 0877454191 0877454388 (pbk)
Anderson moved to a 12,000 acre research forest hoping to find solitude and silence, but clearcutting plans forced him to become actively involved in protest and public education.

408. Banuri, Tariq, and Marglin, Frederique Apffel. eds. *Who will save the forests? : knowledge, power, and environmental destruction*. Atlantic Highlands, NJ: Zed, 1993. 195 p. 1856491595 1856491609
Explains how the relationships between people and the environment differ depending on the knowledge systems of respective societies. Scientific knowledge has been used to marginalize the practices of age-old communities which have sustainably managed the forests for centuries. India, Finland, Maine, and the Himalayas are considered.

409. Booth, Douglas E. *Valuing nature : the decline and preservation of old-growth forests*. Lanham, MD: Rowman & Littlefield, 1994. 287 p. 0847678598 0847678601 (pbk)
Investigates historical attitudes toward old-growth forests over time, the effect of attitudes on forest use, whether the cost-benefit evaluation of old growth is legitimate, and ethical standards for the question of whether old growth should be exploited or preserved.

410. Bowden, Charles. *The secret forest*. Albuquerque: Univ. of New Mexico, 1993. 141 p. : photos. 0826314031
Remnants of a unique tropical forest extend from New Mexico to the Andes in what Bowden calls a "wonderful collision of desert, and tropics, and sierra." He describes its natural history and ecology, details how the indigenous dwellers use native plants for food and medicine, and provides philosophical and preservationist perspectives.

411. Connor, Sheila. *New England natives*. Cambridge: Harvard, 1994. 274 p.

: ill. 0674613503
A geological and botanical history of New England from the last Ice Age to
the present and an economic history of every industry which has used
wood. HR.

412. Dennis, John V. *The great cypress swamps.* Baton Rouge: LSU, 1988. 222
p. : col. photos, maps. 0807115010
Describes ten southern swamps in detail. Using cypress trees as a backdrop, Dennis outlines the strange geological formations of the cypress
swamp and uncovers its distinctive plant and animal life. He also considers
the swamp's vanishing species and hidden dangers.

413. Devall, Bill. ed. *Clearcut : the tragedy of industrial forestry.* SF: Sierra Club,
1994. 291 p. : col. photos. 0871564947
This collection of revealing photos and incisive essays by Dregson, Foreman, Lansky, Maser and others depicts many forms of extreme damage
caused by clearcutting: massive erosion, habitat and species loss, economic boom and bust, etc. HR.

414. Dietrich, William. *The final forest : the battle for the last great trees of the
Pacific Northwest.* NY. Simon & Schuster, 1992. 303 p. : maps. 0671729675
A Pulitzer Prize winning journalist's story of the struggle between the forest
products industry and environmental groups. Depicts people examining
new solutions: bioregionalists, ecologists trying to improve tree growth
while maintaining a healthy forest ecology, and a minority of timber companies practicing sustainable forestry.

415. Engbeck, Joseph H. *The enduring giants: the giant sequoias, their place in
evolution and in the Sierra Nevada forest community; History of the Calaveras big
trees; The story of Calaveras Big Trees State Park.* 3d ed Sacramento: Cal. Dept.
of Parks & Rec., 1988, c1973. 120 p. : ill. 094192503X
Covers the history and natural history of the Giant Sequoia and the
Calveras Big Trees, their use by foresters and tourists, and both historic
and modern efforts to preserve them.

416. Fitzharris, Tim. *Forest : a National Audubon Society book.* Washington:
Starwood, 1991. 174 p. : col. photos. 0912347953
A review of and tribute to the many forests of North America from the New
England sugar maples to the Olympic rain forest. Covers their history, ecol-
ogy, and importance. Includes observations on the serenity and solitude
offered by undisturbed forests.

417. Flader, Susan. *Thinking like a mountain : Aldo Leopold and the evolution of an
ecological attitude toward deer, wolves, and forests.* Madison: Univ. of Wisconsin,

1994, c1974. 320 p. 0299145042
Delineates how Leopold's ideas and concerns developed over time. His
rejection of the German approach to tree farming and game management
was based on his observation of the destruction of the food chain and natural cycles of growth and regeneration.

418. Fritz, Edward C. *Clearcutting : a crime against nature.* Austin: Eakin, 1989.
160 p. : col. ill. 0890156743
An attorney exposes the extent and effects of Forest Service clearcutting
and attempts to launch a national campaign against it.

419. ___. *Sterile forest : the case against clearcutting.* Austin: Eakin, 1983. 271 p.
: photos. 0890153922
Details his campaign in and out of the courts to halt the Forest Service's
plans to turn the diverse deciduous forests of east Texas into sterile pine
plantations. Reveals corruption within the Forest Service.

420. Guha, Ramachandra. *The unquiet woods : ecological change and peasant
resistance in the Himalaya.* Berkeley: Univ. of California, 1990. 214 p.
0520065018
The Himalyan Chipko (hug the trees) movement has existed for over a century. Guha insists that this is not merely an environmental movement, but
"above all, a peasant struggle in defense of forest rights."

421. Hunter, Malcolm L. *Wildlife, forests, and forestry : principles of managing
forests for biological diversity.* Englewood Cliffs, NJ: Prentice-Hall, 1990. 370 p. :
ill. 0139594795
Answers basic questions and considers species composition, age structure,
spatial heterogerneity, shores, dying and downed trees, vertical structure,
intensive silviculture, forestry planning, and economics. Technical, but
accessible for adults with some science background.

422. Jonas, Gerald. *The living Earth book of North American trees.* Pleasantville,
NY: Reader's Digest, 1993. 216 p. : ill. 0895774887
A study of trees and forest systems of North America, examining them as a
biological system, an endangered species, and a renewable and treasured
resource.

423. Lansky, Mitch. *Beyond the beauty strip : saving what's left of our forests.* Gardiner, Me: Tilbury House, 1992. 453 p. : ill., maps. 0884481034 0884480941
(pbk)
The "beauty strip" is the deceptive practice of leaving stretches of trees
along highways uncut to hide clearcuts. Lansky analyzes the abuses of the
forest products industry and suggests reforms. HR.

424. Maser, Chris. *Forest primeval : the natural history of an ancient forest.* SF: Sierra Club, 1989. 282 p. : ill. 0871566834
This 1,000 year biography of an Oregon old-growth forest traces the intricate web of life woven from the first seeds sown by deer mice to the non-sustainable logging practices that threaten it today.

425. ___. *Sustainable forestry : philosophy, science, and economics.* Delray Beach, FL: St. Lucie, 1994. 400 p. : ill. 1884015166
Discusses what makes up a forest (so much more than trees), the technology of forestry, sustainable forestry, adaptive ecosystem management, and related subjects. Suggests a new role for the Forest Service as a leader in sustainable forestry.

426. McPhee, John. *The Pine Barrens.* NY: Farrar, Straus, Giroux, 1978, c1968. 156 p. : ill. 0374514429
The Pine Barrens which once covered nearly 2,000 square miles of New Jersey have been reduced over 50 percent by industrial and commercial encroachment. Describes the natural, social, and economic history of the largest remaining wilderness in the Eastern industrial corridor.

427. Mitchell, John G. *Dispatches from the deep woods.* Lincoln: Univ. of Nebraska, 1991. 304 p. 0803231466
Describes the various types of forests found in North America, their importance to the environment, the threats to their survival, and what can be done to save them.

428. Olson, Sigurd F. *The hidden forest.* Rev. ed. Stillwater, MN: Voyageur, 1990, c1969. 128 p. : 120 col. photos. 0896581330
This photoessay explores the inner workings of forest ecology.

429. Peattie, Donald Culross. *A natural history of western trees.* Boston: Houghton Mifflin, 1991, c1953. 751 p. : ill. 0395581753
Contains a clear and direct guide to the identification of over 200 kinds of trees. Describes the range, the properties of the woods, the leaf, flower, fruit, and bark of each tree.

430. ___. *A natural history of trees of eastern and central North America.* Boston: Houghton Mifflin, 1991, c1950. 606 p. : ill. 0395581745
See previous record.

431. Perlin, John. *A forest journey : the role of wood in the development of civilization.* Cambridge: Harvard, 1991, c1989. 445 p. 0393026671
Wood has been a primary fuel throughout history, but forests are dwindling, especially in the Third World. Examines the relationship between

forests, deforestation, war, expansionism, and economic and industrial policies.

432. Pyle, Robert Michael. *Wintergreen : listening to the land's heart.* Boston: Houghton Mifflin, 1988, c1986. 303 p. : ill. 0395465591
John Burroughs Medal, 1987.
A brilliant evocation of nature and ecology in the Willapa Hills of western Washington and a powerful indictment of the effects of irresponsible logging practices on the land. HR AMT. #

433. Raphael, Ray. *More tree talk : the people, politics, and economics of timber.* Washington: Island, 1994. 330 p. : ill. 1559632534 1559632542 (pbk) Rev., expanded ed. of *Tree talk.*
AR: *Edges: human ecology of the backcountry.*
Covers the debate over forestry issues, interviewing loggers, foresters, hunters, scholars, and conservationists. Analyzes sustainable forestry, economic factors, and the effect of different types of forest ownership on their management and use.

434. Richards, J. F., and Tucker, Richard P. eds. *World deforestation in the 20th century.* Durham: Duke Univ., 1989. 321 p. 0822310139
AR: *Global deforestation and the 19th-century world economy.*
Pressure for food, natural resources, and profits continues to drive deforestation. These papers discuss deforestation as a process, the conversion of forests to agricultural or mining uses, and forests as a capital resource.

435. Yaffee, Steven Lewis. *The wisdom of the spotted owl : policy lessons for a new century.* Washington: Island, 1994. 430 p. 1559632038 1559632046 (pbk)
Uses the case of the spotted owl as a framework for understanding the realities and idiosyncracies of public policy process. Outlines a set of reforms to re-create natural resource agencies and public policy processes to meet the challenges of the next century.

Books for young adults:

436. George, Jean Craighead. *One day in the woods.* NY: Crowell, 1988. 42 p. : ill. 069004724X
Outstanding Science Trade Book for Children, 1990.
Rebecca discovers many things about plant and animal life when she spends the day in Teatown Woods in the Hudson Highlands of New York looking for the ovenbird. HR GR 4-8.

437. Lauber, Patricia. *Summer of fire : Yellowstone 1988.* NY: Orchard, 1991. 64

p. : col. ill. 053105943X 0531085430
Describes the season of fire that struck Yellowstone in 1988, and examines
the complex ecology that returns plant and animal life to a seemingly barren, ash-covered expanse. GR 5-9.

438. Lerner, Carol. A *forest year.* NY: Morrow, 1987. 48 p. : col. ill.
0688064132 0688064140
Outstanding Science Trade Book for Children, 1987.
Describes how seasonal changes in a forest affect the plants and animals
that live there. HR GR 4-8.

439. Schoonmaker, Peter K. T*he living forest.* Hillside, NJ: Enslow, 1990. 64
p. : ill. 0894902709
Examines the characteristics of the vegetation found in temperate and
boreal (far northern) forests discussing their ecological significance, importance to people, and threats posed by pollution.

Rainforests

Tropical rainforests, considered to be both the lungs and the ethnobotanical medicine chest of the world, are rapidly being cleared for ranching,
mining, logging, and other extractive industries. Indigenous populations
and renewable forest resources are being "cleared away," too. Perhaps
nowhere else is biological diversity and stability in such peril with such
immense worldwide consequences. Several new writers have provided
powerful first-person accounts. They include Allen, Caulfield, Dwyer, Ghinsberg, Meunier and Savarin, O'Hanlon, and Plotkin.

Books for adults:

440. Allen, Roberta. *Amazon dream.* SF: City Lights, 1993. 181 p.
0872862704
Allen travels alone to a conservation camp in the upper Amazon, meets a
variety of shady characters, and discovers a new understanding of the
native people of the area and the complex forces that are destroying both
their cultures and the rain forest itself.

441. Arvigo, Rosita; Epstein, Nadine; and Yaquinto, Marilyn. *Sastun : my
apprenticeship with a Maya healer.* SF: Harper, 1994. 190 p. : photos.
0062502557 006250259X (pbk)
The chronicle of how Elijio Panti, a Mayan healer, influenced the practices
of an American-founded naturopathic medical center in Belize. Demon-

strates that the destruction of native cultures and rain forest vegetation is
a loss of massive magnitude for all of humanity.

442. Balee, William L. *Footprints of the forest : Ka'apor ethnobotany, the historical
ecology of plant utilization by an Amazonian people.* NY: Columbia Univ., 1994.
396 p. : ill., map. 0231074840
Explores the history of the Ka'apor and their present mode of land use, influences on the composition of fragile forests, and forest management practices.
Also discusses the nomenclature and classification of indigenous plants and
the cognitive aspects of magical, medicinal, and poisonous plants.

443. Castner, James L. *Rainforests : a guide to research and tourist facilities at
selected tropical forest sites in Central and South America.* Gainesville: Feline,
1990. 380 p. : ill., maps. 0962515027
Includes reviews of selected facilities for tourists and scientists in seven
countries. Covers locations, facilities, costs, native species, etc.

444. Caufield, Catherine. *In the rainforest : report from a strange, beautiful,
imperiled world.* Chicago: Univ. of Chicago, 1991, c1985. 310 p. : ill.
0226097862
Combines current research with personal experience to show the area's
uniqueness, complexity, and global importance. Chronicles its devastation
by ranchers, miners, foresters, and other developers and argues for its
preservation.

445. Collins, N. Mark. *The Last rain forests : a world conservation atlas.* NY:
Oxford, 1990. 200 p. : col. ill., maps. 0195208366
Describes and depicts what rain forests are, their ecology, why we need
them, pressures from development, indigenous people of the rain forests,
and conservation issues.

446. Cowell, Adrian. *The decade of destruction : the crusade to save the Amazon
rain forest.* NY: Doubleday, 1991, c1990. 215 p : ill. 0385420323
Companion to Cowell's series on the *Frontline* television program. An excellent introduction to rainforest ecology and to the social, economic, and
environmental causes of rainforest destruction. Although Cowell's main
emphasis is on the Amazon, he provides some coverage of other rainforests. HR AOT. #

447. Denslow, Julie Sloan, and Padoch, Christine. eds. *People of the tropical
rain forest.* Berkeley: Univ. of Cal., 1988. 231 p. : ill. 0520062957 0520063511
(pbk)
Contains introductory essays by anthropologists, botanists, conservationists, and ecologists. Covers tropical rain forests in fact and fiction, the pre-

history of tropical forests, the peoples of the forests, big businees in tropical forests, and future prospects. AOT. #

448. Downing, Theodore E. ed. *Development or destruction : the conversion of tropical forest to pasture in Latin America.* Boulder, CO: Westview, 1992. 405 p. : ill. 0813378249
Contains 25 essays by various authors describing economic pressures leading to mass destruction of the rain forests and the ecological, economic, and human costs of development. Anthropological, biological, climatogical, ecological, geographic, and industrial perspectives are considered. Somewhat technical, but accessible for adults with some scientific background.

449. Dwyer, Augusta. *Into the Amazon : Chico Mendes and the struggle for the Rain Forest.* SF: Sierra Club, 1990. 250 p. 1550132237
A personal narrative of people working to save the Amazon Rain Forest and a disturbing account of the devastation that government policies, industrialists, and international exploitation have wreaked upon an area key to planetary survival. AOT. #

450. Forsyth, Adrian, and Miyata, Kenneth. *Tropical nature : life and death in the rainforests of Central and South America.* NY: Scribner, 1987, c1984. 248 p. : ill. 0684187108
Presents the natural history, ecology, diversity, and beauty of tropical rainforests. Good preparation for visiting Central and South American rainforests. HR.

451. Gay, Kathlyn. *Rainforests of the world : a reference handbook.* Santa Barbara: ABC-CLIO, 1993. 219 p. : ill. 0874367123
Covers rain forests worldwide—includes ecology, history, biographical sketches, directory of organizations, print resources, and nonprint resources. The directory of organizations and nonprint resources are the strongest sections.

452. Ghinsberg, Yossi. *Back from Tuichi : the harrowing true-life story of survival in the rainforest.* NY: Random House, 1993. 256 p. 067942458X
Separated from his hiking companions in a rafting accident, Ghinsberg survived alone in the Amazon jungle for three weeks. He describes his ordeal and both the threats and splendors of the rain forest. This exciting adventure is HR AOT. #

453. Gradwohl, Judith, and Greenberg, Russell. *Saving the tropical forests.* Washington: Island, 1988. 214 p. : ill. 0933280815
Another threat to rainforests is that too many people farm too little land

with no time for fallow and recovery, and then cut down forest for more farmland. Shows how forest reserves, natural forest mangement, tropical forest restoration, and sustainable agriculture can be used to counter these trends.

454. Head, Suzanne, and Heinzman, Robert. eds. *Lessons of the rainforest.* SF: Sierra Club, 1990. 275 p. 0871566788
Contains essays by 24 scientists and environmental activists on rainforest conservation.

455. Hecht, Susanna B., and Cockburn, Alexander. *The fate of the forest :developers, destroyers and defenders of the Amazon.* NY: HarperPerennial, 1990. 357 p. : photos. 0060973226
Traces the destruction of the Amazon rainforest from rubber tapping in the 1850s, through post WWII development of a capitalist elite, to the massive national debt and land development that fuel the curent devastation. They call for a "socialist ecology" to save the forest.

456. Hurst, Philip. *Rainforest politics : ecological destruction in South-East Asia.* London: Zed, 1990. 303 p. 0862328381 086232839X (pbk)
Describes the rapid destruction of the world's second largest rainforest; loss of species, habitats, and diversity; the human and economic consequences of deforestation; how the exploitation of resources for Western consumption is a carryover of colonialism; and efforts to preserve this crucial area.

457. Joyce, Christopher. *Earthly goods : medicine-hunting in the rainforest.* Boston: Little, Brown, 1994. 320 p. 0316474088
Reviews the work of ethnobotanists and their descriptions of the rain forest as great botanical and medical treasure troves. Presents a vision of how these resources can be explored and exploited while preserving biodiversity. This book is less ethnobotanical than *Tales of a shaman's apprentice,*(469), examing the role of the international pharmaceutical industry.

458. Le Breton, Binka. *Voices from the Amazon.* West Hartford: Kumarian, 1993. 151 p. 1565490215
LeBreton questioned native peoples, loggers, developers, ranchers, miners, rubber tappers, and activists. The result is an excellent exploration of the complex issues relating to the Brazilian rainforest.

459. Lewis, Scott, and Holing, Dwight. *The rainforest book : how you can save the world's rainforests.* LA: Living Planet, 1990. 112 p. : photos. 0962607215
Provides a concise overview of rain forest ecology and the importance of these forests to the world's biodiversity and environmental quality.

Includes advice on how to help preserve the forests through activism and purchasing environmentally sustainable products. HR. #

460. Mendes, Chico, and Gross, Tony. *Fight for the forest : Chico Mendes in his own words.* NY: Monthly Review, 1992. 118 p. 0906156688
In Mendes' last interview before his murder he discusses his life, the campaign against deforestation, and the stuggle for sustainable agriculture in the tropics. Gross covers the trial of Mendes' assasins and analyzes Brazilian environmental policies.

461. Meunier, Jacques, and Savarin, Anne-Marie. *The Amazonian chronicles.* SF: Mercury House, 1994. 240 p. 1562790536
Compares and contrasts previous anthropological works on this region with their experiences living among the people. Describes how the people integrate dream and imagination into everyday life, and stresses the crucial importance of preserving "primitivism" as part of preserving the environment.

462. Miller, Kenton, and Tangley, Laura. *Trees of life : saving tropical forests and their biological wealth.* Boston: Beacon, 1991. 218 p. 0807085081
Describes the intense connectedness of all life in tropical forests, the value of forests as vast warehouses of medical and other resources, and their destruction.

463. Moffett, Mark W. *The high frontier : exploring the tropical rainforest canopy.* Foreword by E. O. Wilson. Cambridge: Harvard, 1993. 192 p. : col. photos. 0674390385 0674390393 (pbk)
Depicts the natural history and "architecture" of rainforests, their support of rich biotic communties, the relationships between arboreal plants and animals, and the development of rainforest canopy biology. Excellent, nontechnical writing on complex subjects. HR.

464. Myers, Norman. *The primary source : tropical forests and our future.* 2d ed. NY: Norton, 1992. 448 p. : ill. 0393308286
This study of the ecology, economic and environmental value, and destruction of tropical rainforests catalyzed the rainforest preservation movement in the mid-1980's. HR.

465. Nichol, John. *The mighty rainforest.* NY: David & Charles, 1994, c1990. 200 p. : 142 col. photos. 0715302183
AR: *The animal smugglers and other wildlife traders.*
This photoessay is a deep appreciation of the rainforest, a consideration of its economic and inherent value if left in a wild state, and a plea for and primer on preservation. AOT. #

466. O'Hanlon, Redmond. *In trouble again : a journey between the Orinoco and the Amazon*. NY: Vintage, 1990, c1988. 272 p. : ill., map. 0679727140
A very lively and sometimes grisly account of an expedition up the Orinoco River with observations on the forest, the Yanomani Indians, development, and preservation efforts.

467. Perry, Donald R. *Life above the jungle floor*. NY: Simon & Schuster, 1986. 194 p. : col. photos. 0685166551
Perry lived in the Costa Rican rain forest canopy in order to accurately describe and depict its complex plant and animal communities. AOT. #

468. Place, Susan E. ed. *Tropical rain forests : Latin American nature and society in transition*. Wilmington: Scholarly Resources, 1993. 228 p. : maps. 0842024239 0842024271 (pbk)
Contains 30 essays in sections entitled, "Perceptions of the rainforest," "Explanations for deforestation in Latin America," "Why save the rainforest?" and "Prospects for development: alternative futures for Latin America's tropical rainforest." Includes an excellent introduction and an annotated bibliography and list of films.

469. Plotkin, Mark J. *Tales of a shaman's apprentice : an ethnobotanist searches for new medicines in the Amazon rain forest*. NY: Penguin, 1994, c1993. 318 p. : ill. 0670831379
Describes how tribal people have used plants for food, medicine, tools, and religious purposes, how deforestation is destroying nature and cultures, and how preservation of botanic diversity is essential in the development of new drugs and other important products. AOT. #

470. Reiss, Bob. *The road to extrema*. NY: Summit, 1992. 301 p. 067168700X
Explores the relationships between the world's largest natural jungle and one of its largest human ones: New York City. Explains how similar economic and political forces fuel the destruction of both natural and human habitats and demonstrates a more positive link between co-evolutuion and sustainable development.

471. Revkin, Andrew. *The burning season : the murder of Chico Mendes and the fight for the Amazon rain forest*. Boston: Houghton Mifflin, 1990. 317 p. : photos. 039552394x
An impersonal, thorough, and adulatory portrait of the labor leader whose occasionally confrontational demonstrations against ranchers and developers slowed down the forest's rapid destruction and won international support.

472. Rietbergen, Simon. ed. *The Earthscan reader in tropical forestry*. London: Earthscan, 1993. 328 p. : ill. 1853831271

Diagnoses rainforest problems and analyzes options for dealing with them. Topics include forest management, land use and degradation, international trade patterns, forestry development aid, indigenous peoples, and the conservation of biological diversity.

473. Shoumatoff, Alex. *The world is burning*. NY: Avon, 1991. 377 p. : photos. 0380715422
This account is sympathetic to, but not uncritical of, Mendes and his movement. Tells of Mendes' life and death, Brazil's reaction to his assassination, the failure to prosecute the alleged killers, and Hollywood's battle for rights to the story.

474. Watson, Ian. *Fighting over the forests*. Boston: Allen & Unwin, 1990. 173 p. : ill. 0044422083
An account of the conflict over logging in Australia. Interviews over 50 Australian timber workers and preservationists in order to understand their views and search for common ground between them.

Books for young adults:

475. Banks, Martin. *Conserving rain forests*. Austin: Steck-Vaughn, 1990. 48 p. : col. ill. 0811423875
Describes the rapidity with which the rain forests of the world are being destroyed, the harm to plants, animals, and humans, and remedies for preserving them. GR 5-9. 4

476. Forsyth, Adrian. *Journey through a tropical jungle*. NY: Simon & Schuster, 1988. 80 p. : col. ill. 0671662627
Outstanding Science Trade Book, 1989.
Presents some of the plants, animals, and people that live in the the the rainforest of Costa Rica. Emphasizes a respect for nature and the diversity of life in the rain forest, and includes a plea for its preservation. HR GR 4-8.

477. Gallant, Roy A. *Earth's vanishing forests*. NY: Macmillan, 1991. 162 p. : ill. 0027357740
Discusses the ecology of rain forests, the problems posed by the present danger to rain forests through deforestation, and possibilities for the future.

478. George, Jean Craighead. *One day in the tropical rain forest*. NY: Crowell, 1990. 56 p. : ill. 069004769X
Outstanding Science Trade Book for Children, 1990.
The future of the Rain Forest of the Macaw depends on a scientist and a young Indian boy as they search for a nameless butterfly during one day in

the rain forest. HR GR 4-8.

479. Hare, Tony. *Rainforest destruction.* NY: Gloucester, 1990. 32 p. : col. photos. 0531172481
AR: *The ozone layer.*
Examines the crisis the world may face as a result of rainforest destruction and efforts being made to preserve them. Brief and concise.

480. Lourie, Peter. *Amazon : a young reader's look at the last frontier.* Honesdale, PA: Caroline House, 1991. 46 p. : ill. 1878093002
Recounts a journey through the Amazon River Valley, describing some of the traditional ways of life that still exist there and the encroachment of colonists that threatens the rain forests. GR 4-7.

481. Miller, Christina G., and Berry, Louise A. *Jungle rescue : saving the New World tropical rain forests.* NY: Atheneum, 1991. 118 p. : photos. 0689314876
AR: *Wastes; Acid rain: a sourcebook for young people.*
In order to avoid a global ecological crisis, the alarming destruction of the tropical rain forests of Central and South America must be reversed. Describes rain forest ecology, threats to the forest, and efforts to save it. GR 5-9.

482. Nations, James D. *Tropical rainforests : endangered environment.* NY: Watts, 1988. 143 p. : ill. 0531106047
Describes the oldest and most complex of ecosystems, the tropical rainforest: its wonders, its importance, its people, and the threats against which we must protect it.

483. Rowland-Entwistle, Theodore. *Jungles and rainforests.* Morristown, NJ: Silver Burdett, 1987. 48 p. : col. ill. 0382095006
Describes the characteristics of tropical rain forests and other kinds of jungles and examines their effect on the plants, animals, and people living in or near them. GR 5-9.

484. Stone, Lynn M. *Rain forests.* Vero Beach, FL: Rourke, 1989. 48 p. : col. ill. 0865924376
Examines the rain forest as an ecological niche and describes the plant and animal life supported there. GR 5-9.

Freshwater

This section includes books on rivers, streams, lakes, ponds, freshwater marshes and wetlands, and the Everglades. All are threatened by pollution, soil erosion, toxic contamination, ill-advised dams and water projects, loss of biodiversity, and drought. Also see the Water Supply section. Important riparian naturalists include Douglas Graves, Matthiessen, Powell, Stoneman, and Zwinger.

Books for adults:

485. Ashworth, William. *The late, Great Lakes : an environmental history.* NY: Knopf, 1986. 274 p. : maps. 0394551516
Combines natural history, meteorology, ecology, economics, and history to demonstrate the environmental degradation of the Great Lakes.

486. Brown, David E. *Arizona wetlands and waterfowl.* Tucson: Univ. of Arizona, 1985. 169 p. : ill., maps. 0816509042
Describes the history and ecology of six varied wetlands, waterfowl biology, resource management, and various species. Details both intentional and thoughtless destruction of waterfowl and wetlands in an attempt to raise public consciousnes and concern.

487. Budd, Ken. *Tatshenshini river wild.* Photos by Ric Careless. Englewood, CO: Westcliffe, 1993. 128 p. : col. photos. 1565790405
Depicts the beauty and biotic wealth of North America's wildest river, which is currently being threatened by development of a huge open pit copper mine. #

488. Carrier, Jim. *Down the Colorado : travels on a western waterway.* Boulder, CO: Rinehart, 1989. 141 p. : photos, maps. 0911797556
A reporter's journal of his trip from the Colorado headwaters to the Gulf of Mexico includes observations on nature, history, and the struggle to preserve the river, which is a mere trickle as it enters the Gulf. #

489. Cassuto, David N. *Cold running river.* Ann Arbor: Univ. of Michigan, 1993. 141 p. 0472104748 0472082388 (pbk)
Provides an ecological and economic history of a Michigan river, from the mixed impact of Native American settlements, through the logging of its banks, to the increasingly heavy demands made on its resources today, including conflicts between anglers and other river users.

490. Douglas, Marjory Stoneman. *The Everglades : river of grass.* Sarasota, FL: Pineapple, 1988, c1947. 447 p. : maps. 0910923388
This natural history of the Everglades details its destruction and includes a strong plea for preservation. Very influential on the early conservation movement in Florida. HR.

491. Douglas, Marjory Stoneman, and Rothchild, John. *Marjory Stoneman Douglas, voice of the river : an autobiography.* Sarasota, FL: Pineapple, 1990, c1987. 267 p. 0910923949
This biography of a key conservationist provides historic, naturalistic, and political perspectives.

492. Dugan, Patrick. *Wetlands in danger : a world conservation atlas.* NY: Oxford, 1993. 187 p. : col. photos, maps. 0195209427
Library Journal Best Reference Source, 1994.
In addition to furnishing detailed maps and data on worldwide wetlands, Dugan considers what wetlands are, their biotic and climatic importance, plant and animal adaptations to wetlands, and the challenges of conservation. HR.

493. Finlayson, C. M., and Moser, Mike. eds. *Wetlands.* NY: Facts on File, 1991. 224 p. : col. photos, maps. 0816025568
Contains seven essays describing the functions and importance of wetlands, their worldwide distribution, threats to their continued existence, negative consequences of wetland loss, and measures to save them.

494. Furtman, Michael. *On the wings of a north wind : the waterfowl and wetlands of America's inland flyways.* Harrisburg, PA: Stackpole, 1991. 161 p. : col. photos. 0811717879
Furtman followed duck migrations for three months, discovering that many wetlands have been plowed over for agriculture or dried by drought, and noting increases in predator populations and game poaching. He also met biologists working with farmers and ranchers to expand wetlands.

495. Graves, John. *Goodbye to a river : a narrative.* Houston: Gulf, 1991, c1960. 306 p. 0932012752
AR: *Hard scrabble, Texas heartland, The water hustlers.*
Graves was a professor, farmer, and amateur naturalist in north Texas. This highly personal elegy to the Brazos River considers the plight of the Brazos in particular and nature in general while encouraging readers to strengthen their bonds with nature.

496. Harris, Tom. *Death in the marsh.* Washington: Island, 1991. 245 p. 1559630701 1559630698 (pbk)
A winner of the Polk and Clarion awards for environmental reporting describes his eight year effort to determine why so many marsh animals were becoming extinct or deformed and how the government, in collusion with ranchers and agribusiness, has deceived investigators and resisted reforms. HR.

497. Johnsgard, Paul A. *Ducks in the wild : conserving waterfowl and their habitats.*
NY: Prentice Hall, 1992. 160 p. : col. photos. 0671850075
Covers the natural history of ducks and other waterfowl, wetlands ecology,
and conservation.

498. Lanting, Frans, and Eckstrom, Christine K. *Okavango : Africa's last Eden.*
SF: Chronicle, 1993. 168 p. : col. ill. 0811805271
The Okavango is a large and ecologically crucial wetland in the Kalahari
desert which is threatened by encroaching development. This photoessay
captures the wildlife, landforms, and natural cycles of the area. HR AOT. #

499. Maser, Chris, and Sedell, James R. *From the forest to the sea : the ecology
of wood in streams, rivers, estuaries, and oceans.* Delray Beach, FL: St. Lucie,
1994. 200 p. : ill. 1884015174
Describes processes by which wood in the water provides habitats, enrich-
es the food chain, stabilizes channels and coastlines, forms floating surface
communities, etc. Warns of potential ecological and economic consequence of the loss of wood in aquatic ecosystems.

500. Matthiessen, Peter. *Baikal : sacred sea of Siberia.* SF: Sierra Club, 1992.
89 p. : col. photos. 0871565846
Baikal is the largest freshwater lake in the world; a spiritual center for
Siberia and Mongolia; a unique ecosystem that is seriously polluted; and
the center of attention for the emerging Russian and international environmental communities. AOT.

501. McNamee, Gregory. *Gila : the life and death of an American river.* NY:
Orion, 1993. 240 p. : ill. 0517591634
Presents an environmental history of how and why a major southwestern
river has been destroyed by overexploitation. These lessons can be applied
to saving other rivers from similar fates. HR.

502. McNulty, Tim; O'Hara, Pat; and North, Douglass A. *Washington's wild
rivers : the unfinished work.* Seattle: Mountaineers, 1990. 144 p. : col. ill.,
maps. 0898861705
More than a dozen rivers are profiled to reaffirm the vital connection
between clean, free-flowing rivers, healthy wildlife population, and biotic
diversity.

503. Meloy, Ellen Ditzler. *Raven's exile : a season on the Green River.* NY: Holt,
1994. 256 p. : ill. 0805024972
Provides a first-person account of the annual use and abuse of the Colorado's largest tributary. Covers both natural and cultural history.

504. Crump, Donald J. ed. *America's wild and scenic rivers*. Washington: Nat'l Geographic Soc., 1983. 199 p. : col. photos. 0870444409 087044445X
This photoessay on over 25 wild and scenic rivers offers personal accounts of river adventures. Also covers the rapid loss of wild rivers and riparian habitat as well as efforts to preserve remaining wild rivers.

505. Palmer, Tim. *Lifelines : the case for river conservation*. Washington: Island, 1994. 200 p. : photos. 1559632194 1559632208 (pbk)
AR: *Endangered rivers and the conservation movement*.
Covers issues such as the salmon population, water quality, hydropower, riparian habitats, and ecosystems.

506. ___. *The Snake River : window to the West*. Washington: Island, 1991. 322 p. : map., photos. 0933280599
Describes the wide variety of natural wonders and habitats along the Snake; its natural history and ecology; its use for irrigation, hydropower, and recreation; and controversy over its present and future uses.

507. ___. *The wild and scenic rivers of America*. Washington: Island, 1993. 338 p. : ill. 1559631457 1559631449 (pbk)
Covers the importance of protecting river ecosystems, state and local protection systems, the National Wild and Scenic Rivers system, continuing threats to rivers, and how to improve preservation efforts. HR.

508. Powell, John Wesley. *The exploration of the Colorado River and its canyons*. Intro. by Wallace Stegner. NY: Penguin, 1987, c1875. 397 p. : ill., maps. 0140170006
Original title: *Report on the exploration of the Colorado River of the West and its tributaries*. An account of the first major scientific expedition down the the Colorado River through the Grand Canyon. HR.

509. Schafer, Jim, and Sajna, Mike. *The Allegheny River : watershed of the nation*. University Park: Penn. State Univ., 1992. 304 p. : 220 ill., maps 0271008369
This mix of historical, travel, nature and environmental writing paints a portrait of this important river yesterday and today. Washington, Carnegie, Johnny Appleseed, Mike Fink, Rachel Carson, Ida Tarbell, and other real and mythical figures are considered.

510. Stranahan, Susan Q. *Susquehanna, river of dreams*. Baltimore: Johns Hopkins Univ., 1993. 322 p. : ill., maps. 0801846021
A Pulitzer Award winning journalist describes how canal builders, loggers, miners, and industrialists nearly destroyed the source of their wealth. Describes the area's geology, floods, economic development, nuclear

development, farming, devastation in some areas, restoration in others, and many colorful characters.

511. Webb, Roy. *Call of the Colorado*. Moscow: Univ. of Idaho, 1994. 175 p. : ill., maps. 0893011614
AR: *Riverman*; *If we had a boat*.
Prospectors, photographers, scientists, surveyors, outfitters, and adventurers are profiled. Also presents a historical survey of river running on the Green and Colorado and geographical notes about the river system.

512. Weller, Milton Webster. *Freshwater marshes : ecology and wildlife management*. 3d ed Minneapolis: Univ. of Minnesota, 1994. 154 p. : col. photos, maps. 0816624062 0816624070 (pbk)
Covers marsh basins, hydrology, and diversity; the marsh as a system; habitat and behavior patterns; dominant animals; habitat dynamics; management and restoration; and marshes and people. .

513. Zwinger, Ann. *Run, river, run : a naturalist's journey down one of the great rivers of the American West*. Tucson: Univ. of Arizona, 1984, c1975. 317 p. : ill., maps. 0816508852
AR: *A conscious stillness*, *Plants in danger*.
Ride the rapids, study the wildlife, explore the canyons, and relive the prehistory and history of the wild and scenic Green River with one of America's pre-eminent nature writers. #

Books for young adults:

514. Gay, Kathlyn. *Water pollution*. NY: Watts, 1990. 144 p. : photos. 0531109496
Discusses the problem of contaminated rivers, lakes, and oceans, and proposes ways to purify them. Effectively promotes personal involvement in preservation and conservation.

515. Hoff, Mary King, and Rodgers, Mary M. *Our endangered planet : Rivers and lakes*. Minneapolis: Lerner, 1991. 64 p. : ill. 0822525011
Alerts the reader to the dangers of surface water pollution and the global imperative to keep these waters fresh. GR 4-7.

516. Parker, Steve. *Pond & river*. NY: Knopf, 1988. 63 p. : photos. 0394996151
A photoessay about the range of plants and animals found in fresh water throughout the year. Examines living conditions and survival mechanisms of creatures dwelling at the edge of the water, on its surface, or under the mud. British orientation. Primarily for GR 5-9, but also of interest to older readers. #

517. Stone, Lynn M. *Wetlands*. Vero Beach, FL: Rourke, 1989. 48 p. : col. photos. 0865924473
Examines the wetland as an ecological niche and describes the plant and animal life supported there. Emphasizes the ecological balance and fragility of wetlands, and threats to them. GR 5-7.

Marine Environments

Oceans, seas, saltwater estuaries, coral reefs, and coastlines are covered in these books. The world's largest single biome is being contaminated by solid, toxic, and radioactive wastes, and by spills from offshore oil drilling and shipping. Related works can be found in the Wastes section. The possibility of rising sea levels due to global warming is potentially a serious threat to the world's coastlines, lowlands, and harbors. Famous writers wandering the seashore include Carr, Carson, Steinbeck, and Ricketts. Others can be found in Bennett's *Seaside Reader*.

Books for adults:

518. Bennett, D. W. *The Seaside reader*. NY: Lyons & Burford, 1992. 311 p. 1558211977
Includes selections on seaside biology, ecology, and natural history by Rachel Carson, John Steinbeck and Ed Ricketts, Jacques Cousteau, John Updike, Norman Mailer, William Least Heat Moon, Peter Matthiessen, Izaak Walton, John Hay, and 16 others. HR.

519. Brower, Kenneth. *Realms of the sea*. Washington: Nat'l Geographic Soc., 1991. 278 p. : col. photos. 0870448552 0870448560 (deluxe)
A photoessay exploring ocean dynamics, their role in the creation of life and biodiversity, and how oceans effect daily lives. Covers the deep ocean, polar seas, open ocean, temperate seas, and tropical seas.

520. Bulloch, David K. *The wasted ocean*. NY: Lyons & Burford, 1989. 150 p. 1558210199 1558210342 (pbk)
Covers the many causes of marine pollution, takes the reader on a systematic tour of America's troubled coastlines, and provides suggestions for action to mitigate existing damage and minimize further damage.

521. Carr, Archie Fairly. *The windward road : adventures of a naturalist on remote Caribbean shores*. Tallahassee: Univ. of Florida, 1979, c1955. 313 p. : photos. 0813006392
A noted conservationist and biologist offers a very interesting and readable

account of his work in the Caribbean, focusing particularly on sea turtles, but also reflecting on nature in general. HR.

522. Carson, Rachel. *The edge of the sea*. Boston: Houghton Mifflin, 1979, c1955. 276 p. : ill. 0395285194
Covers all aspects of seashore biology and ecology. Includes a guide to identification of flora and fauna on shore rocks, sand beaches, and coral reefs. HR

523. ___. *Under the sea wind*. 50th anniversary ed. NY: Plume, 1992, c1941. 304 p. : ill. 0452269180
Written almost as a series of nature adventure stories, Carson follows several marine species through their natural cycles. Presents biological, ecological, conservationist and philosophical observations. HR AOT. #

524. ___. *The sea around us*. NY: Oxford, 1989, c1950. 250 p. : ill., maps. 0195061861
National Book Award, 1951. John Burroughs Medal, 1952.
An account of the sea's origins, dynamics, and relationships to humanity. Presents scientific research in an almost lyrical manner. HR AOT. #

525. Clark, R. B. *Marine pollution*. 3d ed. NY: Oxford, 1992. 172 p. 0198546858
In the 1986 ed. Clark was virtually an apologist for industry, stating that ocean dumping is neccesary and can be done safely. By 1992 he recognized a need to reduce dumping, suggesting that a balance be struck between the cost of environmental standards and industrial activity. Readers should consult Gourlay's *Poisoners of the seas* for balance.

526. Edgerton, Lynn T. *The rising tide : global warming and world sea levels*. Washington: Island, 1991. 139 p. 155963068X
Describes the consequences of rising sea levels caused by global warming: loss of coastal ecosystems, estuaries, natural and constructed coastal protection systems; flooding of coastal cities; increased turbidity, river salinization, etc. Outlines state, national, and international responses to these threats.

527. Gorman, Martha. *Environmental hazards : marine pollution*. Santa Barbara: ABC-CLIO, 1993. 252 p. 0874366410
Covers the history of marine pollution, types of pollution, effects on marine life, and potential planetary consequences. Includes a directory of organizations and libraries, bibliography, audiovisual resources, and computer networks. Essential for activists. HR.

528. Gray, William. *Coral reefs & islands : the natural history of a threatened par-*

adise. Newton Abbot, Eng.: David & Charles, 1993. 192 p. : col. ill. 0715300776
Provides a survey of the world's major coral reefs, describing their natural history, ecology, beauty, complexity, and fragility. Not quite as strong as the *Greenpeace book of coral reefs*, but still recommended AOT. #

529. Kruckeberg, Arthur R. *The natural history of Puget Sound country*. Seattle: Univ. of Washington, 1991. 468 p. : ill. 0295970197
Covers landforms and geology; climate and weather; marine life; lowland forests; other lowland habitats; animal life; montane natural history; water and quality of life; aboriginal Indians; impact of settlement. For both general readers and specialists.

530. Murray, John A. compiler. *A thousand leagues of blue : the Sierra Club book of the Pacific, a literary voyage*. SF: Sierra Club, 1993. 426 p. 0871564521
Contains 26 nonfiction and fiction selections written from 1768 to 1993. Authors include Dillard, Lopez, Melville, Darwin, Matthiessen, E. O. Wilson, London, Abbey, Twain, D. R. Wallace and others.

531. Ricketts, Edward Flanders; Calvin, Jack; and Hedgpeth, Joel Walker. *Between Pacific tides*. 5th ed. rev. by D. W. Phillips. Stanford: Stanford Univ., 1985. 652 p. : ill. 0804712298 0804712441 (student ed.)
A classic work on west coast marine biology and shoreline ecology covering the protected outer coast, the open coast, bays and estuaries, marine communities around wharf pilings, intertidal zonation, and principles of intertidal ecology. For adults with some background in biology. HR.

532. Rudloe, Jack. *The living dock*. Ill. by Walter Inglis Anderson. Golden, CO: Fulcrum, 1988. 273 p. : ill. 155591036X
An impassioned, respected naturalist reveals the infinite variety of marine life on Florida's Gulf Coast. HR.

533. Schultz, Stewart T. *The Northwest Coast : a natural history*. Portland: Timber, 1990. 89 p. : col. ill. 0881921424
Covers the natural history, ecology, geologic and oceanic forces; plant and animal interactions; relation to estuarine and forest ecosystems; and impact from human activity of the Pacific Northwest coast from Cape Mendocino, California, to Cape Flattery, Washington.

534. Steinbeck, John, and Ricketts, Edward Flanders. *Sea of Cortez : a leisurely journal of travel and research, with a scientific appendix comprising materials for a source book on the marine animals of the Panamic faunal province*. Mount Vernon, NY: Appel, 1982, c1941. 598 p. : ill. 0911858083
This unique and fascinating combination of natural history, travelogue, and

philosophy is a seamless collaboration by a great novelist and a great marine biologist. For adults interested in either marine biology or Mexican culture. HR.

535. Teal, John, and Teal, Mildred. *Life and death of the salt marsh.* NY: Ballantine, 1983, c1969. 274 p. : ill. 0345310276
A string of complex salt marshes edges the Atlantic from Newfoundland to Florida. This book shows how the marshes are developed, what kinds of life inhabit them, how they have contributed to civilization, and how they are being destroyed.

536. Thorne-Miller, Boyce, and Gritz, James. *Ocean : photographs from the worlds greatest underwater photographers.* SF: Collins, 1993. 240 p. : col. photos. 000255156X
This photoessay depicts many aspects of marine biology and ecology such as photosynthesis, migration, symbiosis, and the techniques of underwater photography.

537. Wells, Sue, and Hanna, Nick. *The Greenpeace book of coral reefs.* NY: Sterling, 1992. 160 p. : col. photos. 0806987952
Provides information on the uniqueness, variety, beauty, and fragility of reefs; damage from tourism, industry, pollution, and commercial coral harvesting; and efforts to conserve reefs. HR. #

Books for young adults:

538. Baines, John D. *Protecting the oceans.* Austin: Steck Vaughn, 1991. 48 p. : ill. 0811423913
Discusses the importance of the oceans, the sources and effects of their pollution and misuse, and ways to protect them. GR 4-9.

539. Fine, John Christopher. *Oceans in peril.* NY: Atheneum, 1987. 141 p. : photos. 0689313284
Examines plant and animal resources found in the sea, the effects of human intrusion and pollution, and possible solutions to the threats posed to our oceans. GR 5-8.

540. Hare, Tony. *Polluting the sea.* NY: Gloucester, 1991. 32 p. : col. photos. 0531172902
Examines the benefits we reap from the oceans and the damage that oil spills, metal poisoning, and sewage dumping cause. GR 4-8.

541. Hecht, Jeff. *Shifting shores : rising seas, retreating coastlines.* NY: Scribner,

1990. 151 p. : photos. 0684190877
Describes the various factors that change the shape of coastlines including storms, natural erosion, and rising sea levels. Also discusses future implications of these changes on coastal cities and what can be done to protect the coastlines and slow the process of change.

542. Lye, Keith. *Coasts*. Englewood Cliffs, NJ: Silver Burdett, 1988. 48 p. : col. photos. 0382097904
Discusses the characteristics of coasts around the world, including how coastlines recede and advance, climatic conditions, hurricanes and tsunamis, seashore plants and animals, and how people live in coastal communities. GR 5-8.

543. Mattson, Robert A. *The living ocean*. Hillside, NJ: Enslow, 1991. 64 p. : ill. 0894902776
Explores the ocean environment as a habitat for life, the kinds of organisms found in the ocean, and the relations of people to the sea. GR 6-9.

544. Miller, Christina G., and Berry, Louise A. *Coastal rescue : preserving our seashores*. NY: Atheneum, 1989. 120 p. : photos. 0689312881
Examines different types of coasts, how they are shaped by nature, how the development of coasts has destroyed plant and animal life, beaches, and marshes; threats to seacoasts, and ways to use coastal resources and still preserve them. Includes projects. GR 5-9.

545. Parker, Steve. *Seashore*. NY: Knopf, 1989. 63 p. : col. photos. 0394822544 0394922549
Brief text and photos introduce the animal inhabitants of the seashore, including fish, crustaceans, snails, and shorebirds. Contains good biological information, but weak on environmental problems of seashores. GR 5-9.

546. Tesar, Jenny E. *Threatened oceans*. NY: Facts on File, 1992. 112 p. : photos 0816024944
Explores the vital importance of ocean ecosystems and resources, and focuses on how fishing, dumping, oil spills, and other human influences endanger the marine environment. #

Mountains and Alpine Tundra

Mountainous areas are threatened by forest loss from acid rain, overlogging, soil erosion, and stream pollution. Many of these fragile areas are overused for mining, grazing, housing, and recreational development. Books by pre-eminent authors such as Abbey, Bowden, Burroughs, Matthiessen, Muir, Schaller, Wallace, and Zwinger can be found here. Other works about mountains by King (590), the Muries (601), and Muir (596-600) are in the Wilderness section

Books for adults:

547. Abbey, Edward. *Appalachian wilderness : the Great Smoky Mountains.* Photos by Eliot Porter. NY: Arrowood, 1988, c1970. 123 p. : col. photos. 0884860124
Contrasts Porter's beautiful photos with Abbey's hardscrabble account of the removal of the Cherokee Indians and the degradation of the area by mining and motorized tourism. HR.

548. Bowden, Charles. *Frog Mountain blues.* Photos by Jack Dykinga. Tucson: Univ. of Arizona, 1994, c1987. 165 p. : col. photos. 0816515018
Describes the geology, ecology, natural history, and cultural history of these dramatic mountains just outside Tucson. Considers their spiritual importance to Indians, their biologic significance, and efforts to either develop or preserve them.

549. Burroughs, John. *In the Catskills : selections from the writings of John Burroughs.* Marietta, Ga: Cherokee, 1990, c1910. 263 p. : ill. 0877971846
Contains much of his work about living with and learning from nature around his home in the Catskills, which contained a great deal of real wilderness in the early 1990s.

550. Carey, Ken. *Flat rock journal : a day in the Ozark mountains.* SF: Harper, 1994. 230 p. 0062510061 0062502751 (pbk)
This memoir of 20 years of walks in the Ozarks provides vivid descriptions of nature and its effect on the author's consciousness, the effects of clearcutting, struggles to preserve nature, and family life on a remote farm. The implicit message is the importance of preservation.

551. Hubler, Clark. *America's mountains : an exploration of their origins and influences from the Alaska Range to the Appalachians.* NY: Facts on File, 1994. 272 p. : photos. 0816026610
Explores orogeny, a mountain building process of earthquakes and eruptions; the influence of mountains on plants, animals, people, and the general environment; the origins of every major mountain range in America; plains and plateaus; and glaciers.

552. Matthiessen, Peter. *The snow leopard.* NY: Penguin, 1987, c1978. 338 p. 0140102663

An account of Matthiessen's and George Schaller's expedition to the Himalayas to locate the rare snow leopard and other endangered wildlife. It also became a deep spiritual oddyssey for Matthiessen who describes his deep experiences at remote lamaseries. HR AMT. #

553. Muir, John. *The mountains of California.* Ed. by Robert. C. Baron. Golden, CO: Fulcrum, 1988, c1894. 291 p. : ill. 1555910343
AR: *Mountaineering essays, Northwest passages.*
In his first complete book, Muir describes mountaineering adventures in the Sierra Nevada, and the natural history of the area from foothills and forests, to alpine meadows, glaciers, and mountaintops. This inspiring work is HR. #

554. ___. *My first summer in the Sierra.* SF: Sierra Club, 1990, c1911. 188 p. 0871565358
A 1910 reworking of Muir's early journals including a powerful account of his "conversion experience" as a young man in the Sierra Nevada. Many consider this Muir's finest work. HR AOT. #

555. ___. *South of Yosemite : selected writings of John Muir.* Berkeley: Wilderness, 1988. 220 p. : photos. 0899970958
A collection of newspaper and magazine articles, journal entries, and miscellaneous writings about the southern Sierra Nevada. Muir's love for the creatures, forests, alpine regions, and seasons of the mountains and his clarion call to conservation are present throughout. AOT. #

556. ___. *Steep trails.* SF: Sierra Club, 1994, c1918. 391 p. : ill. 0871565358
A collection of journal entries, letters, and articles written in the field in mountainous locations throughout the west over the course of 29 years. Includes his famous essay on Mount Shasta. HR. #

557. ___. *The story of my boyhood and youth.* SF: Sierra Club, 1988, c1913. 162 p. 0871567490
Muir reminisces on his boyhood in Scotland, adolescence on a Wisconsin farm, the grueling work regimen dictated by his father, his early interest in nature, discoveries and inventions, and university education. AMT. #

558. ___. *The Yosemite : the original John Muir text.* SF: Sierra Club, 1988, c1912. 288 p. : col. ill. 0871567822
Muir presents a detailed and delightful description of the geology, climate, forests, flowers, birds, animals, and weather of the Park. He also describes Hetch Hetchy Valley and makes a powerful but unsuccessful plea to save it from damming. HR. #

559. Schaller, George B. *Stones of silence : journeys in the Himalaya.* Chicago:

Univ. of Chicago, 1988, c1980. 292 p. : ill. 0226736466
Describes Schaller's six year study of the wildlife of the Himalyas, especially the endangered sheep and goats. Includes specific recommendations on wildlife preservation and observations on the effects of social change in the area. HR.

560. Vale, Thomas R., and Vale, Geraldine R. *Time and the Tuolumne landscape : continuity and change in the Yosemite high country.* Salt Lake City: Univ. of Utah, 1994. 212 p. : photos. 0874804299
Uses scenes in old photos precisely rephotographed over time to document continuity and change in Yosemite for over a century.

561. Van Dyke, John Charles. *The mountain : renewed studies in impressions and appearances.* Salt Lake City: Univ. of Utah, 1992, c1916. 234 p. 087480387X
Van Dyke's ride across the high plains of Montana into the Rockies becomes a journey into the soul of nature and the human soul as they become one. HR.

562. Wallace, David Rains. *The Klamath knot : explorations of myth and evolution.* SF: Sierra Club, 1983. 149 p. 0871568179
An almost poetic account of evolution, ecological change, and the relationship between geology, climate, biology, and mythology in the Klamath Mountains. This book has widespread appeal for environmentalists, literary authors, folklorists, philosophers, and general readers. HR.

563. Zwinger, Ann. *Beyond the aspen grove.* Tucson: Univ. of Arizona, 1988, c1970. 345 p. 0816510547
The author reflects on the years spent living on forty acres of Colorado mountain land.

564. Zwinger, Ann, and Willard, Beatrice E. *Land above the trees : a guide to American alpine tundra.* NY: Perennial, 1986, c1972. 489 p. : ill. 0060913657
Describes the natural history, geology, and ecology of the alpine tundra of seven western mountain ranges and New Hampshire's White Mountains. Highly detailed, but accessible to most adults.

Books for young adults:

565. George, Jean Craighead. *One day in the alpine tundra.* NY: Crowell, 1984. 44 p. : ill. 0690043260
Outstanding Science Trade Book for Children, 1984.
Relates a boy's adventure when he is alone on the alpine tundra on a stormy day. Describes the ecological relations between plants and animals

such as the marmot, pika, and pocket gopher, and how animals prepare for winter. HR GR 5-7.

Prairies

Prairies and the complex ecosystems they support are shrinking rapidly due to agricultural use, desertification, and urban expansion. Prominent writers associated with the prairies include Derleth, Hasselstrom, Heat Moon, and Janovy.

Books for adults:

566. Derleth, August William. *Walden West*. Madison: Univ. of Wisconsin, 1992, c1961. 262 p. : ill. 0299135942 (pbk) 09913590X
AR: *The only place we live.*
Uses descriptions of life and nature in the Wisconsin prairies to elucidate three related themes: the persistance of memory, the sounds and odors of the country, and Thoreau's observation that most people live quiet lives of desperation.

567. Fitzharris, Tim. *The wild prairie : a natural history of the Western Plains*. NY: Oxford, 1984. 143 p. : col. ill. 0195404386
The wild prairie once extended from Edmonton to Mexico City, supporting 45,000,000 bison and huge populations of other animals, but has shrunk to a fraction of its original size and animal population. Describes these animals and their relation to one another.

568. Hasselstrom, Linda M. *Going over east : reflections of a woman rancher*. Golden, CO: Fulcrum, 1993, c1987. 206 p. 1555911412
A trip into the author's past, focusing on the family cow-calf operation and a celebration of the wilderness. HR AOT. #

569. ___. *Land circle : writings collected from the land*. Golden, CO: Fulcrum, 1993, c1991. 369 p. 1555911420
Explores the responsibilty to the land and her own love of it.. Mountain and Plains Booksellers Regional Book Award, 1992. HR AOT. #

570. ___. *Windbreak : a woman rancher on the northern plains*. Berkeley: Barn Owl, 1987. 233 p. 0960962638
A year in the life of a rancher-poet-conservationist. This fine evocation of the Great Plains contains observations about the its ecology and preservation. AOT. #

571. Heat Moon, William Least*. *PrairyErth : a deep map*. Boston: Houghton Mifflin, 1991. 624 p. : maps. 0395486025
Focuses on one of the last true tallgrass prairies, debunking myths about Kansas and reaching deep into its natural and human history. Filled with anecdotes, Native American lore, and comments on human ecology. AOT. #

572. Matthews, Anne. *Where the buffalo roam*. NY: Grove Weidenfeld, 1992. 193 p. : maps. 0802114083
Describes the efforts of Frank and Deborah Popper to return millions of devastated acres to their natural state and reintroduce huge buffalo herds.

573. Shirley, Shirley. *Restoring the tallgrass prairie : an illustrated manual for Iowa and the upper Midwest*. Iowa City: Univ. of Iowa, 1994. 344 p. : ill. 0877454698 (pbk) 087745468X
A step-by-step guide for recreating prairies from cropland and other uses. Covers site preparation, planting, maintenance, and over 100 species of grasses and wildflowers. Includes a directory of relevant organizations and seed and equipment suppliers.

574. Thompson, Janette R. *Prairies, forests, and wetlands : the restoration of natural landscape communities in Iowa*. Iowa City: Univ. of Iowa, 1992. 139 p. : maps, photos. 0877453713 (pbk) 0877453721
Over 95% of Iowa's landscape was converted to agricultural purposes in the past century. Combines a nontechnical natural history of each native community with a how-to manual for lay restorationists dedicated to reconstructing prairies, forests, and wetlands.

Books for young adults:

575. George, Jean Craighead. *One day in the prairie*. NY: HarperCollins, 1986. 42 p. : ill. 0690045662
Outstanding Science Trade Book for Children, 1986.
The animals on a prairie wildlife refuge sense an approaching tornado and seek protection before it touches down and destroys everything in its path. Excellent descriptions of buffalo, elk, coyote, prairie dogs, and animal interdependence. HR GR 5-12.

576. Lambert, David. *Grasslands*. Englewood Cliffs, NJ: Silver Burdett, 1988. 48 p. : ill. 0382097890
Explores different types of grasslands located around the world, describing their development, climate, plant and animal life, exploitation, and preservation. GR 5-8.

Wilderness

This section covers wilderness areas and national parks. Wilderness areas around the world are threatened by development, recreational overuse, grazing, mining, clearcutting, and the division of wilderness and parklands into artificial managerial fiefdoms. Wilderness areas often support large migratory animals which require large feeding areas and migration zones. Bears, wolves, and other species understand the unitary nature of wilderness far better than bureaucrats. Important advocates of wilderness include Abbey, King, Matthiessen, Muir, the Muries, Olson, and Simpson. Their work and that of many others can also be found in *Words for the Wild* and *The Wilderness Reader*.

Books for adults:

577. Abbey, Edward. *The best of Edward Abbey*. SF: Sierra Club, 1988, c1984. 383 p. 0871567865
Original title: *Slumgullion stew*. Abbey's personal selection of excerpts from his novels and nonfiction works written between 1954 and 1982. HR AMT. #

578. ___. *The journey home : some words in defense of the American West*. NY: Dutton, 1991, c1977. 256 p. 0452265622
Abbey offers a ringing testimonial for and defense of wilderness in the canyonlands and deserts of the Southwest.

579. ___. *One life at a time, please*. NY: Holt, 1988. 225 p. 0805006036
Contains an interview with Joseph Wood Krutch, an essay about San Francisco, a diatribe against cowboys, a study of Ralph Waldo Emerson, a television script, and the controversial, racist essay "Immigration and liberal taboos."

580. ___. *A voice crying in the wilderness = Vox clamantis in deserto : notes from a secret journal*. NY: St. Martin's, 1990. 112 p. : ill. 0312041470
Abbey's last work was selecting short quotations from a 21 volume journal. In addition to nature and conservation, he offers ripostes about government, politics, sex, science, literature, and music. A great source for spicing up papers or speeches. HR AMT. #

581. Abbey, Edward, and Guthrie, A. B. *Images from the Great West*. La Canada, CA: Chaco, 1990. 133 p. : col. photos. 0961601949
Photos depicting the landscape and natural history of the west by Mark Gaede accompanied by prose selections by Guthrie and Abbey. HR AOT. #

582. Bergon, Frank. ed. *The Wilderness reader*. NY: New American Library, 1980. 372 p. 0451625897

Considers relations between civilization and wilderness. Authors include William Byrd, William Bartram, Meriwether Lewis, George Catlin, John James Audubon, Francis Parkman, and John Muir. HR.

583. Chase, Alston. *Playing God in Yellowstone : the destruction of America's first national park.* San Diego: Harcourt Brace Jovanovich, 1987. 464 p. 0156720361
Chase criticizes both Yellowstone Park management and the environmental movement for debating philosophical points, while losing touch with the realities of environmental politics and the destruction of the environment.

584. Crump, Donald J. ed. *America's hidden wilderness : lands of seclusion.* Washington: Nat'l Geographic Soc., 1988. 199 p. : col. photos, maps. 0870446665
Describes the natural history and efforts to preserve the Arctic National Wildlife Refuge, Riviere a l'Eau Claire, Maine's Baxter Sate Park, Montana's Great Burn, Utah's Grand Gulch, the Mojave Desert, and Mexico's Lacandon Wilderness. #

585. Few, Roger. *The atlas of wild places : in search of the Earth's last wildernesses.* NY: Facts on File, 1994. 240 p. : col. photos, maps. 0816031681
Offers a visual and narrative journey to 40 of the world's last pockets of nature that have gone untouched by human development. #

586. Foreman, Dave, and Wolke, Howie. *The big outside : a descriptive inventory of the big wilderness areas of the United States.* Rev. ed. NY: Harmony, 1992. 500 p. : maps. 0517587378
It is crucial that wilderness areas be large enough to maintain proper habitats for wildlife diversity. Includes detailed descriptions of all major American roadless areas.

587. Gruchow, Paul. *The necessity of empty places.* NY: St. Martin's, 1988. 291 p. : map. 0312021984
Landscapes of Western Europe presented from a philosophical standpoint.. Similar to the Erlichs' *The Solace of empty spaces.* AOT. #

588. Harte, John. *The green fuse : an ecological odyssey.* Berkeley: Univ. of Cal., 1993. 156 p. 0520082079
Harte takes us on a fact-based but passionate journey through true wilderness in the Everglades, Tibet, Alaska, Vermont, Costa Rica, Indonesia, and elsewhere so that we might experience "the green fuse" of "the oneness or unity of nature."

589. Hess, Karl. *Rocky times in Rocky Mountain National Park : an unnatural his-*

tory. Niwot: Univ. of Colorado, 1993. 167 p. : ill. 0870813099
A consultant describes environmental pressures on the park from without and within, the impact of a growing elk herd, the effects of fire suppression, and the Park Service's preservation failures as a result of "predatory politics." Calls for major changes in how national parks are managed.

590. King, Clarence. *Mountaineering in the Sierra Nevada*. NY: Penguin, 1989, c1872. 282 p. 0140170154
Covers the relationship between wilderness and humanity.

591. Lowry, William R. T*he capacity for wonder : preserving national parks*. Washington: Brookings Institution, 1994. 280 p. 0815752989
The national parks of the U.S. and Canada are being degraded by overuse, rising crime, pollution, and contention over preservation vs. use. Whereas the US National Park system has suffered budget cuts, Canada has made preservation the highest priority. Lowry considers proposals to change the park system.

592. Martin, Vance. *For the conservation of Earth*. Golden, CO: Fulcrum, 1988. 418 p. : ill. 1555910262
Contains over 80 papers on wilderness; the global challenge of environmental problems; science and management; tropical forests; development and the environment; and culture and the environment. Includes national case studies from around the world.

593. Matthews, Rupert. T*he atlas of natural wonders*. NY: Facts on File, 1988. 240 p. : col. maps, photos. 0816019932
Provides information on the geography, geology, flora and fauna, climate, ecology, history, and use of over 50 of the world's primary wildernesses and natural wonders. #

594. Matthiessen, Peter. T*he cloud forest : a chronicle of the South American wilderness*. NY: Penguin, 1987, c1961. 322 p. : photos. 0140095497
A personal account of explorations in, and the natural and human ecology of the Amazon forests, Sargasso Sea, Andes, Tierra del Fuego, Matto Grosso and other areas.

595. McPhee, John. E*ncounters with the archdruid*. NY: Noonday, 1990, c1971. 245 p. 0374514313
Originally appeared in T*he New Yorker*. A personal account of three separate confrontations and lively discussions between David Brower and a mineral engineer, a dam builder, and a resort developer. Takes place on the Georgia Sea Islands, Washington's Glacier Peak Wilderness, and the Colorado River. HR AMT. #

596. Muir, John. *The American wilderness*. Photos by Ansel Adams. NY: Barnes & Noble, 1993. 223 p. : photos. 1566191033
Includes articles by Muir on Hetch Hetchy Valley; Sequoia, Yosemite, General Grant, and Yellowstone National Parks; animals; the mountains in winter and spring; and parks, forests, and nature in general. HR. #

597. ___. *The eight wilderness discovery books*. Seattle: Mountaineers, 1992. 1030 p. : ill., maps. 089886335X
Contents: *The story of my boyhood and youth — A thousand mile walk to the Gulf — My first summer in the Sierra — The mountains of California — Our national parks — The Yosemite — Travels in Alaska — Steep trails.*
This excellent anthology includes an introduction by Terry Gifford; the first two chapters of the rare novel *Zanita : Tale of the Yosemite* by Therese Yelverton; notes on the making of the eight wilderness discovery books; a chronology of Muir's life; and an article on the John Muir Trust. HR. #

598. ___. *Muir among the animals: the wildlife writings of John Muir*. SF: Sierra Club, 1989, c1986. 196 p. 0871566079
Anthology of essays and articles which present Muir's studies of and attitudes about animals and wildlife conservation. Includes The Wild Sheep of California, The Passenger Pigeon, Stickeen, The Bee-Pastures of Oregon, and the Coyote. HR AOT. #

599. ___. *Our national parks*. SF: Sierra Club, 1991, c1901. 278 p. 0871566265
Contains ten articles written for *Atlantic Monthly* which were instrumental in the founding of the national park system and the Sierra Club. An excellent example of scientifically accurate, highly literate, politically effective conservation writing. HR, particularly for activists. #

600. ___. *Wilderness essays*. Salt Lake City: Peregrine Smith, 1988, c1980. 263 p. 0879053011
An anthology of some of Muir's best-known work on wilderness. HR AOT. #

601. Murie, Margaret E., and Murie, Olaus Johan. *Wapiti wilderness*. Boulder: Colorado Assoc. Univ., 1985, c1963. 302 p. : ill. 087081155X
Olaus describes his work as a field biologist and Margaret offers observations on their life together in the Wyoming wilderness during all four seasons.

602. Nash, Roderick. *Wilderness and the American mind*. 3d ed. New Haven, CT: Yale, 1982, c1967. 425 p. 0300029055
Traces the evolution of wilderness concepts from the earliest settlers to Thoreau, Muir, and Aldo Leopold, to modern ecologists. Possibly the most important study of the effect of wilderness on human consciousness. HR.

603. Oelschlaeger, Max. *The idea of wilderness : from prehistory to the age of ecology.* New Haven, CT: Yale, 1991. 477 p. 0300053703
Spanning prehistoric hunter-gatherer societies to modern civilization, Oelschlaeger surveys how the ideas of wilderness and humanity's place in nature have changed over time. He critiques the work of Thoreau, Leopold, Muir and other leaders. HR.

604. ___. ed. *The wilderness condition : essays on environment and civilization.* SF: Sierra Club, 1992. 345 p. 0871566427
Contains papers on various aspects of the relationship between wilderness and civilization by Oelschlager, Snyder, Shepard, Sessions, LaChapelle and others.

605. Olson, Sigurd F. *The singing wilderness.* NY: Knopf, 1990, c1956. 244 p. 0394445600
Describes 30 years of studying and enjoying this wilderness area along the U.S.-Canadian border. Includes the seasons, natural history, and a plea for its preservation. HR.

606. Robbins, Jim. *Last refuge : the environmental showdown in Yellowstone and the American West.* NY: Morrow, 1993. 285 p. : map. 0688111785
Examines the conflict between loggers-ranchers-miners-developers who want immediate access to more resources and environmentalists who wish to protect resources. Presents results of historical and field research and proposes a solution that would allow people to make a living off the land without destroying it.

607. Ronald, Ann. ed. *Words for the wild : the Sierra Club trailside reader.* SF: Sierra Club, 1987. 365 p. 0871567091
Includes nearly 40 essays by Emerson, Thoreau, Muir, Burroughs, Leopold, Krutch, Teale, Eiseley, Stegner, Abbey, McPhee, Dillard, Austin, Lopez, Powell, and others. HR for the trail or home. #

608. Sholly, Dan R., and Newman, Steven M. *Guardians of Yellowstone : an intimate look at the challenges of protecting America's foremost wilderness park.* NY: Quill, 1993. 317 p. : maps, photos. 0688125743
Yellowstone's chief ranger describes the challenge of balancing the needs of preservation with the demands of tourism. Covers such controversial subjects as the grizzly bear, reintroducing wolves to Yellowstone, and the fires of 1988.

609. Simpson, Bob. *Wilderness is where you find it.* Chapel Hill: Algonquin, 1988. 195 p.: col. photos. 091269789X
Describes Simpson's wilderness adventures across North America.

Although it consists mostly of hunting, fishing, and hiking stories, it conveys a strong sense of reverence for nature and pleas for conservation.

610. Stegner, Wallace Earle. ed. *This is Dinosaur : Echo Park country and its magic rivers.* Boulder, CO: Roberts Rinehart, 1985, c1955. 93 p. : photos. 0911797114
This book helped save Dinosaur Canyon from damming and flooding, and establish it as a national monument. The new edition calls for renewed conservation and preservation efforts.

611. Swanson, Timothy M., and Barbier, Edward. eds. *Economics for the wilds : wildlife, diversity, and development.* Washington: Island, 1992. 226 p. 1559632119 1559632127 (pbk)
AR: *Economics, natural-resource scarcity and development.*
Includes an explanation of the economic distortions that work against sustainable management, an analysis of the implications of resource loss, a consideration of the value of natural resources for developing societies, and methods of designing appropriate policies for integrating wildlife and wildland utilization in economic development.

612. Thoreau, Henry David. *A yearning toward wildness : environmentalism and inspiration.* Ed. by Tim Homan. Atlanta: Peachtree, 1991. 166 p. 1561450359
Includes short quotations, aphorisms, and some longer excerpts from Thoreau's complete works. An excellent source of quotations for speakers, writers, and activists. AOT. #

613. Turner, Tom. *Wild by law : the Sierra Club Legal Defense Fund and the places it has saved.* SF: Sierra Club, 1990. 154 p. : col. photos. 0871566273
Covers the establishment of environmental law in general, the SCLDF in particular, and the lands it has protected. Contains little known facts about litigation concerning landmark cases.

614. Wallace, David Rains. *The Quetzal and the Macaw : the story of Costa Rica's national parks.* SF: Sierra Club, 1992. 222 p. 0871565854
Over half of the Costa Rican rainforest has been cleared for cattle grazing in the last 40 years, but in the last 20 this trend has been countered by the development of one of the best systems of national preserves on Earth. Offers valuable lessons to other countries.

615. Wolke, Howie. *Wilderness on the rocks.* Foreword by Edward Abbey. Tucson: Ned Ludd, 1991. 246 p. : photos. 0933285051
Contends that the National Wilderness Preservation System is a system of very scenic areas that lack ecological integrity and do not begin to provide the necessary habitat for sensitive species. Proposes major reforms. HR.

Books for young adults:

616. Ansell, Rod, and Percy, Rachel. *To fight the wild*. NY: Harcourt Brace Jovanovich, 1986, c1980. 151 p. : ill. 0152890688
A man struggles to survive on his own in one of the most isolated corners of Australia for two months before he is rescued.

617. Patent, Dorothy Hinshaw. *Yellowstone fires : flames and rebirth*. NY: Holiday House, 1990. 40 p. : col. photos. 0823408078
Outstanding Science Trade Book for Children, 1990.
Describes the massive forest fires that burned almost one million acres of Yellowstone National Park in 1988 and the effects, both positive and negative, on the ecology of the forest there. HR GR 5-8.

618. Young, Donald, and Bix, Cynthia Overbeck. *The Sierra Club book of our national parks*. SF: Sierra Club, 1990. 64 p. : ill. 0316977446
Describes the history, attractions, uses, and diversity of, as well as threats to, a number of national parks, including the Grand Canyon, Maine's Acadia National Park, and Alaska's Gates of the Arctic National Park and Preserve. GR 4-8.

Urban and Suburban Areas

Urban areas have unique environmental problems, and their growth presents threats to other biomes. Urban environmental issues include land use and zoning, pollution, waste disposal, slums, and generally overstripping the carrying capacity of the land. Urban centers and their surrounding land and water supplies are considered in these books. Related works can be found in the Wastes and Water Supply sections. Many noted naturalists may work for urban universities or institutes, but almost all seem to focus their attention outside the city. Pyle and Wallace are probably the best known writers in this section.

Books for adults:

619. Adams, Lowell W. *Urban wildlife habitats : a landscape perspective*. Minneapolis: Univ. of Minnesota, 1994. 199 p. : ill. 0816622124 0816622132 (pbk)
Provides both scientific and practical information on wildlife habitats in urban and suburban areas and the conservation and management of urban wildlife.

620. Garber, Steven D. *The urban naturalist.* NY: Wiley, 1987. 242 p. : ill.
0471857939
Practical advice on how to locate, identify, study, and preserve grasses,
wildflowers, trees, insects, fish, amphibians, reptiles, birds and mammals
in urban settings. AOT. #

621. Garoogian, Rhoda, and Garoogian, Andrew. eds. *Health and environment in America's top-rated cities : a statistical profile.* Boca Raton: Universal Reference, bi-annual. 1881220125 (1994)
Arranged by city. Includes statistical and other data on health care reform,
AIDS, infant mortality, children's health, death/suicide rates, hospitals,
physicians, health care costs, crime, air and water quality, toxic releases,
Superfund sites, pollen calendar, climate, recycling, etc.

622. Goldstein, Eric A., and Izeman, Mark A. *The New York environment book.*
Washington: Island, 1990. 267 p. : photos. 1559630191
Describes the immense amount of pollution and waste generated in New
York City in five categories (solid waste, air pollution, water, toxics, and
waterways) and provides suggestions for citizen action to help alleviate
these problems.

623. Gordon, David. ed. *Green cities : ecologically sound approaches to urban
space.* Montreal: Black Rose, 1990. 299 p. : ill. 0921689551 0921689543
(pbk)
Includes 24 essays by various authors defining the many aspects of the
"green city," the naturalization of urban environments, and practical methods for overcoming existing barriers and effecting change.

624. Lipkis, Andy, and Lipkis, Katie. *The simple act of planting a tree : a citizen
forester's guide to healing your neighborhood, your city, and your world.* LA: Tarcher,
1990. 237 p. : ill. 0874776023
Demonstrates the crucial importance of trees to the quality of life. Shows
how to form local tree planting groups, raise funds, work with local government, and where, when, and how to plant trees. Brimming with good sense
and practical advice. HR AOT. #

625. McHarg, Ian L. *Design with nature.* NY: Wiley, 1991, c1969. 197 p. : ill.
0471557978
How does one arrive at plans to harmonize the "built environment" with
the natural one? McHarg recommends the "overlay method" which overlays
maps depicting environmental and socioeconomic factors. Especially for
architects and planners. HR.

626. Pick, Maritza. *How to save your neighborhood, city, or town : the Sierra Club*

guide to community organizing. SF: Sierra Club, 1993. 213 p. : ill. 0871565226
"Think locally, act vocally" by organizing grassroots efforts for preservation of wildlife, conservation of wildlife, and mitigation of environmental problems. Covers advertising, public relations, recruitment, organization, and related topics. HR.

627. Platt, Rutherford H.; Rowntree, Rowan A.; and Muick, Pamela C. eds. *The Ecological city : preserving and restoring urban biodiversity.* Amherst: Univ. of Mass., 1994. 291 p. : ill., maps. 0870238833 0870238841 (pbk)
Contains 17 essays by various authors on evolving relations between cities and their natural environments. Includes perspectives from geography, ecology, landscape architecture, urban forestry, law, and environmental education. Applies the concept of sustainability to the urban environment.

628. Pyle, Robert Michael. *The thunder tree : lessons from an urban wildland.* Boston: Houghton Mifflin, 1993. 220 p. 0395466318
Pyle describes how a "wasteland" on the edge of Denver became his boyhood sanctuary, wilderness area, and teacher. Makes a strong case for preserving nature in and around urban areas. HR. #

629. Rockland, Michael Aaron. *Snowshoeing through sewers : adventures in New York City, New Jersey, and Philadelphia.* New Brunswick, NJ: Rutgers, 1994. 170 p 0813521157
Rockland reverses Greeley's advice and decides to "go east, middle-age America," exploring both industrial wasteland and pockets of wilderness by foot, bike, and canoe. Includes observations on nature, human ecology, and urban naturalism.

630. Seamon, David. ed. *Dwelling, seeing, and designing : toward a phenomenological ecology.* Albany: SUNY, 1993. 363 p. : ill., maps. 0791412776
Contains 14 papers by architects, planners, geographers, and ecologists on three major themes: modernity and the built environment, architectural and landscape meaning, and relations between living and design.

631. Temple, Judy Nolte. *Open spaces, city places : contemporary writers on the changing Southwest.* Tucson: Univ. of Arizona, 1994. 144 p. 0816511659 0816514402 (pbk)
Contains 14 essays by Southwestern writers contrasting the area's burgeoning urbanism and environmental decay with their nature and conservation oriented writing. Authors include Rudolfo Anaya, Charles Bowden, Lawrence Clark Powell, Luci Tapahonso, Frederick Turner, and Ann Zwinger.

632. Tuan, Yi-fu. *Topophilia : a study of environmental perception, attitudes, and values.* NY: Columbia Univ., 1990, c1974. 260 p. 0231073941

AR: *Passing strange and wonderful.*
Topophilia means love of place. Considers how to dwell ecologically in natural places by creating structures that maintain the integrity of topographic domains.

633. Wallace, David Rains. *Idle weeds : the life of a sandstone ridge.* SF: Sierra Club, 1980. 183 p. 0871562715
Uses a detailed description of plant and animal life on a suburban ridge in Ohio as a springboard to consider a variety of ecological and philosophical issues including evolution, the relations between nature and civilization, and the place of the individual within nature and society. AOT. #

634. Whipple, A. B. C. *Critters : adventures in wildest suburbia.* NY: St. Martin's, 1994. 208 p. 0312104456
A personalized and often amusing account of wild animals that the author has encountered in his Connecticut home and yard over 40 years, including skunks, rabbits, raccoons, deer, Canada geese, bats, and swans. Includes natural history information for each animal encountered.

Books for young adults:

635. Herberman, Ethan. *The city kid's field guide.* NY: Simon & Schuster, 1989. 48 p. : col. photos. 0671677497
Companion book to a *Nova* TV program. Outstanding Science Trade Book for Children, 1989.
Describes the wildlife commonly found in a variety of urban environments including backyards, vacant lots, parks, and city margins. Takes a strong ecological approach, describing natural communities and processes. HR GR 5-9.

4. Activities and Issues

This chapter includes books on human activities which affect the entire global environment and the impacts and issues surrounding them. How we grow and distribute food, for example, affects virtually all biomes, is dependent upon water supply, and generates wastes. Energy required for industry and for personal use creates pollution and poses waste disposal problems. One controversial form of energy, nuclear power, is also linked to the production of nuclear weapons and is a source of radioactive waste. Just as the science of ecology stresses the interconnectedness of all life processes, the study of human activities demonstrates that they, too, are all linked and have inevitable consequences. The Environmental Action chapter stresses how to limit or mitigate such activities or replace them with less harmful alternatives

Agriculture

Berry, Kittredge, Logan, Logsdon, Osborn, and Russell provide philosophical and social commentary on agriculture. Problems of contemporary agriculture include pollution from pestcides and herbicides; cropland loss to development or desertification, soil erosion, and groundwater depletion, as revealed in books by Fitchen, Graham, Grainger, Jacobs, Opie, Sauder, Symanski, and Wright. The promise and performance of organic agriculture, integrated pest management, and sustainable agriculture are explored by Altieri, Jackson, Lappe, Mollison, Savory, and Fukuoka. Nabhan's books on Native American ethnobotany are listed here, while those on tropical ethnobotany can be found in Rainforests (441-2, 457, 469) Home gardening books are in Greener Homes and Gardens. Ancona and Anderson's photoessay on *The American Family Farm* is highly recommended for readers of all ages.

Books for adults:

636. Altieri, Miguel A. *Agroecology : the scientific basis of alternative agriculture.* Boulder, CO: Westview, 1987. 227 p. 0813372844
Describes how ecologically sound agricultural methods can yield large,

sustainable harvests. Provides a historical and theoretical basis and compares various methods and systems. HR.

637. Ausubel, Ken. *Seeds of Change : the living treasure : the passionate story of the growing movement to restore biodiversity and revolutionize the way we think about food.* SF: Harper, 1994. 232 p. : col. photos. 0062500082
Documents and demonstrates how biodiversity and public health can be improved through heirloom seeds, heritage vegetables, and organic farming. He provides a tour of Seeds of Change Farm and recipes from celebrated chefs Alice Waters, Charley Trotter, and Mark Miller.

638. Berry, Wendell. *A continuous harmony; essays cultural and agricultural.* NY: Harcourt Brace Jovanovich, 1975, c1972. 182 p. 0516225751
AR: *The long-legged house.*
Berry calls for a less materialistic but more spiritually rich lifestyle, exposes the devastation and dangers of strip mining, and proposes greater unity between humankind, other animals, and the land. HR.

639. ___. *The gift of good land : further essays, cultural and agricultural.* SF: North Point, 1981. 281 p. 0865470529
Berry continues his crusade for holistic agriculture and economics as presented in *The unsettling of America*, arguing that the natural and the cultural should be combined and sustained, not consumed. HR.

640. ___. *The unsettling of America : culture & agriculture.* SF: Sierra Club, 1986, c1977. 228 p. 0865470529
Advocates a return to family farming, arguing that the environmental crisis is a moral consequence of humanity's separation from the land. HR.

641. Callaway, M. Brett, and Francis, Charles A. eds. *Crop improvement for sustainable agriculture.* Lincoln: Univ. of Nebraska, 1993. 261 p. 0803214626
Contains 18 essays on how to change the adaptation of plants to environmental conditions; genetic improvement of crop plants for stress conditions; and future sustainable agricultural systems. Somewhat technical, but accessible for adults with some background in agronomy.

642. Clark, Robert. *Our sustainable table.* SF: North Point, 1990. 176 p. 0865474451
These 12 essays by Gary Nabhan, Gretel Ehrlich, Alice Waters, Wendell Berry, Wes Jackson, Frances Moore Lappe, and others explore the connections between sustainable agriculture and cooking based on fresh, organically grown, regional foods. Considers the political, social, and economic aspects of farming, cooking and eating.

643. Fitchen, Janet M. *Endangered spaces, enduring places : change, identity, and survival in rural America*. Boulder: Westview, 1991. 314 p. : ill. 0813311144 0813311152 (pbk)
Focusing on upstate New York, Fitchen depicts farmers trying to hold on to their land; communities whose water has been despoiled by toxic waste; and citizens trying to maintain a decent quality of environmental and economic life.

644. Fukuoka, Masanobu. *The natural way of farming : the theory and practice of green philosophy*. NY: Japan, 1985. 280 p. : photos. 0870406132
AR: *The one-straw revolution*.
Fukuoka's approach to farming the natural way, the theory and practice of working with nature, and living better for it. Demonstrates the practical advantages of organic farming and no-tillage and explains how we can and must change our way of doing things if we are to make lasting peace with the Earth and ourselves.

645. Jackson, Wes. *Altars of unhewn stone : science and the Earth*. SF: North Point, 1987. 158 p. 0865472874
Considers how environmentally conscious farmers can simulate the natural environment as closely as possible to increase genetic variation. This classic study of sustainable agriculture is HR.

646. ___. *Becoming native to this place*. Lexington: Univ. of Kentucky, 1994. 121 p. 0813118468
Jackson urges us to base our culture and agriculture on nature's principles, to recycle. Calls for "a new generation of homecomers" to establish their roots in the soil, and local communities to bring about a sustainable ecological transformation of America. Similar to Berry's work. HR.

647. ___. *New roots for agriculture*. New ed. Lincoln: Univ. of Nebraska, 1985, c1980. 150 p. 0803275625
Demonstrates how massive plowing and other industrial agricultural practices have damaged and eroded the soil. Recommends the use of perennial crops and other alternatives. HR.

648. Jackson, Wes; Berry, Wendell; and Colman, Bruce. eds. *Meeting the expectations of the land : essays in sustainable agriculture and stewardship*. SF: North Point, 1984. 250 p. 086547172X
Contains 17 essays by the editors, Donald Worster, Amory Lovins, Gary Nabhan and others on a new agriculture which combines modern science and ancient wisdom related to ecology, conservation, pollution, economics, etc. For adults with a basic knowledge of agriculture and ecology. HR.

649. Jacobs, Lynn. *Waste of the West : public lands ranching*. Tucson: L. Jacobs, 1991. 602 p. : ill., maps. 0962938602
A well-researched, comprehensive consideration of the many negative consequences of ranching on the western U.S. (desertification, deforestation, pollution, contamination, species loss, economic and social consequences, etc.) HR for adults, especially for activists.

650. Kittredge, William. *Hole in the sky : a memoir*. NY: Knopf, 1992. 238 p. 0679411666
The story of how three generations bought a large ranching claim, developed it beyond its carrying capacity, and diminished many existing wildlife species. Tells how the loss of harmony with the land and between family members as they pursued economic goals is a metaphor for the ecological destruction of the West. HR.

651. ___. *Owning it all : essays*. Saint Paul, MN: Graywolf, 1987. 182 p. 0915308967
Autobiographical essays that delve into those myths of land, manhood, and manifest destiny and their effect on our lives today. Kittredge left the farm and began to think about how the myths of the West no longer serve us.

652. Langer, Richard W. *Grow it! : the beginner's complete in-harmony-with-nature small farm guide, from vegetable and grain growing to livestock care*. Rev. ed. NY: Noonday, 1994. 360 p. : ill. 0374523908
An introductory to intermediate level guide to organic farming including planning, soils, composting, organic pest management and fertilization, livestock, stocking ponds with fish, simple living, etc. This practical and philosophical book is for both backyard gardeners and professional farmers.

653. Lappe, Frances Moore. *Diet for a small planet*. NY: Ballantine, 1991, c1971. 479 p. : ill. 0345373669
AR:*Recipes for a small planet*, *World hunger: twelve myths*.
This is primarily a consideration of how much food is used to produce livestock, the environmental consequences, how grains could be used more effectively to feed the human population directly, and how a grain-based diet is healthier than a meat-based one. It is secondarily a renowned vegetarian cookbook HR.

654. Lappe, Frances Moore; Collins, Joseph; and Fowler, Cary. *Food first : beyond the myth of scarcity*. Rev. ed. NY: Ballantine, 1980, c1978. 619 p. 0345298187
Contends that although overpopulation and drought effect food scarcity, the main cause is the replacement of localized, self-reliant agricultural systems with those designed for cash-crop export. HR.

655. Leslie, Anne R., and Cuperus, Gerrit W. eds. *Successful implementation of integrated pest management for agricultural crops.* Boca Raton, FL: Lewis, 1993. 193 p. 0873715020
Pest control through cultural practices allowing less dependence on chemicals.

656. Logan, Ben. *The land remembers : the story of a farm and its people.* Madison: Stanton & Lee, 1986 printing, c1975. 281 p. 0883610957
The memoir of a young man growing up on a farm in southwestern Wisconsin, what the land meant to him and his family, and how they never abused the land with destructive agricultural techniques.

657. Logsdon, Gene. *At nature's pace : farming and the American dream.* Foreword by Wendell Berry. NY: Pantheon, 1994. 208 p. 0679427414
The authors vision of nature and farming expressed through a series of essays.

658. ___. *The contrary farmer.* Post Mills, VT: Chelsea Green, 1994. 236 p. 0930031679
Recommends cottage farming: small-scale, part-time growing that aims to reduce food expenses and increase pleasure in living. Logsdon passionately combines scientific pragmatism and idealism. HR.

659. Mollison, B. C. *Permaculture : a practical guide for a sustainable future.* Washington: Island, 1990. 615 p. : col. photos. 1559630485
Permaculture is the conscious design and maintenance of economical, agriculturally productive ecosystems that have the diversity, stability, and resilience of natural ecosystems. Both a practical manual and a consideration of integrating landscape and human activity. HR.

660. Nabhan, Gary Paul. *The desert smells like rain : a naturalist in Papago Indian country.* SF: North Point, 1987, c1982. 192 p. 0865470502
Provides insights into the natural history of desert plants and animals as it documents a dying agricultural tradition that has enriched the biological diversity of the Papago's seemingly harsh desert environment.

661. ___. *Enduring seeds : native American agriculture and wild plant conservation.* SF: North Point, 1989. 225 p. 0865473439
Outlines rediscovered Native American farming practices and shows how they may ensure the food supply of the future.

662. ___. *Gathering the desert.* Tucson: Univ. of Arizona, 1985. 209 p. 0816509352
Focuses on 12 of the over 425 edible Sonoran wild plants. Demonstrates

how traditional plant-human interaction has sustained both land and culture. Tells how, when these patterns are disrupted, co-evolved desert wildlife is as dramatically affected as the plants themselves.

663. U. S. Nat'l Research Council. Comm. on the Role of Alternative Farming Methods in Modern Production Agriculture. *Alternative agriculture.* Washington: Nat'l Academy, 1989. 448 p. 0309039878 0309039851 (pbk) Describes the increased use of alternative practices to reduce pesticide and chemical fertilizer use, cut costs, increase profits, and minimize negative impacts. Considers federal policy and economic factors in relation to these alternatives.

664. Opie, John. *Ogallala : water for a dry land.* Lincoln: Univ. of Nebraska, 1993. 412 p. 0803235577
The massive Ogallala Aquifer has been crucial in supplying irrigation water from South Dakota to Texas, but its overuse could lead to a return to mere subsistence level farming.

665. Russell, Sharman Apt. *Songs of the fluteplayer : seasons of life in the Southwest.* Reading MS: Addison-Wesley, 1991. 160 p. 0201570939
A couple learns to adapt to and appreciate the natural hardships and splendors of the Mimbres River Valley in New Mexico. The annual water cycle of drought and flash floods is particularly well depicted.

666. Ruttan, Vernon W. ed. *Agriculture, environment, and health : sustainable development in the 21st century.* Minneapolis: Univ. of Minnesota, 1994. 401 p. 0816622914 0816622922 (pbk)
Contains 14 papers examining biological and technical constraints on crop and animal production, resource and environmental constraints on sustainable agriculture, and health constraints on agricultural production.

667. Sargent, Frederic O.; Lusk, Paul; Rivera, Jose; and Varela, Maria. *Rural environmental planning for sustainable communities.* Washington: Island, 1991. 254 p. : photos. 1559630248 (pbk) 1559630256
REP is a method for citizens in small towns and rural areas to participate in planning. This hands-on guide is designed to help create growth plans that are environmentally sound, politically acceptable, and economically feasible.

668. Sauder, Robert A. *The lost frontier : water diversion in the growth and destruction of Owens Valley agriculture.* Tucson: Univ. of Arizona, 1994. 208 p. : ill. 0816513813
Covers the agricultural settlement and communities of the valley prior to 1913, the building of a huge water system to benefit Los Angeles industries, and the subsequent destruction of agriculture in the valley. This valu-

able case history is particularly recommended to water resources activists.

669. Savory, Allan. *Holistic resource management*. Washington: Island, 1988.
564 p. : ill. 0933280629 0933280610 (pbk)
AR: *Holistic resource management workbook*.
A comprehensive manual for integrating environmental practices into
farming, ranching, and livestock management. Demonstrates the practicali-
ty and profitability of this approach. Recommended for ranchers, farmers,
and land management activists.

670. Vail, David J.; Hasund, Knut Per; and Drake, Lars. *The greening of agricultural reform in industrial societies : Swedish reforms in comparative perspective*.
Ithaca, NY: Cornell Univ., 1994. 300 p. : ill. 0801427509
Swedish progress in sustainable agriculture, farm animal welfare, conservation, pesticide taxes, organic farming subsidies, etc. is described and compared to other industrialized nations. For political economists, rural
sociologists, agricultural scientists, policy analysts, and activists.

671. Whynott, Douglas. *Following the bloom : across America with the migratory
beekeepers*. Harrisburg, PA: Stackpole, 1991. 214 p. 0811719448
Interweaves four themes: the lives and work of migratory beekeepers; the
importance of bees as the primary pollinators in agriculture; the natural
history and social orders of bees; and the potential effects of the African-
ized bee on American ecology and agriculture.

672. Wolf, Edward C. *Beyond the green revolution : new approaches for Third
World agriculture*. Washington: Worldwatch Inst., 1986. 46 p. 0916468747
Although the Green Revolution was successful in a few cases, in others it
has had negative effects on the land and local farmers. Crop failures are
frequent. Suggests ways to combine the advantages of traditional farming
with appropriate modern methods.

673. Wright, Angus Lindsay. *The death of Ramon Gonzalez : the modern agricultural dilemma*. Austin: Univ. of Texas, 1990. 337 p. 0292715609
Uses the case of one Mexican farmworker who died from pesticides to illu-
minate the collusion of industries and governments in the U.S. and the
Third World, to continue using carcinogenic and environmentally hostile
agrichemicals. HR.

Books for young adults:

674. Aaseng, Nathan. *Ending world hunger*. NY: Watts, 1991. 143 p. : photos.
0531110079

AR: *Overpopulation*.
Studies the political, social, and scientific-technical reasons for world food shortages, and possible approaches to increasing food production. Makes a strong argument for sustainable agriculture.

675. Ancona, George, and Anderson, Joan. *The American family farm : a photo essay*. San Diego: Harcourt Brace Jovanovich, 1989. 94 p. : photos. 0152030255
ALA Notable Book Award, Notable Children's Trade Book in Social Studies, 1989.
Focuses on the daily lives of three families in Massachusetts, Georgia, and Iowa. All three work hard to preserve the land, and one uses no chemical pesticides or fertilizers. HR GR 4-8 and for family reading. #

676. Fine, John Christopher. *The hunger road*. NY: Atheneum, 1988. 148 p. : photos. 0689313616
Discusses world starvation; its escalation because of overpopulation, poverty, habitat destruction, development, and inequities in distributing a limited food supply; and past and present efforts to alleviate the problem.

677. Lee, Sally. *Pesticides*. NY: Watts, 1991. 143 p. : ill. 0531130177
Discusses the uses of pesticides, the dangers they may pose to our food supply, and governmental restrictions imposed on those considered harmful.

678. McCoy, J. J. *How safe is our food supply?* NY: Watts, 1990. 159 p. : ill. 0531109356
Outlines the various procedures by which foods are preserved, grown, and stored, demonstrating how some of these methods may be detrimental to human health.

Human-Animal Relations

Humans and other animals interact at all levels in hunting, fishing, agriculture, science, industry, and the home. A range of positions on hunting is presented in books by Anderson, Bass, Bourjaily, Burroughs, Dizard, Dobie, Gish, Hemingway, Roosevelt, Rue, and Wright. Although the humane treatment of animals has been a concern of many people and cultures for millenia, the use of animals in factory farming and research has generated great controversy during the past twenty years, thus generating the animal rights movement. Clark, Fox, Hearne, Mason, Mighetto, Shepard, Sanders, and Sorabji explore philosophical and moral aspects of human-animal relations. Shepard and Sanders' *Sacred Paw* is particulary fascinating. Newkirk, Regan, Singer, and Welsh promote the animal rights movement while Howard and Marquardt provide critiques of it. Pringle's *The Animal Rights Controversy* is particularly recommended for teens.

Books for adults:

679. Anderson, Dennis. *An hour before dawn : so some cranes may dance : stories of the outdoors*. Stillwater, MN: Voyageur, 1993. 221 p. 0896581802
The first section consists of hunting, fishing, and nature stories with a common theme of threats to the North American duck population from over-hunting and habitat loss. The second describes largely successful efforts to save 15 crane species.

680. Bass, Rick. *The deer pasture*. NY: Norton, 1989, c1985. 122 p. 0393305899
This account of a family's annual deer hunt considers the role of hunting as part of the ecological balance, and some philosophical and moral aspects of hunting. #

681. Bourjaily, Vance Nye. *The unnatural enemy : essays on hunting*. Tucson: Univ. of Arizona, 1984, c1963. 180 p. : ill. 0816508844
A consideration of how great hunters of the past have looked upon the sport and of how their views apply and fail to apply to contemporary behavior.

682. Bourjaily, Vance Nye, and Bourjaily, Philip. *Fishing by mail : the outdoor life of a father and son*. NY: Atlantic Monthly, 1993. 200 p. 0871135566 0802112633
This exchange of highly literate and often amusing letters between a father and son describes their hunting and fishing activities along with asides on what hunting, fishing, and nature mean to them.

683. Burroughs, Franklin. *Billy Watson's croker sack : essays*. Boston: Houghton Mifflin, 1991. 149 p. 0395619009
AR: *The river home*.
Whether waiting in a duck blind, fishing in a stream, gathering flora and fauna, or talking to his neighbors, Burrough's central theme in these essays is the intimate connection between the natural world and the self.

684. Clark, Stephen R. L. *The moral status of animals*. NY: Oxford, 1984, c1977. 221 p. 0192830406
A relatively early rejection of speciesism (the concept that some species, particularly humans, are superior to others) and summary of earlier think-

ing on animal rights. Clark calls for vegetarianism and an end to animal experimentation.

685. ___. *The nature of the beast : are animals moral?* NY: Oxford, 1982. 127 p. 0192830414
In this continuation of *The moral status of animals*, Clark proposes that people and other animals have similar evolutionary intellectual roots and that animals are not driven entirely by instinct, but think and act intentionally as well.

686. Dizard, Jan E. *Going wild : hunting, animal rights, and the contested meaning of nature.* Amherst: Univ. of Mass., 1994. 200 p. 0870239082
Dizard uses the case a special nine day deer hunt in which 576 deer were killed to examine several questions: Should nature be seen as a self-balancing harmony or a challenge to use and dominate? Who decides what is wild? Is "wildlife management" an oxymoron?

687. Dobie, J. Frank. *The Ben Lilly legend.* Austin: Univ. of Texas, 1981, c1950. 237 p. 0292707282
The chronicle of an aging hunter who was noted for his understanding of wildlife and ecology in the Southwest. To Dobie, Lilly was a personification of his ideal of a white person living by a Native American nature-oriented cosmology.

688. Fitzgerald, Sarah. *International wildlife trade : whose business is it?* Washington: World Wildlife Fund, 1989. 494 p. : ill. 0942635108 (pbk) 0942635132
Explains laws regulating wildlife trade, the problems of poaching and smuggling wildlife, and the consequences of such trade, particularly on rare and endangered species.

689. Fox, Michael W. *Agricide : the hidden crisis that affects us all.* NY: Schocken, 1986. 229 p. : photos. 0805240136 0805208186 (pbk)
AR: *Returning to Eden, The way of the dolphin, Whitepaws.*
Demonstrates how modern agribusiness' bottom-line orientation has created practices which are both environmentally and economically unsound and are "stealing from the future." He analyzes the abuses of factory farming and offers alternatives.

690. Gish, Robert. *Songs of my hunter heart : a western kinship.* Albuquerque: Univ. of New Mexico, 1994, c1992. 168 p. 0862315240
Lyrically traces the evolution of Gish's changing attitudes on hunting, masculinity, ecology, and the human relation to nature. HR.

691. Hearne, Vicki. *Animal happiness*. NY: HarperCollins, 1994. 238 p. 0060190167
Contains 40 short selections about animals and animal trainers, showing that each species experiences happiness in its own ways, and that people can be happier and more aware through their own interactions with animals.

692. Hemingway, Ernest. *Green hills of Africa*. NY: Collier, 1987, c1935. 304 p. 0020519303
This account of one of his hunting expeditions reads almost like fiction, and although soaked in macho bloodlust, contains a few lyrical passages vividly describing African nature and wildlife. Not his best work, but of interest to Hemingway buffs.

693. Howard, Walter E. *Nature and animal welfare : both are misunderstood*. Pompano Beach, FL: Exposition, 1986. 67 p. : ill. 0682403148
Considers animal welfare and rights, vested interests of environmental organizations, the biological role of hunters and trappers, wildlife management, and nature's main problem: human procreation.

694. Kerasote, Ted. *Bloodties : nature, culture, and the hunt*. NY: Random House, 1993. 277 p. 0394576098
Contrasts the attitudes and activities of subsistence hunters in Greenland with those of "trophy" hunters. He rejects the antihunting movement and examines a philosophical middle ground of responsible hunting in a context of respect for nature.

695. Linzey, Andrew. *Christianity and the rights of animals*. NY: Crossroad, 1987. 197 p. 0824508769 0824508750 (pbk)
Contends that other animals are the creation of and have inherent value to God and should be treated as such. Includes 16 church statements on animals, issued between 1956-86.

696. Marquardt, Kathleen; Levine, Herbert M.; and LaRochelle, Mark. *AnimalScam : the beastly abuse of human rights*. Washington: Regnery, 1993. 221 p. 0895264986
A critique of the animal rights movement by the founders of Putting People First. Although it contains a great deal of hyperbole, this book includes documentation on abuses perpetrated by animal rights extremists.

697. Mason, Jim. *An unnatural order : uncovering the roots of our domination of nature and each other*. NY: Simon & Schuster, 1993. 319 p. 0671769235
Analyzes the links between the "dominationist" view toward animals (traced to the transition from feminist paganism to patriarchal monothe-

ism), fears of our own animal nature, sexual repression, misogyny, racism, and colonialism. HR.

698. Mighetto, Lisa. *Wild animals and American environmental ethics*. Tucson: Univ. of Arizona, 1991. 177 p. : photos. 0816511608 0816512663 (pbk)
Identifies the roles played by popular writers, hunters, humanitarians, and ecologists in influencing attitudes toward animals. Discusses predator control, biocentrism, elimination of exotic species, and other current topics.

699. Newkirk, Ingrid. *Free the animals!* : *the untold story of the* Animal Liberation Front *and its founder, "Valerie"*. Chicago: Noble, 1992. 372 p. 187936011X
The founder of the Animal Liberation Front describes the motivations, origins, history, tactics, and goals of this radical group. Newkirk provides additional information and attempts to make a balanced assessment of ALF.

700. ___. *Save the animals!* : 101 *easy things you can do*. NY: Warner, 1990. 192 p. 0446392340
Advice on cruelty-free products, vegetarianism, lobbying, and community action. Each chapter includes sections describing the problem, the solution, resources, and what you can do. HR. #

701. Regan, Tom. *The case for animal rights*. Berkeley: Univ. of Cal., 1983. 425 p. 0520049047
AR: *All that dwell therein*.
Argues that animals have memory, beliefs, emotions, a sense of the future, and conscious intentionality. He calls not only for more humane treatment of animals, but a more inclusive way of looking at their role and ours in nature and society.

702. ___. *The struggle for animal rights*. Clarks Summit, PA: Int'l Soc. for Animal Rights, 1987. 197 p. 0960263217
Provides a history and analysis of the animal rights movement and a consideration of experimentation on animals, and related issues.

703. Regan, Tom, and Singer, Peter. eds. *Animal rights and human obligations*. 2d rev. ed. Englewood Cliffs, NJ: Prentice Hall, 1989, c1976. 280 p. 0130368644
Philosophers and ethologists argue for and against animal rights. Issues include vivisection, other types of laboratory testing, vegetarianism, and the similarities and differences between humans and other animals.

704. Reisner, Marc. *Game wars* : *the undercover pursuit of wildlife poachers*. NY: Viking, 1991. 352 p. 0670814865
Describes how undercover game wardens detect poachers, create covers,

impersonate smugglers, perform stings, and infiltrate crime rings. Focuses on Dave Hall, a legendary agent who has cracked poaching rings from Louisiana to Alaska.

705. Roosevelt, Theodore. *Ranch life and the hunting-trail.* NY: St. Martin's, 1985, c1896. 186 p. 031266365X
AR: *Hunting trips of a ranchman, The wilderness hunter.*
Considers how animals, humans, and methods must adapt to succeed in the harshness of North Dakota; how nature can be both appreciated and exploited; and how frontier justice serves the ends of a larger morality. His later conservationist values (wise, long term use rather than preservation for nature's sake) can be traced to this book.

706. Rue, Leonard Lee. *The deer of North America.* 2d exp. ed. Danbury, CT: Grolier, 1989. 588 p. : photos. 1556540515
An extensive study with three sections: "The animal and its behavior," "How the year goes," and "Toward sound deer management." Provides an historical view and current approaches to deer management. Makes a case for controlled hunting as an important tool in population management.

707. Shepard, Paul, and Sanders, Barry. *The sacred paw : the bear in nature, myth, and literature.* NY: Viking, 1985. 270 p. : ill. 0670151335
By relating both the natural and the mythic history of bears as real animals and as symbols, the authors also provide a context for understanding the correct human place in the natural world. This fascinating, multifaceted work is HR.

708. Singer, Peter. *Animal liberation.* 2d rev. ed. NY: Avon, 1991. 352 p. : ill. 0380713330
Presents a strong case against speciesism. Encourages vegetarianism and opposes the use of animals in scientific experimentation. This ed. also covers the rise of the animal rights movement.

709. Singer, Peter, and Cavalieri, Peter. eds. *The great ape project : equality beyond humanity.* NY: St. Martin's, 1994. 312 p. 0312104731
A collection of 30 essays by various scientists supporting the "Declaration on Great Apes" which demands the right to life, protection of individual liberty, and prohibition of torture for chimpanzees, gorillas, and orangutans. The biological and behavioral links between apes suggest that humans are "the third chimpanzee."

710. Sorabji, Richard. *Animal minds and human morals : the origins of the Western debate.* Ithaca, NY: Cornell, 1993. 267 p. 080142948X
Traces the roots of current thinking about animals and the relationship

between animals and humans to Aristotle and other classical philosophers. Considers animal intelligence, animal welfare, and comparative psychology.

711. Welsh, Heidi J. *Animal testing and consumer products.* Washington: Investor Responsibility Research Center, 1990. 167 p. 0931035392
Covers the evolution of the animal protection movement, legislation, current testing methods, alternative methods, shareholder resolutions in nine companies, and testing in universities, hospitals and corporations. Includes a directory of 47 animal protection groups.

712. Wright, William H. *The grizzly bear : the narrative of a hunter-naturalist.* Lincoln: Univ. of Nebraska, 1977, c1909. 304 p. : ill. 0803209274
An informative and amusing account of the habits and habitat of the grizzly based on a quarter century of observation.

Books for young adults:

713. Bloyd, Sunni. *Animal rights.* San Diego: Lucent, 1990. 127 p. : ill. 1560061146
Discusses current animal rights issues, their underlying philosophies, and ways to curb animal abuse.

714. Burroughs, Nigel. *Nature's chicken : a book for animal lovers.* Summertown, TN: Book Co., 1992. 35 p. : ill. 0913990922
Describes the adverse effects of factory poultry farming on human health and the chickens themselves, and discusses the advantages of becoming a vegan, someone who does not eat meat or dairy products. GR 4-8.

715. Fox, Michael W. *Animals have rights, too : a primer for parents, teachers and young people.* NY: Crossroad, 1991. 144 p. : ill. 0824512855
This informal, conversational approach to a controversial subject sometimes suffers from overemotionalism, but is a good, elementary overview of the uses and abuses of animals. #

716. Goodman, Billy. *A kid's guide to how to save the animals.* NY: Avon, 1991. 136 p. : ill. 0380766515
Discusses endangered animals, non-threatened species of animals that suffer through mistreatment, and what children can do to protect animals. GR 4-8.

717. Patterson, Charles. *Animal rights.* Hillside, NJ: Enslow, 1993. 112 p. : photos. 089490468X

Discusses the ways in which animals are used for medical research, food, education, and entertainment, and presents the views of people concerned with the inhumane treatment of animals. GR 7-10.

718. Pringle, Laurence P. *The animal rights controversy*. San Diego: Harcourt Brace Jovanovich, 1989. 103 p. : photos. 0152035591
Outstanding Science Trade Book for Children, Notable Children's Trade Book in Social Studies, 1989.
Presents viewpoints of scientists and animal rights advocates on the use of animals for scientific research, for food, and for human enjoyment. Considers the moral and ethical issues. Lists relevant organizations. The best book on the subject for teens. HR.

Endangered Species

Many animal and plant species have been eliminated, and many others face extinction. This section includes books which emphasize the endangerment factor. Books emphasizing the natural history of individual species can be found in Natural History. Adams and Carwardine, Bergman, and Matthiesen offer philosophical outlooks. Books on biodiversity and/or conservation and species restoration efforts include those written or edited by Barbour, Beatley, Day, DeBlieu, DiSilvestro, Durrell, Grumbine, Hudson, Jensen, James, Lowe, Norton, Schaller, Tudge, Wallace, Ward, and Wilson. Related books can be found in the Conservation section. The books by the Facklams, Patent, and Pringle are highly recommended for teens.

Books for adults:

719. Adams, Douglas, and Carwardine, Mark. *Last chance to see*. NY: Ballantine, 1992, c1990. 262 p. : col. photos. 0517571195I
A novelist and a zoologist describe unusual endangered species such as New Zealand kakapos, Komodo dragons, blind river dolphins, and white rhinos. A highly personalized account of their travels and discoveries, told with both humor and poignancy. AOT. #

720. Allaby, Michael, and Lovelock, J. E. *The great extinction*. London: Paladin, 1985. 192 p. : col. photos. 0586085017
When the dinosaurs became extinct, so did nearly three quarters of all living species. Allaby and Lovelock propose that an accumulation of terrestrial and extraterrestrial events created a level of air pollution that modern industry could not create on its own, even through the greenhouse effect.

721. Barbour, Michael G.; Pavlik, Bruce; Drysdale, Frank; and Lindstrom, Susan. *California's changing landscapes : diversity and conservation of California vegetation*. Sacramento: Cal. Native Plant Soc., 1993. 244 p. : ill. 0943460174
Describes historic landscapes and plant resources, how they came to be consumed, what replaced them, and what might remain in the future. Portraits of devastation are balanced by the delight that the authors find in the plant kingdom. Three potential scenarios are examined.

722. Barker, Jerry R., and Tingey, David T. eds. *Air pollution effects on biodiversity*. NY: Van Nostrand Reinhold, 1992. 322 p. 0442007485
Contains fourteen papers from an EPA/Fish and Wildlife Service workshop on biodiversity, air pollution exposure and effects, consequences of air pollution on biodiversity, and policy issues and research needs.

723. Barker, Rocky. *Saving all the parts : reconciling economics and the Endangered Species Act*. Washington: Island, 1993. 268 p. : ill. 155963202X 1559632011 (pbk)
Presents an overview of current endangered species controversies, why protection is important, how we know species are in trouble, and the implications of human activities. Analyzes the "jobs vs. nature" controversy, trends in resource mangement, and the movement toward sustainability.

724. Beatley, Timothy. *Habitat conservation planning : endangered species and urban growth*. Austin: Univ. of Texas, 1994. 234 p. 0292707991 0292708068 (pbk)
AR: *Ethical Land use : principles of policy and planning*.
Examines the success of habitat conservation plans and their political and practical issues and constraints.

725. Bergman, Charles. *Wild echoes : encounters with the most endangered animals in North America*. NY: McGraw-Hill, 1990. 342 p. : ill. 007004922X
Combines natural history, intellectual history and highly personalized-possibly embellished-accounts of his encounters with endangered species. Somewhat similar to Durrell.

726. Brown, David E. ed. *The grizzly in the Southwest : documentary of an extinction*. Norman: Univ. of Oklahoma, 1985. 274 p. : ill., maps. 0806119306
A detailed account of the extinction of the grizzly from Arizona, New Mexico, southern Colorado, and northern Mexico. Includes implications for other species, other areas, and wildlife reintroduction. HR.

727. Brown, David E., and Gish, Dan Miles. eds. *The wolf in the Southwest : the making of an endangered species*. Tucson: Univ. of Arizona, 1983. 195 p. : ill., maps. 0816507821 0816507961 (pbk)
Since the turn of the century this once plentiful species has been systemat-

ically eradicated under pressure from the livestock industry. This book details the eradication, and although somewhat technical, is accessible to adults with some scientific background.

728. Day, David. *Noah's choice : true stories of extinction and survival.* London: Viking, 1990. 169 p. 0670806692
Details the extinction of hundreds of species and the success of saving some species from the brink of extinction. HR AMT. #

729. DeBlieu, Jan. *Meant to be wild : the struggle to save endangered species through captive breeding.* Golden, CO: Fulcrum, 1991. 302 p. : ill. 1555910742 1555911668 (pbk)
Library Journal Best Sci-Tech Book 1991.
Describes efforts to preserve the endangered peregrine falcon, California condor, blackfooted ferret, Puerto Rican parrot, Florida panther, and other species through captive breeding and release. Also considers the pros and cons of this controversial practice. HR AOT. #

730. Dietz, Tim. *The call of the siren : manatees and dugongs.* Golden, CO: Fulcrum, 1992. 196 p. : ill. 1555911048
Manatees and dugongs are legendary, large, gentle, endangered marine mammals. Dietz describes their place in myth, natural history; their habits and habitat; decimation through hunting and environmental threats; and efforts to save them. AMT. #

731. DiSilvestro, Roger L. *Reclaiming the last wild places : a new agenda for biodiversity.* NY: Wiley, 1993. 266 p. 0471572446
Library Journal Best Sci-Tech Book Award, 1994.
AR: *Audubon perspectives, birth of nature; Audubon perspectives, fight for survival.*
Boundaries of a species' territory are natural, but those of wildlife refuges are not. Claiming that isolating species artificially cannot preserve them, DiSilvestro advocates an 'applied biodiversity' approach instead. HR.

732. Durrell, Gerald Malcolm. *The aye-aye and I : a rescue mission in Madagascar.* NY: Simon & Schuster, 1994, c1993. 175 p. 0671884395
AR: *Ark on the move, A bevy of beasts, The stationary ark, The talking parcel.*
Durrell captures a variety of exotic endangered animals including lemurs, tortoises, leaping rats, and fosa for captive breeding. He also describes the people, language, landscape, and rapid deforestation of this fascinating island. #

733. Ehrlich, Paul R.; Dobkin, David S.; and Wheye, Darryl. *Birds in jeopardy : the imperiled and extinct birds of the United States and Canada including Hawaii and Puerto Rico.* Stanford: Stanford Univ., 1992. 259 p. : col. ill. 0804719675

Covers nesting, feeding, breeding, migration, wintering, population estimates, relevant behavioral factors and threats, status of imperilment, conservation efforts, and recovery plans for 60 species. Also covers birds which
have become extinct, drawing lessons from their demise. Highly detailed,
but nontechnical.

734. Ehrlich, Paul R., and Ehrlich, Anne H. *Extinction : the causes and consequences of the disappearance of species.* NY: Ballantine, 1985, c1981. 305 p.
0345330943
Empirically documents the rate of species loss, warns of the social and
environmental consequences, and considers political and ethical aspects
of the problem. HR.

735. Grumbine, R. Edward. *Ghost bears : exploring the biodiversity crisis.* Intro.
by Michael Soule. Washington: Island, 1992. 294 p. 155963152X
1559631511 (pbk)
Explores the concept of biodiversity, focusing on the decline of grizzly bear
populations as an example of biodiversity loss. Describes the connections
between conservation biology, environmental laws, land management
practices, and environmental values. HR.

736. Hudson, Wendy E. ed. *Landscape linkages and biodiversity.* Washington:
Island, 1991. 196 p. 1559631082 1559631090 (pbk)
Contains ten essays by various authors. They propose that fragmentary
approaches to wildlife protection be replaced with a unified approach of
habitat preservaton, maintenance of land corridors, and integrating biodiversity conservation with human activities.

737. James, Valentine Udoh. *Africa's ecology : sustaining the biological and environmental diversity of a continent.* Jefferson, NC: McFarland, 1993. 293 p.
0899508596
Disputes the dominance of technical issues in wildlife protection, arguing
that social and cultural aspects must be considered. Reviews the status of
the African environment and critiques policies for developing its resources.

738. Jensen, Deborah B.; Torn, Margaret S.; and Harte, John. *In our own
hands : a strategy for conserving California's biological diversity.* Berkeley: Univ. of
Cal., 1993. 302 p. 0520080157
Discusses the threats to biodiversity posed by increased demand due to
overpopulation and inefficient use of resources, and proposes a plan to
conserve it. Particularly recommended for Californians.

739. LaBastille, Anne. *Mama Poc : an ecologist's account of the extinction of a
species.* NY: Norton, 1990. 320 p. : photos. 0393308006

AR: *Assignment wildlife.*
A leading conservationist presents a 20 year study of the extinction of the Guatemalan giant grebe and considers the implications for other species and for preservation strategies.

740. Lowe, David W.; Matthews, John R.; and Moseley, Charles J. eds. *The Official World Wildlife Fund guide to endangered species of North America.* Washington: Beacham, 1990-1992. 3 v. (1670 p.) : col. photos, maps 0933833172
A compendium of information on hundreds of endangered animals and plants. Encourages preservation efforts. HR.

741. Marshall Cavendish Corp. *Endangered wildlife of the world.* NY: Cavendish, 1993. 13 volumes : col. photos, maps. 1854354892
American Libraries Best Reference Source, 1994.
Describes various endangered or threatened species around the world, covering their habitat, behavior, and efforts to protect them. HR. #

742. Matthiessen, Peter. *African silences.* NY: Vintage, 1992. 225 p. : map. 067931024
Matthiessen's views on endangered creatures in the wild.

743. Mech, L. David. *The wolf : the ecology and behavior of an endangered species.* Minneapolis: Univ. of Minnesota, 1981, c1970. 384 p. : photos, maps. 0816610266
Describes the behavior and ecological role of the wolves, demonstrating that they are harmless to humans, but have an extremely beneficial impact on other species, including prey. HR.

744. Mowat, Farley. *Sea of slaughter.* NY: Bantam, 1989, c1984. 448 p. 0553302694
Covers both the extinction of undersea species and those that still survive despite the constant threat of man.

745. Norton, Bryan G. ed. *The preservation of species : the value of biological diversity.* Princeton, NJ: Princeton Univ., 1986. 305 p. 0691083894
Contains 11 essays on the value of biodiversity, problems associated with species loss, and managing natural resources to maintain biodiversity. Recommended for interested laypersons, legislators, activists, policymakers, and higher-level managers.

746. Peterson, Roger Tory; Schreiber, Rudolf L.; and Cronkite, Walter. *Save the birds.* Boston: Houghton Mifflin, 1989. 384 p. : col. ill., maps. 0395511720
A lavishly illustrated compendium of information on bird habitats and

species with background information on bird preservation and a guide to preservation efforts around the world. HR. #

747. Phillips, Kathryn. *Tracking the vanishing frogs*. NY: St. Martin's, 1994. 256 p. : photos. 0312109733
For three years Phillips interviewed scientists and accompanied them on expeditions to determine why so many amphibian species are disappearing. Likely causes include acid rain, UV radiation, drought, habitat destruction from over-grazing and logging, dams, and non-native predatory fish.

748. Quinn, John R. *Wildlife survivors : the flora and fauna of tomorrow*. Blue Ridge Summit, PA: TAB, 1993. 208 p. : ill. 083064346X 0830643451 (pbk)
Discusses the forces that work against biodiversity and predicts which plants and animals will survive into the increasingly urbanized 21st century. AOT. #

749. Schaller, George B. *The last panda*. Chicago: Univ. of Chicago, 1993. 291 p. : col. photos, maps. 0226736288
Library Journal Best Sci-Tech Book, ALA Notable Book Award, 1994.
A husband-wife team spent nearly five years in China's Wolong Reserve studying the lives, life-cycle, and habitat of pandas, golden monkeys, red pandas, and takin. Also presents a panda preservation plan. HR.

750. Tudge, Colin. *Last animals at the zoo : how mass extinction can be stopped*. Washington: Island, 1992. 266 p. 1559631570
Defends zoo breeding programs and provides a lucid account of a genetic strategy to save animals from extinction. Tudge places the debate about conservation in a wide ecological context and urges us to think in terms of centuries instead of years.

751. U.S. Nat'l Research Council Comm. on the Formation of the Nat'l Biological Survey. *A biological survey for the nation*. Washington: Nat'l Academy, 1993. 205 p. 0309049849
Presents the planning for a comprehensive survey intended to gather, analyze, and disseminate information required for the wise stewardship of America's natural resources, and to foster an understanding of our natural systems and the benefits they provide to society.

752. Wallace, David Rains. *Life in the balance : companion to the Audubon television specials*. San Diego: Harcourt Brace Jovanovich, 1987. 309 p. : col. photos. 0151515611
Describes the probability of mass extinctions of wildlife, plants, rain forests, and insects. Presents some preventive and restorative methods. #

753. Ward, Geoffrey C., and Ward, Diane Raines. *Tiger Wallahs : encounters with the men who tried to save the greatest of the great cats.* NY: HarperCollins, 1993. 174 p. : ill. 0060167955
Describes the natural history of the tiger while emphasizing the efforts of dedicated conservationists to preserve it despite an atmosphere of government corruption, poaching, and population pressures for development of natural areas.

754. Ward, Peter Douglas. *The end of evolution : on mass extinctions and the preservation of biodiversity.* NY: Bantam, 1994. 301 p. : ill. 0553088122
AR: *On Methuselah's trail : living fossils and the great extinctions.*
Presents research on past mass extinctions and asserts that a "third event," largely human-made, began with the end of the last Ice Age.

755. Wilson, Edward Osborne. *The diversity of life.* NY: Norton, 1993. 443 p. : ill. 0393035387
Describes the evolutionary process of speciation that created abundant biodiversity. Presents the concepts of conservation biology in a non-technical way. Considers the struggle between developmental and conservation efforts. HR.

756. Wilson, Edward Osborne, and Peter, Frances M. eds. *Biodiversity.* Washington: Nat'l. Academy, 1988. 521 p. 0309037395 (pbk) 0309037832
Contains essays exploring the crucial importance of biodiversity, threats to it, and strategies for preservation. For adults with some background in biology and economics. HR.

Books for young adults:

757. Bloyd, Sunni. *Endangered species.* San Diego: Lucent, 1989. 96 p. : photos. 1560061065
Describes how and why various species of animals and the tropical rain forests are threatened with extinction. Discusses the importance of ensuring their survival. GR 5-9.

758. Facklam, Howard, and Facklam, Margery. *Plants : extinction or survival?* Hillside, NJ: Enslow, 1990. 96 p. : photos. 0894902482
Outstanding Science Trade Book for Children, 1990.
The importance of diversity of plant life and threats to diversity are examined. Describes the importance of germ plasm and how storage banks around the world are preserving plant seeds so that plant species will not be lost through disease and environmental conditions. HR GR 6-10.

759. Facklam, Margery. *And then there was one : the mysteries of extinction*. Ill. by Pamela Johnson. SF: Sierra Club, 1990. 56 p. : ill. 0316259829
Examines many reasons for the extinction and near-extinction of animal species. Discusses how some near-extinctions have been reversed through special breeding programs and legislation to save endangered species. GR 4-8.

760. Few, Roger. *Children's guide to endangered animals*. NY: Macmillan, 1993. 96 p. : col. ill., maps. 0027345459
A.k.a. *The MacMillan children's guide to endangered animals*.
Describes 150 endangered species throughout the world, the factors that threaten their survival, and current efforts to save them from extinction.

761. Kraus, Scott D., and Mallory, Kenneth. *The search for the right whale*. NY: Crown, 1993. 36 p. : col. photos. 051757845X
Follows a team of New England Aquarium scientists as they follow and study migrating North Atlantic right whales and speculates about the future survival of this endangered species. GR 5-9.

762. Lampton, Christopher. *Endangered species*. NY: Watts, 1988. 128 p. : ill. 0531105105
Explains what species are, how they become extinct, and the effect of extinction on the ecology. Looks at endangered species of plants and animals and offers possible solutions.

763. McClung, Robert M. *Lost wild America : the story of our extinct and vanishing wildlife*. Rev. ed. Hamden, CT: Linnet, 1993. Over 300 p. : ill. 0208023593
AR: *Lost wild worlds*.
Traces the history of wildlife conservation and environmental politics in America up to 1992 and describes various extinct or endangered species. #

764. Patent, Dorothy Hinshaw. *The whooping crane : a comeback story*. NY: Clarion, 1988. 88 p. : col. photos. 0899194559
Oustanding Science Trade Book for Children, 1988.
Traces the 40 year attempt to save the endangered whooping crane from extinction, focusing on efforts at wildlife refuges and the captive breeding program. HR GR 4-8.

765. Pringle, Laurence P. *Living treasure : saving Earth's threatened biodiversity*. NY: Morrow, 1991. 64 p. : ill. 0688077102
Discusses the rich variety of life on Earth, the origins and importance of such diversity, its rapid loss, and how to save organisms from extinction. HR GR 4-8.

766. Silverstein, Alvin; Silverstein, Virginia B.; and Silverstein, Robert A.

Saving endangered animals. Hillside, NJ: Enslow, 1993. 128 p. : photos.
0894904027
Discusses the endangerment and extinction of different species of wildlife,
and conservation efforts now underway. GR 7-10.

Energy

Energy is required for almost all industrial activity. Cartledge and Gates
provide general and comparative information on various forms of energy
while Davis, Lovins, and Tygiel consider the social and economic aspects of
those choices. Civilization's overreliance on nonrenewable fossil fuels and
the environmental hazards of oil drilling and shipping are covered by
Freudenburg, Gever, Hawley, Holing, Keeble, Nadler, Strohmeyer, and Wills.
Golob, Hills, Kreith, and Lovins propose alternative and renewable
sources. For other books on personal and industrial energy conservation
see Capone (1603), Hines (1666), and Westerman (1680)

Books for adults:

767. Brodeur, Paul. *Currents of death : power lines, computer terminals, and the
attempt to cover up their threat to your health.* NY: Simon & Schuster, 1989. 333
p. 0671678450
AR: *The great power-line cover-up, The zapping of America.*
Demonstrates the incremental negative health effects of low-level, alter-
nating-current electromagnetic fields from a variety of sources.

768. Cartledge, Bryan. ed. *Energy and the environment.* NY: Oxford, 1993. 170
p. 019858413X 0198584199 (pbk)
Representatives of various types of energy-coal and oil, gas, nuclear power,
and alternative technologies-set out the pros and cons of these sources
and discuss their environmental consequences. The ommission of conservation and underemphasis of "the soft path" are major flaws, but information on other types of energy is useful.

769. Davis, David Howard. *Energy politics.* 4th ed. NY: St. Martin's, 1993.
321 p. 0312072325
Organized around different sources-coal, oil, natural gas, electricity,
nuclear, and newer energy sources-the text shows how each form of energy
involves distinct policy and mangement issues. Expanded coverage of environmental impacts in this edition.

770. Freudenburg, William R., and Gramling, Robert. *Oil in troubled waters* :

perceptions, politics, and the battle over offshore drilling. Albany: SUNY, 1994. 179 p. : photos, maps. 0791418820 (pbk) 0791418812
Why is offshore oil drilling welcome in some places, such as Louisiana, when it is vehemently opposed in others, such as California? The authors provide an analysis of the social, historical, economic and environmental factors that cause such a paradox.

771. Gates, David Murray. *Energy and ecology.* Sunderland, MS: Sinauer Associates, 1985. 377 p. : ill., maps. 0878932305 0878932313 (pbk)
Covers the ecological and environmental consequences of energy extraction, conversion, use and discharge. Contains accounts of world energy resources and reserves, their rates of use and expected lifetimes, and environmental consequences. Somewhat technical, but accessible to adults with some science background.

772. Gever, John. *Beyond oil : the threat to food and fuel in the coming decades.* 3d ed. Boulder: Univ. of Colorado, 1991. 304 p. 0870812424
Presents the results of a computer projection of American energy in the year 2005: it will take more energy to explore for oil than the wells will produce; agriculture is so oil-dependent that the U.S. may cease to be a major food exporter; U.S. oil reserves will be dangerously low. Particularly useful to promote alternative energy, energy conservation, and sustainable agriculture. HR.

773. Golob, Richard, and Brus, Eric. *The almanac of renewable energy : the complete guide to emerging energy technologies.* NY: Holt, 1993. 348 p. : ill. 0805019480
Choice Outstanding Academic Book Award, 1993.
Offers detailed information on which emerging sources are available, how they are developed, and their economic, ecological, political, and personal impacts. Covers hydroelectric, biomass, geothermal, solar thermal, photovoltaic, wind, and ocean energy, energy storage, and energy efficiency. HR.

774. Hawley, T. M. *Against the fires of hell : the environmental disaster of the Gulf War.* NY: Harcourt Brace Jovanovich, 1992. 208 p. 0151039690
The immense oil fires, their environmental impact, and the efforts to extinguish them are described in detail. Also covers the inadequate cleanup, the Kuwaiti government's failure to protect its population or cooperate with international organizations, and the continuing suffering of people and wildlife after the war.

775. Hills, Richard Leslie. *Power from wind : a history of windmill technology.* NY: Cambridge Univ., 1994. 324 p. : ill., photos. 0521413982
Covers the history of the windmill from the year 1000, focusing primarily on

Europe. Traces the declining use of windmills in industry, land drainage and irrigation, and the current interest in windmills as a source of electrical power.

776. Holing, Dwight. *Coastal alert : ecosystems, energy, and offshore oil drilling.* Washington: Island, 1990. 126 p. : photos. 1559630507
A comprehensive treatment of the ecological problems caused by offshore oil drilling, alternatives for reducing our dependency on fossil fuels, and guidelines for citizen action.

777. Keeble, John. *Out of the channel : the* Exxon Valdez *oil spill in* Prince William Sound. NY: HarperCollins, 1991. 290 p. 0060163348
Provides a detailed description of the spill, its effect on wildlife and human and animal habitats, and Exxon's shoddy mitigation effort. He places blame not on the captain, but on collusion between big oil and government in response to insatiable demands of American consumers.

778. Kreith, Frank, and Burmeister, George. *Energy management & conservation.* Denver: Nat'l Conf. of State Legislatures, 1993. 363 p. 1555163750
Provides examples of innovative energy conservation tactics, an in-depth anlysis of American energy consumption, and alternative fuels. Analyzes the Energy Policy Act of 1992.

779. Lovins, Amory B. *Soft energy paths : toward a durable peace.* NY: Harper, 1979, c1977. 231 p. 0060906537
Demonstrates the wastefulness inherent in current industrial practices. Presents an alternative similar to Commoner's "soft path" that relies on solar power, conservation, and the development of local, renewable resources. Still an influential book despite its relative age. HR.

780. Nalder, Eric. *Tankers full of trouble : the perilous journey of Alaskan crude.* NY: Grove, 1994. 320 p. 080211458X
Goes beyond the Exxon Valdez and other major spills to consider the tanker industry, politics, regulations, and greed. Calmer and less one-sided than some books on the subject. HR.

781. National Audubon Society. *Audubon House : building the environmentally responsible, energy efficient office.* NY: Wiley, 1994. 207 p. : photos, plans. 0471024961
Demonstrates how a 19th century building was restored and retrofitted using low-toxic natural materials, high efficiency heating and cooling with almost no CFCs, the use of daylighting, and other techniques.

782. Strohmeyer, John. *Extreme conditions : big oil and the transformation of Alas-*

ka. NY: Simon & Schuster, 1993. 287 p. : maps. 067176697X
A Pulitzer Prize winning journalist chronicles the history of abuse of the oil industry in Alaska from the days of the early boosters, through turf battles between companies, government corruption, the swindling of Native cultures, pipeline construction, and the development of practices that made a disaster like the *Exxon-Valdez* inevitable.

783. Tygiel, Jules. *The great Los Angeles swindle : oil, stocks, and scandal during the Roaring Twenties.* NY: Oxford, 1994. 398 p. 019505489X
The story of C.C. Julian, who used bribery, deceit, environmental devastation, corrupt stock practices, and even murder to create a $150,000,000 fraud that destroyed many honest businesses while it raped the land. This book reads like a detective novel, but it's all too true.

784. Wills, Jonathan, and Warner, Karen. *Innocent passage : the wreck of the tanker Braer.* London: Mainstream, 1993. 192 p. : col. photos. 1851585427
Describes the grounding of a Liberian-registered oil tanker. Covers environmental impacts; the collusion of government, the oil industry, and insurance firms in creating conditions that make spills inevitable; and possible actions to prevent future spills.

Books for young adults:

785. Carr, Terry. *Spill! : the story of the Exxon Valdez.* NY: Watts, 1991. 64 p. : col. photos, maps. 0531109984
Discusses the carelessness, neglect, and the industry-wide intentional noncompliance with safety regulations that led to the oil spill in Prince William Sound; the cleanup effort; and the long-term consequences of this disaster. HR GR 4-8.

786. Cozic, Charles P., and Polesetsky, Matthew. *Energy alternatives.* San Diego: Greenhaven, 1991. 176 p. 0899085776 0899085830 (pbk)
A collection of articles offering opinions for and against such energy issues as whether fossil fuels should be replaced, the uses of nuclear power, alternatives to gasoline-powered cars, and the need for a national energy policy. Good for term papers. #

787. Gardiner, Brian. *Energy demands.* NY: Gloucester, 1990. 36 p. : ill. 0531171973
A succinct look at how we supply our energy demands and why alternative energy sources must be developed. GR 4-8.

788. Godman, Arthur. *Energy supply A-Z.* Hillside, NJ: Enslow, 1991. 144 p.

0894902628
Defines energy related terms and provides an overview of energy issues. GR
7-10.

789. Goldin, Augusta R. *Small energy sources : choices that work.* San Diego:
Harcourt Brace Jovanovich, 1988. 178 p. : ill. 0152762159
Acknowledging the problems of meeting increasing energy demands,
Goldin describes some intriguing means of production that do not depend
on nuclear power or fossil fuels.

790. Gutnik, Martin J., and Browne-Gutnik, Natalie. *The energy question :
thinking about tomorrow.* Hillside, NJ: Enslow, 1993. 104 p. : photos.
0894904000
Examines the limited supplies of fossil fuels left and discusses alternative
sources of energy. GR 7-10.

791. Herda, D. J., and Madden, Margaret L. *Energy resources : towards a
renewable future.* NY: Watts, 1991. 140 p. : ill. 0531110052
Chronicles the story of natural resources exploited by humanity for energy
and considers the prospect of renewable resources for the future. GR 7-10.

792. Keeler, Barbara. *Energy alternatives.* San Diego: Lucent, 1990. 112 p. :
photos. 1560061189
Discusses the present energy crisis, types of fuels now in use, alternative
energy sources, and the conservation of energy. Provides advice on individual energy conservation. GR 5-9.

793. Kerrod, Robin. *Future energy and resources.* NY: Gloucester, 1990. 36 p. :
col. ill. 053117221X (lib. bdg.) 053117221X
Discusses our dwindling energy sources and alternative sources of energy
for the future. Presents the pros and cons of various sources including conservation. GR 4-8.

794. Yanda, Bill. *Rads, ergs, and cheeseburgers : the kids' guide to energy and the
environment.* Santa Fe: Muir, 1991. 100 p. : ill. 0945465750
Ergon, a magical being, discusses the generation of various forms of energy, its transportation, uses in everyday life, conservation, and development
of alternative sources. GR 5-8.

Nuclear Power and Weapons

Nuclear power is clearly a very controversial form of energy. Megaw's *How Safe?* weighs the pluses and minuses. Church, O'Neill, and Waters' novel, *The Woman at Otowi Crossing* (2037), provides enlightening historical views. Bertell, Fradkin, Gallagher, Gould, May, and Williams explore the health hazards of radiation. Williams' *Refuge* is particularly powerful. D'Antonio, Gould, May, Robinson, Schell, and Udall explore the deadly connection between nuclear power and weapons. Also see Gyorgy's *No Nukes* (1431). The Millers provide an excellent handbook on radioactive waste. Pringle's *Nuclear Energy* is particularly recommended for young adults.

Books for adults:

795. Bertell, Rosalie. *No immediate danger : prognosis for a radioactive Earth*. Summertown, TN: Book Co., 1986. 435 p. 0913990256
Details the many abuses and dangers of nuclear development and operations. Reveals how industry and government have colluded to cover the problems up while promoting nuclear energy and weapons. Includes a strong call to action. HR.

796. Church, Peggy Pond. *The house at Otowi Bridge : the story of Edith Warner and Los Alamos*. Albuquerque: Univ. of New Mexico, 1973, c1960. 149 p. 0826302815
The story of the "lizard woman" who lived between the two worlds of the Indians of San Ildefonso Pueblo and the scientists who developed the atomic bomb. HR.

797. D'Antonio, Michael. *Atomic harvest : Hanford and the lethal toll of America's nuclear arsenal*. NY: Crown, 1993. 304 p. : ill. 0517589818
Details the development of Hanford, its role in creating nuclear power and weapons, the alarming incidence of cancers and other diseases in the area, efforts to expose these problems, and the propaganda program which is still being used to defend operations at other nuclear sites. HR.

798. Fradkin, Philip L. *Fallout : an American nuclear tragedy*. Tucson: Univ. of Arizona, 1989. 300 p. 0816510865
An exhaustively researched case study of a lawsuit brought against the government by families whose members had been killed or diseased by radiation from nuclear testing. HR.

799. Gallagher, Carole. *American ground zero : the secret nuclear war*. Cambridge: MIT, 1993. 427 p. : photos. 0262071460

Library Journal Best Sci-Tech Book Award, 1994.
This study of the United States government's intentional exposure of unknowing citizens to radiation from nuclear weapons testing and its long coverup includes personal accounts from cancer victims. HR.

800. Gould, Jay M.; Goldman, Benjamin A.; and Millpointer, Kate. *Deadly deceit : low-level radiation, high-level cover-up.* 2d rev ed. NY: Four Walls Eight Windows, 1991. 266 p. : maps. 0941423352 0941423565 (pbk)
An expose of the unrecognized hazards of low level radiation from nuclear power and weapons and the role of federal agencies in falsifying or suppressing data to protect corporations. HR.

801. May, John. *The Greenpeace book of the nuclear age : the hidden history, the human cost.* NY: Pantheon, 1989. 378 p. : ill. 0394585534 0679729631 (pbk)
A highly detailed and documented account of hundreds of nuclear accidents and radiation incidents from the 1940s through the 80s. HR.

802. Megaw, W. J. *How safe? : Three Mile Island, Chernobyl and beyond.* Toronto: Stoddart, 1987. 240 p. : ill. 0773721118
A nuclear scientist examines the civil nuclear industry worldwide: its history, people, mismanagement and political manipulation. Provides an understanding of the benefits and the potential hazards, including that of a terrorist attack on a nuclear installation.

803. Miller, E. Willard, and Miller, Ruby M. *Environmental hazards : radioactive materials and wastes : a reference handbook.* Santa Barbara: ABC-CLIO, 1990. 298 p. 0874362342
Covers the growth of the nuclear industry and subsequent proliferation of nuclear waste, waste handling and storage techniques, government policy, and the controversy over nuclear waste. AOT. #

804. O'Neill, Daniel T. *The firecracker boys.* NY: St. Martin's, 1994. 384 p. 0312110863
In the late 1950s the AEC devised a plan to use nuclear weapons to excavate a deepwater port in Alaska near the oldest continually occupied settlement in the Americas. This is the story of government deception and opposition by local residents and scientists.

805. Robinson, Marilynne. *Mother country.* NY: Farrar, Straus, Giroux, 1989. 261 p. 0374213615
Great Britain, not the United States or the former Soviet Union, is the largest commercial source of plutonium and presents the greatest radioactive threat to the global environment. Novelist Robinson provides factual information and a powerful plea for radical reform.

806. Schell, Jonathan. *The fate of the Earth*. NY: Knopf, 1988, c1982. 244 p. 0394525590
Vividly describes the devastating environmental consequences of nuclear war and nuclear winter. Also analyzes the political aspects of the arms race and considers how it can be stopped. HR.

807. Udall, Stewart L. *The myths of August : a personal exploration of our tragic Cold War affair with the atom*. NY: Pantheon, 1994. 399 p. 0679433643
AR:*The quiet crisis and the next generation*, *The energy balloon*, *Agenda for tomorrow*. Summarizes the negative environmental, health, political, and personal consequences of nuclear weapons testing and "the peaceful atom."

808. Williams, Terry Tempest. *Refuge : an unnatural history of family and place*. NY: Vintage, 1992, c1991. 304 p. : ill. 0679740244
AR: *Pieces of a white shell: a journey to Navajoland*.
A powerful, highly personal account of a waterfowl refuge in Utah, her mother's ovarian cancer, and the discovery that the unusually high incidence of ovarian and breast cancer in the area was caused by radioactive fallout from bomb testing. HR.

Books for young adults:

809. Hawkes, Nigel. *Nuclear power*. Vero Beach, FL: Rourke, 1990. 48 p. : ill. 0865920982
Describes nuclear fuel, its sources, how it is processed, and its uses. Also examines opposing viewpoints concerning nuclear power as a source of energy. GR 5-9.

810. Pringle, Laurence P. *Nuclear energy : troubled past, uncertain future*. NY: Macmillan, 1989. 124 p. : photos. 0027753913
Outstanding Science Trade Book for Children, 1989.
Surveys the history and development of nuclear power and presents both sides of the controversy surrounding its use. HR GR 5-10.

Solid and Toxic Wastes

Toxic wastes pose similar problems to those of nuclear wastes. More than half of both the adult and young adult books in the Wastes section are about toxics, while the rest are about garbage and recycling. Most of the adult books I could find on these subjects were written for scientists and engineers and are thus excluded. *The Recyclers Handbook* and the *McGraw-Hill*

Recycling Handbook are in this section, but several books on locating and reducing toxics in the home are described in the Greener Homes and Gardens section (1600, 1608, 1614, 1647) Zipko's *Toxic Threat* is listed with the young adult books, but it is also one of the best books on toxics for adults.

Books for adults:

811. Barnett, Harold C. *Toxic debts and the Superfund dilemma.* Chapel Hill: Univ. of North Carolina, 1994. 350 p. 0807821241 0807844357 (pbk)
The Superfund has failed because of conflict over who will pay the toxic debt and the impact of this conflict on interdependent funding and enforcement decisions at all levels of government. Excoriates the Reagan-Bush legacy of deregulation and nonenforcement.

812. Brown, Michael Harold. *Laying waste : the poisoning of America by toxic chemicals.* Updated ed. NY: Pocket, 1981. 363 p. 0671422634
The reader is given a firsthand look at some of the countries most horrendous toxic waste cesspools.

813. Earth Works Group. *The recycler's handbook.* Berkeley: Earth Works, 1990. 132 p. : ill. 092963408X
Provides practical information on how to recycle glass, paper, plastic, and other recyclables. Also includes information on products made of recycled materials and a guide to relevant state agencies and national organizations. HR.

814. Lappe, Marc. *Chemical deception : the toxic threat to health and the environment.* SF: Sierra Club, 1991. 360 p. 0871566036 (pbk) 0871565110
Explains the threats posed by toxic production methods, the vast extent of use of toxics in manufacuring, deceptive pubic relations practices, and the importance of reform by both producers and consumers.

815. Lund, Herbert F. ed. *The McGraw-Hill recycling handbook.* NY: McGraw-Hill, 1993. 992 p. : ill. 0070390967
American Libraries Outstanding Reference Source, 1994.
Covers the basics of recycling eleven major types of waste, setting recycling goals and priorities, processing sites, methods and equipment, financial planning, the psychology of recycling, public awareness, employee training, quality monitoring, and related topics. HR.

816. Lynch, Kevin, and Southworth, Michael. *Wasting away.* SF: Sierra Club, 1990. 270 p. 0871566753
Presents waste and waste disposal as an essential part of life and growth, but overconsumption has created a "cacatopia." Examines all types of

waste, including wasteful thinking, suggesting that in addition to cutting down on waste we need to consider it a potential resource and reuse it.

817. Rathje, William L., and Murphy, Cullen. *Rubbish! : the archaeology of garbage.* NY: Harper, 1993, c1992. 263 p. 0060922281
This factual but humorous book analyzes America's consumption and disposal habits, waste-to-energy options, reuse and recycling, landfills, incineration, and related topics. Includes "The ten commandments of garbage."

818. Schweitzer, Glenn E. *Borrowed Earth, borrowed time : healing America's chemical wounds.* NY: Plenum, 1991. 298 p. 030643766X
Provides a frank and insightful look at pesticide and toxic chemical issues.

819. Shulman, Seth. *The threat at home : confronting the toxic legacy of the* U.S. *military.* Boston: Beacon, 1994, c1992. 272 p. 0807004170
Makes a strong case that the military is the most dangerous polluter, shielded from public view and environmental regulations by its special status. Thousands of toxic sites exist, with some "national sacrifice zones" located near major cities.

820. Whitaker, Jennifer Seymour. *Salvaging a land of plenty : garbage and the American dream.* NY: Morrow, 1994. 288 p. 0688101305
Examines the causes and consequences of the huge amount of waste generated in America. Recommends shifting the cost of waste disposal to product pricing, arguing that an improved standard of living and a clean environment are posssible in an economy based on conservation.

Books for young adults:

821. Arneson, D. J. *Toxic cops.* NY: Watts, 1991. 125 p. : ill. 0531125254
Discusses several threats to the environment, laws that have been enacted to protect it, and the branch of the Environmental Protection Agency and local groups that enforce such laws. GR 7-10.

822. Donnelly, Judy, and Kramer, Sydelle. *Space junk : pollution beyond the Earth.* NY: Morrow, 1990. 106 p. : ill. 0688086780 0688086799 (lib. bdg.)
Discusses the miscellaneous junk floating in space, how it got there, what effects it can have, and what can be done about it. GR 5-8.

823. Gay, Kathlyn. *Garbage and recycling.* Hillside, NJ: Enslow, 1991. 128 p. : photos. 0894903217
Examines the problem of garbage accumulation in America and different recycling solutions which may prevent the situation from getting worse in

the future. Includes advice on how to get involved in recycling.

824. ___. *Global garbage : exporting trash and toxic waste.* NY: Watts, 1992. 144 p. : ill. 0531130096
Examines the increasing problems of toxic waste disposal, such as dumping in poor nations, military dumping, and waste disposal in space. Includes treaties on hazardous waste and plans for global protection.

825. ___. *Silent killers : radon and other hazards.* NY: Watts, 1988. 128 p. : ill. 0531105989
Examines potentially dangerous chemicals, gases, and metals that can unknowingly endanger people in their homes, schools, and workplaces and destroy the environment.

826. Hadingham, Evan, and Hadingham, Janet. *Garbage! : where it comes from, where it goes.* NY: Simon & Schuster, 1990. 48 p. : ill., maps. 0671694243 067169426X (pbk)
Companion book for the *Nova* TV series on . Documents the ever-increasing problem of what can be done to dispose of our garbage. GR 5-9.

827. Kronenwetter, Michael. *Managing toxic wastes.* Englewood Cliffs, NJ: Messner, 1989. 118 p. : photos. 0671690515
Discusses toxic wastes, their effects on the environment, their handling and disposal, and government regulation of such pollutants.

828. Lee, Sally. *The throwaway society.* NY: Watts, 1990. 128 p. : ill. 053110947X
Examines the growing problem of how to handle solid wastes, exploring such areas as collecting and transporting waste, sanitary landfills, incineration, recycling, and ocean dumping.

829. Tesar, Jenny E. *The waste crisis.* NY: Facts on File, 1991. 112 p. : photos. 081602491X
Examines all kinds of waste, including commercial, industrial, toxic, and radioactive waste, and discusses the problems and possible solutions connected with the existence and management of such pollutants.

830. Zipko, Stephen James. *Toxic threat : how hazardous substances poison our lives.* Rev. ed. Englewood Cliffs, NJ: Messner, 1990. 249 p. : ill. 0671693301 067169331X (pbk)
Describes hazardous substances in our environment, how they get there, and the problems they cause. Explains how food webs and other natural cycles are disrupted by toxic wastes. Projects, activities, and ways to get involved are provided. HR. #

Water Supply

Water, like energy, is essential to virtually all human endeavors, but the quality of groundwater is declining, aquifers are being depleted, and demands for safe water continue to increase. Powledge provides a good general overview of these issues while the rest of the authors consider the interlocking economic, social, and political aspects of water use. Bowden, Fradkin, and Resiner provide particularly pointed commentary on these issues for adults while Cossi demystifies them for teens. Also see the Freshwater section and McGuire's *Indian Water in the New West* (991)

Books for adults:

831. Bates, Sarah F.; Getches, David H.; MacDonnell, Lawrence J.; and Wilkinson, Charles F. *Searching out the headwaters : change and rediscovery in western water policy.* Washington: Island, 1993. 241 p. 1559632178 1559632186 (pbk)
Traces the history of western water use, describes demographic and economic changes, considers the emerging reform movement, and outlines a new water policy designed to protect the environment while effectively serving the water needs of the new West.

832. Bowden, Charles. *Killing the hidden waters : the slow destruction of water resources in the American southwest.* Austin: Univ. of Texas, 1985, c1977. 174 p. : ill. 0292743068
Examines groundwater depletion in Arizona and the southern Great Plains by settlers, contrasting their behavior with that of the Pima, Papago, and Comanche Indians.

833. Fradkin, Philip L. *A river no more : the Colorado River and the West.* Tucson: Univ. of Arizona, 1984, c1981. 360 p. : ill. 0816508232
Fradkin describes his explorations of the river, its history, how politics and economics now play a greater role than natural forces, its overuse and pollution, and the "bust" consequences which are starting to follow the artificial economic "boom."

834. Martin, Russell. *A story that stands like a dam : Glen Canyon and the struggle for the soul of the West.* NY: Holt, 1990. 354 p. 0805008225
Covers the demand for power and water that resulted in the building of this dam, the controversy over its construction, its environmental and economic consequences, and the irony of how its construction raised public consciousness and transformed the environmental movement.

835. Powledge, Fred. *Water : the nature, uses, and future of our most precious and*

abused resource. NY: Farrar Straus Giroux, 1982. 423 p. 0374286604
Traces the use and abuse of American water resources, the development of parallel water quality and supply crises, how water projects have been used as pork by politicians, the rise of conservation consciousness, and possible future scenarios. HR.

836. Reisner, Marc. *Cadillac desert : the American West and its disappearing water*. Rev. ed. NY: Penguin, 1993. 606 p. : ill. 0140178244
Traces over a century of "water wars" between and among government, farmers, ranchers, industry, developers, and conservationists. Demonstrates how corruption, political favoritism, and greed rather than science or common sense have depleted supplies and contributed to desertification. HR.

837. Reisner, Marc, and Bates, Sarah F. *Overtapped oasis : reform or revolution for western water*. Washington: Island, 1990. 200 p. 0933280769
Presents an accounting of the history of western water issues.

838. Stegner, Wallace Earle. *Beyond the hundredth meridian : John Wesley Powell and the second opening of the West*. NY: Penguin, 1992, c1954. 481 p. : ill. 0140159940
While detailing the life and achievements of this explorer, government official, and conservationist, Stegner also provides a lively history of the use and abuse of western lands, irrigation and other water projects, and the rise of the conservation movement.

839. Wilkinson, Charles F. *Crossing the next meridian : land, water, and the future of the West*. Washington: Island, 1992. 376 p. 155963149X (pbk) 1559631503
Focuses on practices which have evolved over the past 150 years in five areas: mining, lumber, grazing, dams, and water storage and diversion. Demonstrates how Indians, Mexican Americans, and small farmers have been affected by water policies. Proposes that further development be based on the principle of sustainability.

840. Worster, Donald. *Rivers of empire : water, aridity, and the growth of the American West*. NY: Oxford, 1992, c1985. 402 p. 0195078063
Places the history of water development in the arid West in the context of the concept of "hydraulic civilization." Criticizes this concept and resultant developments as being ill-conceived and unsustainable. HR.

Books for young adults:

841. Cossi, Olga. *Water wars : the fight to control and conserve nature's most pre-*

cious resource. NY: New Discovery, 1993. 127 p. 0027245950
Discusses water's sources, uses, and shortages; water quality problems and management programs; and what individuals can do to ensure cleaner water. Includes a chronology of legislation and a list of relevant oganizations.

842. Hoff, Mary King, and Rodgers, Mary M. *Our endangered planet : Groundwater.* Minneapolis: Lerner, 1991. 64 p. : ill. 0822525003
Describes the global uses and abuses of groundwater and suggests ways to preserve this valuable resource. GR 4-8.

5. Cultural Factors

This chapter covers books about the political, social, and economic aspects of the environment. Related books can be found in the sections on Conservation (political and social aspects of international conservation), Ecofeminism, Green Movement, Business and Industry, and Environmental Action.

Politics

Because the environment does not recognize human-created boundaries, many environmental problems have an international dimension. Adamson, Gore, Baden, Israelson, Lieferinck, Rogers, Sagoff, and Zaelke explore this dimension. Heaton, MacDonnell, Meiners, O'Leary, Ophuls, and the Rifkins focus on the political scene in the U.S. *Whose Common Future?* from the British journal, *The Ecologist*, and Meiners and Yandle's *Taking the Environment Seriously* reject national controls for local ones or market-oriented solutions respectively, while Paehlke, Schnaiberg, and Switzer advocate moving from a geopolitical to an ecopolitical consciousness. Books specifically about Third World ecopolitics are included in section 5.3. Related books in other chapters include Biehl (1139), Mellor (1164), Bregman (1408), John (1437) , Fogelman (1426), Gorczynski (1430), Jessup (1437), and Jorgenson (1439).

Books for adults:

843. Adamson, David. *Defending the world : the politics and diplomacy of the environment.* NY: St. Martin's, 1991. 246 p. 185043302X
Analyzes the international and political aspects of the environmental movement, such as linking economic aid with environmental policies; climate change and world agriculture practices; increased use of coal in India and China; pollution in Eastern Europe; and the commons of the oceans and Antarctica.

844. Baden, John. ed. *Environmental Gore : a constructive response to Earth in the balance.* SF: Pacific Research Inst. for Public Policy, 1994. 288 p. 0936488786
Essays purporting to offer an alternative vision of the state of the global environment to that of VP Gore, but actually little more than a political hit

piece by industrial apologists. Included only as an example of a cleverly disguised anti-environmental book.

845. Ecologist (Journal). *Whose common future? : reclaiming the commons.* Philadelphia: New Society, 1993. 216 p. 086571276X 0865712778 (pbk)
Traces the environmental crisis to federal and state governments and businesses which overexploit resources for short term economic or political gain. Debunks "mainstream" solutions. Proposes the imposition of local controls and highlights successful efforts to do so.

846. Gore, Albert. *Earth in the balance : ecology and the human spirit.* Boston: Houghton Mifflin, 1992. 407 p. 0452269350
Examines how to improve the Earth's environment and renew the human spirit given the practical limitiations of economics and politics. Proposes a Global Marshall Plan and that the strategic defense initiative be supplanted by a strategic environmental initiative.

847. Heaton, George R.; Repetto, Robert C.; and Sobin, Rodney. *Backs to the future* : U.S. *government policy toward environmentally critical technology.* Washington: World Resources Inst., 1992. 41 p. 0915825759
Recommends shifting the 60% of federal r&d dollars spent on the military to technologies that markedly increase the efficiency of energy and raw material use, or that eliminate most of the pollution from agricultural and industrial processes. Includes a list of these technologies.

848. Israelson, David. *Silent Earth : the politics of our survival.* Toronto: Penguin, 1991, c1990. 278 p. 0140129227
Covers environmental politics and behavior from the national to the personal level. Demonstrates how environmental devastation is the result of activities at all levels, and that any solutions must incorporate widespread changes in values and behavior. Contains a section on environmental terrorism and the Gulf War.

849. Liefferink, J. D.; Lowe, Philip; and Mol, A. P. J. eds. *European integration and environmental policy.* NY: Bellhaven, 1993. 241 p. 0470221054
The process of environmental policy making in the EEC is characterized by a delicate interplay of political agendas, institutional relations and regulatory cultures at national and transnational levels. Contains 12 essays by various authors.

850. MacDonnell, Lawrence J., and Bates, Sarah F. eds. *Natural resources policy and law : trends and directions.* Washington: Island, 1993. 241 p. 1559632453 1559632461 (pbk)
Contains ten essays by various authors on fundamental changes taking

place in environmental and natural resource laws. Also considers the history and evolution of resource law and policy; laws on mining, drilling, and public lands; how environmentalism affects environmental law; and future directions.

851. Meiners, Roger E., and Yandle, Bruce. eds. *Taking the environment seriously*. Lanham, MD: Rowman & Littlefield, 1993. 270 p. 0847678733
Contains ten essays by various authors about how present federal efforts to protect the environment have failed and how market-oriented solutions could be implemented.

852. O'Leary, Rosemary. *Environmental change : federal courts and the* EPA. Philadelphia: Temple Univ., 1993. 256 p. 1566390958
O'Leary contends that the EPA's most important policy areas are water quality, pesticides, toxic substances, air quality, and hazardous waste; that court decisions have effected both law and agencies; that a new partnership is forming between the courts and EPA; and that court decisions are crucial given competing interests.

853. Ophuls, William, and Boyan, A. Stephen. *Ecology and the politics of scarcity revisited : the unraveling of the American dream*. NY: Freeman, 1992. 379 p. 0716723131
Argues that unless we change present political, social, and economic structures, we will be unable to adequately deal with scarce resources. Calls for the creation of a survivalist class of 'ecological guardians' even at the expense of social liberties and personal freedom. Controversial and HR.

854. Paehlke, Robert. *Environmentalism and the future of progressive politics*. New Haven, CT: Yale, 1989. 325 p. 0300040210
Attempts to define what environmentalism actually is; how it relates to liberalism, conservatism, socialism, and progressivism; how environmentalists exist across this spectrum; and how a better defined ideology of environmentalism might be constituted.

855. Rifkin, Jeremy, and Rifkin, Carol Grunewald. *Voting green : your complete environmental guide to making political choices in the 90's*. NY: Doubleday, 1992. 390 p. 0385419171
Includes a report card on all members of Congress, a checklist of green positions on many issues, and a platform based on a sustainable economy and global environmental security. The presentation of issues and platform make this book relevant far beyond the 1992 elections. HR.

856. Rogers, Adam. *The Earth Summit : a planetary reckoning*. LA: Global View, 1993. 351 p. 1881294935

Report on and analysis of the Rio Conference on Environment and Development held in 1992. Includes both the issues covered and the "Missing agenda" of the population explosion, how to finance implementation of the treaties, and how to resist pressure from military and industrial polluters.

857. Sagoff, Mark. *The economy of the earth : philosophy, law, and the environment.* NY: Cambridge Univ., 1990, c1988. 271 p. 0521395666
Covers social legislation and its relationship to cultural, ethical, and aesthetic goals.

858. Schnaiberg, Allan, and Gould, Kenneth Alan. *Environment and society : the enduring conflict.* NY: St. Martin's, 1994. 255 p. 0312102666 0312091281 (pbk)
Offers a perspective for solving ecological problems and a systematic approach to how organizations, institutions, and individuals can push for reforms. Analyzes the domestic and global inequalities arising from ecological degradation.

859. Switzer, Jacqueline Vaughn. *Environmental politics : domestic and global dimensions.* NY: St. Martin's, 1994. 380 p. 0312102380 0312083890 (pbk)
Presents the historical framework, the political process, land stewardship, conservation, the politics of energy, managing water resources, air pollution, sick building syndrome, ocean pollution, acid rain, climate change, deforestation, endangered species, polar regions, overpopulation, and moving from geopolitics to ecopolitics.

860. Zaelke, Durwood, Orbuch, Paul, and Housman, Robert F. eds. *Trade and the environment : law, economics, and policy.* Washington: Island, 1993. 318 p. 1559632674 1559632682 (pbk)
Twenty industrial leaders, trade advocates, environmentalists, and policymakers explore issues and concerns surrounding the complex interactions between trade agreements and environmental protection.

Social Factors

This section contains six sub-categories: population, economics, land use, human ecology, ecological history, and factors specific to the Western United States. It should be noted that given the close interrelationship of these topics, few of the books are on a single one.

The Ehrlichs, Meadows, and Turner consider population and its relation to other environmental issues for adults, as do the Becklakes, Gallant, McGraw, and Winkler for young adults.

Economics is the focus of adult books by Boulding, Daly, and Schumacher and Bates' book for teens. Land use and the promising new field of bioregionalism is central to the work of Andruss, Berg, Callicott, Gallagher, Heat Moon, Raphael, and Sale. For other books on bioregionalism see entries 1070, 1360, 1387, 1404, 1407, and 1676. Herda and Newton consider land use issues for young adults.

The multifaceted field of human ecology (including humanity and technology) is addressed by Bates, Berry, Catton, Commmoner, Dubos, Eckholm, Fitzpatrick, Fuller, Matthiessen, Meadows, Redclift, Rifkin, Roszak, Carl Ortwin Sauer, Peter Sauer, Schumacher, Simmons, Ward, Westing, and Worster.

Environmental history is analyzed by Crosby, Mannion, Martin, Melville, Ponting, Sale, Salisbury, Stocking, and Worster. Historical analysis of the Conservation movement can be found in the Conservation section.

Aspects of human ecology and environmental history specific to the West are considered by Conaway, Dayton, DeBuys, Sandoz, Stegner, Turner, Waters, and Wilkinson. Schaefer's *An American Bestiary* combines human ecology, human-animal relations, environmental history, and cosmology in a Western setting.

Books for adults:

861. Andruss, Van; Plant, Christopher; Plant, Judith; and Wright, Eleanor. eds. Home! : *a bioregional reader.* Philadelphia: New Society, 1990. 181 p. : ill. 0865711879 0865711887 (pbk)
Contains 36 essays on defining bioregionalism, living in place, nature-culture and community, reinhabitation and restoration, and self-government. "Required reading" for anyone interested in bioregionalism. HR.

862. Bates, Marston. *The forest and the sea : a look at the economy of nature and the ecology of man.* NY: Lyons, 1988, c1960. 277 p. 1558210091
AR: *Jungle in the house, Man in nature.*
Considers the relationship between nature and human ecology. Includes our relationship to aquatic life; the ecology of streams, rivers, and oceans; threats posed by pollution; and ethical, aesthetic, and utilitarian reasons for preserving natural diversity. HR.

863. Berg, Peter. ed. *Reinhabiting a separate country : a bioregional anthology of northern California.* SF: Planet Drum, 1978. 220 p. : ill. 0937102008
Provides a new synthesis between natural history and culture. Includes essays, stories, and poems on a sense of place in the mountains, river and delta, and cities. Discusses values and exchanges, and developing a multi-species consciousness. HR.

864. Berry, Wendell. *Home economics : fourteen essays.* SF: North Point, 1987. 192 p. 0865472742
AR: *The long-legged house.*
A consideration of the relationships between people and nature in both wilderness areas and populated ones. If we were truly "materialistic" we would value resources and possessions, not waste them. This requires thinking of all creation as our home, hence "home economics." Arguably Berry's finest work. HR.

865. ___. *Recollected essays,* 1965-1980. SF: North Point, 1981. 340 p. 0865470251
An anthology of essays and articles published in magazines and other books.

866. Boulding, Kenneth Ewart. *Ecodynamics : a new theory of societal evolution.* Beverly Hills: Sage, 1981, c1978. 382 p. 0803909454 0803916833
Argues that human economic systems "develop according to the same laws as nature's," that the theory of evolution can be applied to economics, and that combining economic and environmental concerns is a logical evolutionary step.

867. Callicott, J. Baird. *In defense of the land ethic : essays in environmental philosophy.* Albany: SUNY, 1989. 325 p. 0887068995 0887069002 (pbk)
Expands Aldo Leopold's land ethic concept to consider how an evolutionary, ecological way of thinking can "bridge the gap between what is and what ought." Also analyzes animal liberation, holistic environmental ethics, non-anthropocentric value systems, American Indian ethics, education, natural aesthetics, and extraterrestrial life.

868. Catton, William Robert. *Overshoot, the ecological basis of revolutionary change.* Urbana: Univ. of Illinois, 1982. 298 p. 0252009886
Considers carrying capacity as a maximum supportable load on the land, the myth of limitless resources, the drawdown of stealing resources from the future, the delusion that technology will always save us, and overshoot as growth beyond carrying capacity leading to extinction. Proposes conservation and other tactics to counter these conditions.

869. Conaway, James. *The kingdom in the country : wranglers, shepherds, miners & bureaucrats, squatters & gunfighters, Indians & other inhabitants of the land nobody owns.* NY: Avon, 1993, c1987. 293 p. : map. 0380716135
Conaway travels through the vast public lands of the west and chronicles the struggle between exploitation and preservation.

870. Crosby, Alfred W. *Ecological imperialism : the biological expansion of Europe,*

900-1900. NY: Cambridge Univ., 1993, c1986. 368 p. : ill. 0521320097
0521456908 (pbk)
AR: *The Columbian exchange: biological and cultural consequences of* 1492.
Demonstrates how European cultures have used advanced technology, the
military, and expansionist policies to spread a European population
around the Earth, largely for the purpose of dominating the world's food
and other natural resources.

871. DeBuys, William Eno, and Harris, Alex. *River of traps : a village life*. Albuquerque: Univ. of New Mexico, 1990. 238 p. : photos. 0826311822
A personal evocation of the ancient and natural way of life in the mountains of northern New Mexico, where life and death is rooted in and takes
its meaning from the land. The strong implication is that civilization
should learn from and preserve this way of life, rather than destroy it.

872. Duncan, Dayton. *Miles from nowhere : tales from America's contemporary
frontier*. NY: Viking, 1993. 320 p. 0670831956
Describes the unusual challenges, unexpected rewards, independence, and
interdependence of living in America's most remote areas.

873. Ehrlich, Paul R. *The population bomb*. Rev. ed. NY: Ballantine, 1975,
c1971. 201 p. 08191908617
AR: *Ecoscience*.
Makes a strong case for the Malthusian theory that rapid population
growth combined with dwindling resources makes starvation and environmental decay inevitable. HR.

874. Ehrlich, Paul R., and Ehrlich, Anne H. *The population explosion*. NY:
Simon & Schuster, 1990. 320 p. 0671698843
Further elaborates the theories advanced in *The population bomb*, demonstrating that the world is already overpopulated, and arguing in favor of
government and educational efforts to lower the birth rate.

875. Fitzpatrick, Tony. *Signals from the heartland*. NY: Walker, 1993. 230 p. :
photos. 0802712606
Library Journal Best Sci-Tech Book Award, 1994.
A native midwesterner demonstrates how America's heartland is ecologically threatened by human activity, and how human activity is threatened
by natural events such as floods. HR.

876. Fuller, R. Buckminster. *Operating manual for spaceship Earth*. NY: Pen-
guin, 1991, c1979. 143 p. 0140194517
AR: *Earth, inc., Utopia or oblivion, I seem to be a verb* .
Presents "spaceship Earth" as an interconnected body of resources which

can be properly exploited if humanity can evolve mentally to avoid its tendency toward oblivion through war, greed, and misuse of technology. Describes the concepts of general systems theory, synergy, integral functions, and the regenerative landscape. AOT. #

877. Gallagher, Winifred. *The power of place : how our surroundings shape our thoughts, emotions, and actions.* NY: Poseidon, 1993. 240 p. 067172410X
Library Journal Best Sci-Tech Book Award, 1994.
The physical environment in which we live has a profound effect on how we act and think. Draws upon biology, psychology, and environmental sciences to argue that we must cultivate a deeper sense of place in order to live in harmony with the environment. HR.

878. Heat Moon, William Least*. *Blue highways : a journey into America.* 2d pbk ed. Boston: Houghton Mifflin, 1991, c1982. 435 p. : photos, map. 0449211096
The author travels the backroads through America's countrysides and small towns on a voyage of discovery and renewal. He describes a traditional way of life and a positive relation to the land which still persist despite the encroachments of high tech and mass pop culture. AOT. #

879. Luhmann, Niklas. *Ecological communication.* Chicago: Univ. of Chicago, 1989. 187 p. 0226496511
The author probes economy, law, science, politics, religion, and education and their relationship to environmental issues.

880. Mannion, Antoinette M. *Global environmental change : a natural and cultural environmental history.* NY: Halsted, 1991. 404 p. 0470216786
Explores how environmental change takes place through both natural and cultural forces. Begins in prehistoric times, but focuses mainly on changes of the last few centuries due to the agricultural and industrial revolutions. Biotechnology is potentially a major change agent.

881. Martin, Calvin. *In the spirit of the earth : rethinking history and time.* Baltimore Johns Hopkins Univ., 1992. 157 p. 0801843588
Traces human alienation from nature to neolithic times. Suggests how some earlier values can be re-adapted today.

882. Meadows, Donella H. *The global citizen.* Washington, DC: Island, 1991. 300 p. 1559630590
AR: *The limits to growth.*
Considers the relationship between social problems, human ecology, and the destruction of the natural environment. Considers agriculture, population, economics, and growth. HR.

883. Melville, Elinor G. K. *A plague of sheep : environmental consequences of the conquest of Mexico.* NY: Cambridge Univ., 1994. 203 p. 052142061X
Contends that the introduction of new animal species and technologies from the Old World to the New both enabled the conquest of the Americas and made its environmental degradation inevitable.

884. Ponting, Clive. *A green history of the world : the environment and the collapse of great civilizations.* NY: Penguin, 1993, c1991. 432 p. 0140176608
Demonstrates how a series of civilizations have risen through the effective exploitation of natural resources and then reached their demise through overuse. The industrial-based devastation of the Earth's environment has been preceded several times by agricultural-based degradation.

885. Raphael, Ray. *Edges : human ecology of the backcountry.* Lincoln: Univ. of Nebraska, 1986, c1976. 233 p. 0803289227
Depicts a rural way of life in northern California which is based on an intimacy with the land, but is threatened by rising property taxes, highways, and other factors. Advocates the individual right to pursue a simpler, less stressful lifestyle.

886. ___. *An everyday history of somewhere : being the true story of Indians, deer, homesteaders, potatoes, loggers, trees, fishermen, salmon & other living things in the backwoods of northern California.* Redway, CA: Real Books, 1992, c1974. 192 p. : ill. 1881102254
Provides both a historical and a modern perspective on social ecology in this important precursor of bioregionalism. Entertaining, informative, and HR.

887. Redclift, M. R., and Benton, Ted. eds. *Social theory and the global environment.* NY: Routledge, 1994. 256 p. 0415111692 0415111706 (pbk)
Contains essays rejecting the notion that it is solely up to natural scientists to recommend environmental policies, while the only role for social scientists is to study the social effects of those policies. Proposes a more positive, synergistic approach.

888. Rifkin, Jeremy, and Howard, Ted. *Entropy : a new worldview.* Rev. ed. NY: Bantam, 1989. 354 p. 0553347179
AR: *The emerging order: God in the age of scarcity.*
Explains the tendency of economic, social, and environmental systems to move from an ordered to a disordered state (entropy), especially as related to the greenhouse effect and other global environmental problems.

889. Roszak, Theodore. *Where the wasteland ends : politics and transcendence in postindustrial society.* Berkeley: Celestial Arts, 1989, c1972. 492 p. 0890875707

Contends that the repression of religious sensibilities has led to the creation of unnatural simulated environments and technocratic institutions that foster psychic alienation and personal despair.

890. Sale, Kirkpatrick. *The conquest of paradise : Christopher Columbus and the Columbian legacy.* NY: Plume, 1991. 453 p. 0452266696
AR: Barry Lopez's *The rediscovery of North America.*
A detailed examination of the environmental and cultural consequences of the manner in which the Americas were "discovered" and settled. HR.

891. ___. *Dwellers in the land : the bioregional vision.* 2d ed. Philadelphia: New Society, 1991, c1985. 238 p. 0865712255
AR: *Human scale.*
A good introduction to the concept of bioregionalism, i.e. organizing political, social and economic activity along natural, biological lines. Argues that all human activity must be firmly rooted to the land and that politics must be decentralized. HR.

892. Salisbury, Joyce E. ed. *The Medieval world of nature : a book of essays.* NY: Garland, 1993. 265 p. : ill. 0815307527
Twelve modern essays on how medieval Europeans perceived, depicted, used, and related to nature. Topics include animals, gardens, nature mysticism, land tenure, and nature in the work of Chaucer and Dante. Of primary interest to scholars and mystics.

893. Sandoz, Mari. *Hostiles and friendlies : selected short writings of Mari Sandoz.* Lincoln: Univ. of Nebraska, 1992, c1959. 250 p. 0803292082
Includes seven largely autobiographical recollections, three Indian studies, and ten pieces of short fiction.

894. ___. *Old Jules country : a selection from Old Jules and thirty years of writing since the book was published.* Lincoln: Univ. of Nebraska, 1982, c1965. 319 p. 0803291361
A selection of her nonfiction writing excerpted primarily from the Great Plains series: *Beaver men, Crazy Horse, Cheyenne autumn, The buffalo hunters, The Cattlemen,* and *Old Jules.*Three of these books relate the history of the west to the economic exploitation of animals. Also includes excerpts from *These were the Sioux* and other essays.

895. ___. *Sandhill Sundays and other recollections.* Lincoln: Univ. of Nebraska, 1970. 165 p. 0803291485
Essays and stories written 1929-1965 set in the Sandhill country of Nebraska. They deal mainly with nature, homesteading, the Indian way of life, and relations between settlers and Indians.

896. Sauer, Peter H. ed. *Finding home : writing on nature and culture from* Orion *magazine*. Boston: Beacon, 1992. 293 p. 0807085179 (pbk) 0807085189
Contains articles exploring a changing culture's way of living in a changing nature, reflections of our own bewilderment, and a reconsideration of the effort individuals must make to reforge their bonds to nature.

897. Schaefer, Jack. *An American bestiary*. Boston: Houghton Mifflin, 1975. 287 p. : ill. 039520710X
Twelve essays on southwestern animals written in an intentionally anthropomorphic way so their behavior provides a critique of civilization. The implicit message is that humanity must overcome its cultural consciousness in order to understand its effect on nature.

898. Schneider-Hector, Dietmar. *White Sands : the history of a national monument*. Albuquerque: Univ. of New Mexico, 1993. 270 p. : ill. 0826314155
Outlines the history of this unique desert site and issues it faces today. Describes an alliance of environmentalists and ranchers opposed to Air Force and Chamber of Commerce activities which disturb the area.

899. Schumacher, E. F. *Good work*. NY: Harper & Row, 1979. 223 p. 0060138572
Schumacher reiterates that "small is beautiful" and demonstrates how appropriate, small-scale technology can be effectively utilized in both industrialized and Third World countries to correct environmental and economic imbalances. AOT. #

900. ___. *A guide for the perplexed*. NY: Harper, 1978. 147 p. 0060906111
Describes how to live according to the principle of "small is beautiful." Argues that both the narcisssistic "cult of the self" and rigidly hierarchical thinking must be replaced with a point of view which is spiritually based and respectful of nature and other people. AOT. #

901. ___. *Small is beautiful: a study of economics as if people mattered*. NY: Borgo, 1991, c1973. 347 p. 0809591154
An effective critique of why our economic and political systems destroy the natural world and cheapen human experience. Calls for small-scale "technology with a human face" and localized "Buddhist economics." This watershed book has had a profound influence on the environmental, appropriate technology, sustainable development, and new age movements. HR AOT. #

902. Simmons, I. G. *Interpreting nature : cultural constructions of the environment*. NY: Routledge, 1993. 215 p. : ill. 0415097053 0415097061
Analyzes how human attitudes and interactions with the natural world as shaped by psychological processes and cultural beliefs..

903. Stegner, Wallace Earle. *The American West as living space.* Ann Arbor: Univ. of Michigan, 1987. 89 p. : photos. 0472093754 0472063758 (pbk)
AR: *The sound of mountain water.*
These essays on the aridity of the west, environmental and conservation issues, and wilderness also appear in *Where the blue bird sings to the lemonade springs* and are HR.

904. ___. *Wolf willow : a history, a story, and a memory of the last plains frontier.* NY: Penguin, 1990, c1962. 306 p. 0140134395
This fascinating autobiography of Stegner's early years includes a consideration of democracy, an evocation of the northern plains in the early 20th century, the relationship of people and place, and two short stories.

905. Stegner, Wallace Earle, and Etulain, Richard W. *Conversations with Wallace Stegner on Western history and literature.* Rev. ed. Salt Lake City: Univ. of Utah, 1990. 230 p. 0874803535
Stegner comments on environmental innovations, notable transitions in western American literature, his own books, and major contributions to western environmentalism and culture. HR.

906. Stocking, Kathleen. *Letters from the Leelanau : essays of people and place.* Ann Arbor: Univ. of Michigan, 1990. 182 p. 0472094459 0472064452 (pbk)
Argues that the technological transition from hunter-gatherer to agricultural to industrial society proceeded more quickly than the wisdom which could keep us from ecological ruin. She finds renewal in a return to the land and country values, in this case on Michigan's Upper Peninsula.

907. Turner, B. L. ed. *The Earth as transformed by human action : global and regional changes in the biosphere over the past 300 years.* NY: Cambridge Univ., 1990. 713 p. : ill., maps. 0521363578
Contains 42 fairly technical papers on changes in population and society, transformations of the global environment, and understanding these changes in terms of human-nature theory, social relations, and cultural-human ecology. For adults with some scientific background. HR.

908. Turner, Frederick. *Of chiles, cacti, and fighting cocks : notes on the American West.* SF: North Point, 1990. 200 p. 0865474281
Compares and contrasts the legends of the American southwest with the modern realities. Broad in scope, it includes literary, historical, social, and natural perspectives.

909. Turner, Frederick W. *Beyond geography : the western spirit against the wilderness.* New Brunswick, NJ: Rutgers, 1992, c1980. 357 p. 0813519098
Traces western society's obsession for converting "primitive" peoples and

"taming" the land. Looks at the progressive decay of Christianity, from a living mythology to a historically oriented state religion that created a spiritual vacuum in Europe. That vacuum manifested itself in exploration, conquest, and conversion.

910. Waters, Frank*. *To possess the land : a biography of Arthur Rochford Manby.* Athens, OH: Swallow, 1993, c1973. 295 p. 0804009805
A biography of one of the most infamous land-grabbers and despoilers in western history.

911. Westing, Arthur H. *Warfare in a fragile world : military impact on the human environment.* NY: Taylor & Francis, 1980. 249 p. 0856661870
AR: *Ecological consequences of the second Indochina war, Weapons of mass destruction and the environment.*
Military abuses of all major global habitats are evaluated. Concludes that ecological considerations have not weighed heavily in past human affairs, civil or military, and that such neglect is becoming ever more dangerous.

912. Wilson, Edward Osborne. *Sociobiology.* Abr. ed. Cambridge: Harvard, 1980. 366 p. 0674816234
AR: *Naturalist.*
Wilson's claim that all human behavior has a genetic component made this one of the most controversial scientific books of its time. Unabridged ed. entitled *Sociobiology: the new synthesis.* HR.

913. Worster, Donald. ed. *The Ends of the earth : perspectives on modern environmental history.* NY: Cambridge Univ., 1989. 487 p. : ill. 0521343658
AR: *American environmentalism: the formative period, 1860-1915.*
Contains essays on the connections between climate and food supply, demographic pressure and technological innovation, social change and environment in pre-colonial Europe, and the impact of European conquest on the ecosystems and peoples of the rest of the world. HR.

914. ___. *Under western skies : nature and history in the American West.* NY: Oxford, 1992. 292 p. 0195076249
Provides a natural and ecological history of the west, arguing that unlimited personal and corporate freedom untempered by environmental consciousness are destroying the balance of nature.

915. ___. *The wealth of nature : environmental history and the ecological imagination.* NY: Oxford, 1993. 255 p. 0195076249
A book of essays which previously appeared in various journals or books or given as lectures. Reminiscent of the work of Aldo Leopold.

Books for young adults:

916. Becklake, John, and Becklake, Sue. *The population explosion.* NY: Gloucester, 1990. 36 p. : col. photos. 0531171981
Discusses our continually increasing population, its causes and devastating consequences, and efforts by governments and individuals to control its growth. GR 5-9.

917. Gallant, Roy A. *The peopling of planet Earth : human population growth through the ages.* NY: Macmillan, 1990. 163 p. : photos. 0027357724
Outstanding Science Trade Book, 1990.
Examines the impact of human population growth, discussing the origins of the human species, the rise of cities, migration to the new world, population trends, and the scarcity of natural resources. Presents the concept of carrying capacity. HR.

918. Herda, D. J., and Madden, Margaret L. *Land use and abuse.* NY: Watts, 1990. 143 p. 0531109534
Describes the environmental impact of land use, showing how careless development leads to loss of agricultural lands, animal and plant species, and causes deforestation, desertification, and urban sprawl.

919. McGraw, Eric. *Population growth.* Vero Bleach, FL: Rourke, 1987. 46 p. : ill. 0865922764
Covers the history and causes of population growth, world population growth trends, subsequent resource reduction trends, associated problems such as overgrazing and urbanization, and possible ways to curb population growth. GR 5-9.

920. Newton, David E. *Land use A-Z.* Hillside, NJ: Enslow, 1991. 128 p. 0894902601
A dictionary of terms related to the many scientific, technological, and social issues affecting the use of land. GR 7-10.

921. Sauer, Carl Ortwin. *Man in nature.* Berkeley: Turtle Island, 1975, c1939. 267 p. : ill., maps. 0913666017
AR: *Land and life.*
This classic text combines geography, history, anthropology, and natural history, drawing on Native American civilizations to describe our landscape and history. Sauer divides North America into different ecosystems, explaining how Indians related to the land and developed more advanced cultures over time. Despite its age, it remains HR AOT. #

922. Winckler, Suzanne, and Rodgers, Mary M. *Our endangered planet : popu-*

lation growth. Minneapolis: Lerner, 1991. 64 p. : col. ill. 082252502X
Studies the effects of uncontrolled population growth on the global environment, which results in dangerous pressures on natural resources, wildlife, air, water, and living space. GR 5-9.

Third World

Ecopolitics is a contentious struggle in the Third World, where international development efforts frequently impoverish or even eliminate indigenous populations, reduce biodiversity, and destabilize the environment. This struggle is explored by Bull, Chatterjee, Faber, Guimaraes, Hall, Harrison, Little, Ramphal, Rich, Stonich, Weinberg, Weir, and White. Books by Adams, Bonner, Braun, Kempf, and Rabinowitz cover Third World conservation efforts. Related books can be found in the Rainforest and Sustainable Development sections.

Authors who believe that threatened indigenous cultures should be preserved because of their unique knowledge and contributions to the creation of a more ecogical worldview include Cheneviere, Cowan, Davidson, Horton, Maybury-Lewis, Morgan, O'Hanlon, Perkins, Suzuki, Tobias, and Weatherford. Also see Bell's *Daughters of the Dreaming* (1138) The books on endangered cultures by Reynolds are highly recommended for all ages.

Books for adults:

923. Adams, Jonathan S., and McShane, Thomas O. *The myth of wild Africa : conservation without illusion*. NY: Norton, 1992. 266 p. : photos. 0303033961
Argues that well-meaning conservation efforts have helped destroy the balance between people and wildlife in Africa. Many homes, livelihoods, and cultures are being destroyed. Envisions a new conservation in which people and their needs are brought into the equation, but the animals are not abandoned.

924. Agarwal, Anil. ed. *For Earth's sake : a report from the Commission on Developing Countries and Global Change*. Ottawa: IDRC, 1992. 145 p. 0889366225
The goal of the Commission was to raise the profile of Third World environment/development perspectives and concerns within the research community worldwide. Intended to stimulate interest in social studies among a wide range of researchers and activists in environmental and developmental issues.

925. Bonner, Raymond. *At the hand of man : peril and hope for Africa's wildlife*. NY: Knopf, 1993. 322 p. 0679400087
Describes the conflict between the poverty of the African people and efforts

to conserve wildlife. Argues that unless economic and social conditions are improved, wildlife extinctions are inevitable. Also explores controversies within the African conservation movement.

926. Braun, Elisabeth. *Profile in conservation : Africa.* Boulder, CO: North American, 1994. 200 p. : ill. 1555919197
This biography of both famous and little-known conservationists also provides information on the many different natural environments of Africa, endangered species and their habitats, preservation efforts, and the philosophy of village-based conservation projects. #

927. Bull, David. *A growing problem : pesticides and the Third World poor.* Oxford: OXFAM, 1982. 192 p. 0855980648
Examines the complex web of pest and predator, crops, livestock, and people that functions in systems with natural biological controls. Contrasts this with the widespread use of pesticides, which have highly negative impacts on the Third World.

928. Chatterjee, Pratap, and Finger, Matthias. *The Earth brokers : power, politics and world development.* NY: Routledge, 1994. 208 p. 0415109620 0415109639
Was the Rio Conference little more than a green-camoflaged attempt to pursue development schemes in the Third World? Outlines the implications of alliances formed by the summit, suggesting that the "new world order" has inherited a set of management elites who preside over the widespread neglect of the global village. HR.

929. Cheneviere, Alain. *Vanishing tribes : primitive man on Earth.* Garden City, NJ: Doubleday, 1987. 267 p. : col. ill., maps. 0385238975
This photoessay depicts the everyday lives of 20 different "primitive" tribal groups; traditional legends of each tribe; how they live in equilibrium with nature; and developmental, governmental, and environmental threats to their continued existence. HR.

930. Cowan, James. *Letters from a wild state : rediscovering our true relationship to nature.* NY: Bell Tower, 1991. 138 p. 051758770X
This exploration of our essential unity with all life on the planet explores 50,000 years of Australian aboriginal thought..

931. ___. *Messengers of the gods : tribal elders reveal the ancient wisdom of the Earth.* NY: Bell Tower, 1993. 209 p. 0517880784
Similar to *Letters from a wild state,* but with more emphasis on tribal rites and ceremonies.

932. Davidson, Art. *Endangered peoples*. SF: Sierra Club, 1993. 195 p. : col. photos. 0871564572
Thoughtful prose and photos depict the plight of over 250 million people whose cultures are being exterminated and populations decimated by deforestation, development, repression, and assimilation. This grimly realistic but uplifting book is HR.

933. Faber, Daniel J. *Environment under fire : imperialism and the ecological crisis in Central America*. NY: Monthly Review, 1993. 301 p. 0853458391 0853458405 (pbk)
Exploitation of nature is closely tied to the exploitation of people..

934. Guimaraes, Roberto Pereira. *The ecopolitics of development in the Third World : politics & environment in Brazil*. Boulder, CO: L. Rienner, 1991. 271 p. 1555872433
Considers ecopolitics in Brazil before 1964, the expansion of the Brazilian economy and subsequent environmental devastation under military rule, the effect of the UN's Special Secretariat of the Environment on Brazil, and ecological transition and bureaucratic politics in the Third World.

935. Hall, Sam. *The fourth world : the heritage of the Arctic and its destruction*. NY: Vintage, 1988, c1987. 240 p. : maps. 0394756304
Once protected by remoteness and climate, Arctic cultures are now threatened by economic development as the land and water are coveted by oil developers, dam builders, and the military. Describes worldwide environmental threats posed by the melting of the icecap and massive oceanic pollution. HR

936. Harrison, Paul. *Inside the Third World ; the anatomy of poverty*. 3d ed. NY: Penguin, 1993. 529 p. 0140172173
AR: *The greening of Africa*.
Examines how colonialism, superpower rivalry, and economic exploitation with few environmental controls have created or exacerbated poverty, urban slums, ecological catastrophe, malnutrition, disease, overpopulation, and alienation throughout the Third World.

937. Horton, Barbara Curtis. *Tiger bridge : nine days on a bend of the Nauranala*. Santa Barbara: Daniel, 1993. 74 p. 1880284014
While waiting for a tiger to appear for nine days Horton considers the integrity and nobility of wildness and struggles with the mental and cultural encumbrances that prevent people from authentically experiencing a place.

938. Howard, Michael C. ed. *Asia's environmental crisis*. Boulder: Westview, 1993. 293 p. 0813388082

Contains 13 essays on the environmental degradation that has resulted from rapid economic expansion; the connection between politics, economics, and the environment; governmental and nongovernmental initiatives to improve the situation; and possible solutions to interlinked problems.

939. Inter Press Service. *Story Earth : native voices on the environment.* SF: Mercury House, 1993. 200 p. 1562790358
Contains 18 statements from representatives of indigenous cultures on the nature of the global environmental crisis and changes we must make in the way we view the world. Rejects the industrial view of the Earth as a resource to be consumed and recommends listening to the lessons of traditional cultures.

940. Kemf, Elizabeth. *The Law of the mother : protecting indigenous peoples and protected areas.* SF: Sierra Club, 1993. 344 p. : col. photos, maps. 0871564513
Offers a comprehensive vision of how to design and implement conservation projects to provide for the well being of local peoples, wildlife, and the land itself. Controversial issues and innovative solutions are considered.

941. Little, Peter D., and Horowitz, Michael M. eds. *Lands at risk in the Third World : local-level perspectives.* Boulder: Westview, 1987. 416 p. 0813373115
Contains18 papers on the effects of social structure, social progress, and cultural history on practices leading to resource deterioration. Considers why so many well-intentioned programs have gone wrong and why either economic success or failure has led to environmental deterioration. Covers Africa, Asia, and Latin America.

942. Maybury-Lewis, David. *Millennium : tribal wisdom and the modern world.* NY: Viking, 1992. 397 p. : col. photos. 0670829358
Companion volume to the PBS television series.AR: *The savage and the innocent.* Wisdom from "primitive" societies, past and present.

943. Morgan, Sally. *My place.* NY: Arcade, 1990. 360 p. 1559700548
Morgan's physical and psychological journey to find the truth of her origins, which had been hidden from her to protect her from racism. Includes a consideration of the role of women in native cultures and the link between Australian Aborigines and the Earth. HR.

944. O'Hanlon, Redmond. *Into the heart of Borneo : an account of a journey made in 1983 to the mountains of Batu Tiban with James Fenton.* Edinburgh: Salamander, 1984. 191 p. : photos, map. 0907540554
Zany bird-watching Brits with three Iban guides make their way through leeches upriver in Borneo searching for the heart—not of darkness but of wilderness. HR.

945. Perkins, John M. *The world is as you dream it : shamanic teachings from the Amazon and Andes.* Rochester, VT: Destiny, 1994. 144 p. : ill. 0892814594
How the Shuar, Otavalan, and Salasacan Indians use shamanism for establishing unity with the Earth, preservation of the natural world, and self-empowerment. Contends that a mode of materialism and domination must be replaced with a more ecologically and spiritually oriented one.

946. Perrott, John. *Bush for the Bushman : need "the Gods must be crazy" Kalaharipeople die?* Greenville, PA: Beaver Pond, 1992. 227 p. : col. photos. 1881399044
What begins as an adventurous travelogue becomes a cause for the preservation of the !Kung, a highly Earth-centered culture living in the Kalihari. Perrott demonstrates how the preservation of Africa's environment and its native cultures are part of the same struggle.

947. Rabinowitz, Alan. *Chasing the dragon's tail : the struggle to save Thailand's wild cats.* NY: Anchor, 1992. 241 p. : photos, map. 0385415184
This "Indiana Jones" zoologist not only works to preserve Thailand's big cats from extinction by poachers, drug traffickers, and habitat destruction, he also fights floods, fire ant infestations, elephant stampedes, and the unwanted marriage to the daughter of a tribal chief. Entertaining and informative. HR. #

948. Ramphal, S. S. *Our country, the planet : forging a partnership for survival.* Washington: Island, 1992. 291 p. 1559631651 1559631643 (pbk)
Contends that industrial countries must consume less energy so that poor ones may consume more; presses for more equitable trade policies that would allow developing nations to work their way out of poverty; and shows how industrial countries have stymied progress. HR.

949. Rich, Bruce. *Mortgaging the Earth : The World Bank, environmental impoverishment, and the crisis of development.* Boston: Beacon, 1994. 376 p. 080704704X
Demonstrates how World Bank officials have systematically approved projects with disasterous environmental and human rights consequences over the objections of staff members and other experts. Chronicles grassroots efforts by citizens in Third World countries seeking alternatives to ruinous "Bank-style" development.

950. Stonich, Susan C. *"I am destroying the land!" : the political ecology of poverty and environmental destruction in Honduras.* Boulder, CO: Westview, 1993. 191 p. 0813386497
Interconnections involving geographic, historical, demographic, social, economic, and ecological aspects of development.

951. Suzuki, David T., and Knudtson, Peter. *Wisdom of the elders : honoring sacred native visions of nature.* NY: Bantam, 1992. 274 p. 0553088629
A zoologist and a biologist make an effective case that the solutions to the environmental crisis are imbedded in the living mosaic of profound indigenous insights into the workings of the natural world. Scientific and legendary cosmologies converge here.

952. Tobias, Michael. ed. *Mountain people.* Norman: Univ. of Oklahoma, 1986. 219 p. : photos, maps. 0806119764
Although mountain peoples tend to be among the last to be "civilized," industrial development and political and cultural forces are exterminating many remaining groups. Surveys the history, lifestyles, relation to nature, and lessons to be learned from extinct and remaining montane cultures. AOT. #

953. Weatherford, J. McIver. *Savages and civilization : who will survive?* NY: Crown, 1994. 310 p. 0517588609
Argues that the natural and cultural environments of indigenous peoples are being destroyed at the very time that their tribal wisdom, offering us insights into how developed societies can live harmoniously with the Earth, is most needed. HR.

954. Weinberg, Bill. *War on the land : ecology and politics in Central America.* Atlantic Highlands, NJ: Zed, 1991. 203 p. 0862329469 0862329477 (pbk)
Mass agribusiness' "war on the land" has ravaged the environment of Central America, created mass relocations of farmers to slums, and strengthened totalitarian regimes. Makes a strong case for the reversal of US foreign and agricultural policy in the area.

955. Weir, David. *The Bhopal syndrome : pesticides, environment, and health.* SF: Sierra Club, 1988, c1987. 210 p. 0871567970
Uses the 1984 Union Carbide accident in India as a case study of pesticide production and the poor safety records of chemical plants.

956. White, Rodney. *North, South, and the environmental crisis.* Toronto: Univ. of Toronto, 1993. 214 p. 080205952X 0802068855 (pbk)
Analyzes how overpopulation, poverty, inequities between rich and poor nations, the push for material improvements, and industrialization exacerbate the environmental crisis. Seeks to make technical issues more familiar to politicians, journalists, teachers, union leaders, employers, community leaders, and others who shape public opinion.

Books for young adults:

957. Chiasson, John. *African journey.* NY: Bradbury, 1987. 55 p. : col. photos.
0027185303
Notable Childrens Trade Book in Social Studies, ALA Notable Children's
Book Award, 1987.
Depicts how nature dictates the way of life for people in six different
regions of Africa. HR.

958. Cowan, James. *Kun-man-gur the Rainbow Serpent.* Boston: Barefoot,
1994. 31 p. : col. ill. 1569579067
In this retelling of an Aboriginal creation myth, the Rainbow Serpent finds
a home and food for the flying foxes after he punishes a bat for saying that
they smelled bad. Includes a foreword explaining the background and
importance of the myth. GR 4-adult. #

959. Liptak, Karen. *Endangered peoples.* NY: Watts, 1993. 160 p. : ill.
0531109879
Examines five ethnic groups living tribal existences around the world and
shows how assimilation into mainstream society and other factors are
threatening their cultures.

960. Reynolds, Jan. *Vanishing cultures.* San Diego: Harcourt Brace
Jovanovich, 1991-94. 6 v. (32 p. each) : col. photos. Contents: Amazon
Basin (01520228315) — Down under (0152241825) — Far north
(0152271783) — Frozen land (0152387870) — Himalaya (0152344659) —
Mongolia (0152553126) — Sahara (0152699597).
Examines the cultures of tribal people in various areas and the threats to
their environment and continuing existence. Although the primary text is
written for GR 3-6, the photography and the higher level text in the rear of
each volume make these appropriate for all ages. HR. #

Native North Americans

You don't have to leave America to see the Third World, just visit a reservation. Native Americans have a rich ecological tradition which was largely
destroyed during the European conquest of North America but is being
reasserted today. This tradition has been over-romanticized by both native
and non-native authors, a good case in point being a speech attributed to
chief Seattle (Sealth, actually) which was written by a white minister in 1971.
        Fortunately, authentic texts are also available. (An asterisk has been
added to the names of enrolled tribal members. For an explanation, see

the introduction to the Native American Fiction chapter.) Awiakta, the original Black Elk, Fire, Horn, Hughes, Hungry Wolf, Margolin, McLuhan, Nelson, Standing Bear, Stone, Talayesva, the Tedlocks, Thompson, and Trimble all describe traditional values. For collections of legends see entries 966, 969-70, 974-5, 977, 980-81, 983, 985-86, 993, and 998-99.

Analysis of the conquest of North America are made by Cronon, Delage, and Murphy. Personal accounts of white captives who lived among the Indians are provided by Long and Tanner. Beck, Evans-Wentz, Mathews, and Waters attempt to combine the best aspects of native and non-native cosmology. Deloria, Gedicks, Mander, Matthiessen, McGuire, Momaday, Mowat, Vecsey, Wenzel, Wyler, and Whaley describe contemporary native environmental and social problems. Controversial works presenting Native American philosophy and rituals to New Age audiences are provided by Wallace Black Elk, McGaa, Highwater, Summer Rain, and Sun Bear.

For relevant works in other chapters see Arthur (1486), Cajete (1500), Allen (1133), Bell (1138), Jensen (1156), Niethammer (1167), Brower (1033), McLuhan (1106), Modzelewski (1266), Sheldon (393), Heat Moon (571, 878), and Nabhan (660-62)

Many of the legends and works for adults are also appropriate for teens. The books by Benton-Banai, Bruchac, and Jenness in this chapter and by Bruchac in Environmental Education (1494-98) are highly recommended for young adults.

Books for adults:

961. Awiakta, Marilou*. Selu : seeking the Corn-Mother's wisdom. Golden, CO: Fulcrum, 1993. 336 p. : ill. 1555911447
Among Selu's traditional survival wisdoms are strength, respect, balance, adaptability, and cooperation. These common-sense teachings have a direct bearing on preserving the environment, women's issues, cultural diversity, and government.

962. Beck, Peggy V.; Walters, Anna Lee; and Francisco, Nia. The sacred : ways of knowledge, sources of life. Flagstaff: Northland, 1990, c1977. 365 p. : ill. 0912586745 (pbk) 0912586249
This college-level text demonstrates how the reexamination of traditional sacred ways may help in learning about the environment and the interrelationships of all beings. Also shows that although sacred principles and values are eternal, traditional ways change to adapt to modern life.

963. Black Elk*, and Neihardt, John Gneisenau. Black Elk speaks : being the life story of a holy man of the Ogalala Sioux. Alexandria, VA: Time-Life, 1991, c1932. 280 p. : col. ill. 0809489007 0783517505
AR The sacred pipe, The sixth grandfather.

In order to preserve his traditional culture, this Sioux seer related its values and traditions to the wider world. Accounts of his visions and of the tribal dances he carried out according to those visions are particularly vivid and notable. Still very influential today. HR.

964. Black Elk, Wallace H. *., and Lyon, William S. *Black Elk : the sacred ways of a Lakota.* SF: Harper, 1991. 223 p. 0062500740
Describes the Sacred Pipe, Sun Dance, sweat-lodge, vision quest, and other ceremonies. This contemporary populizer of Indian ceremonies (not to be confused with the original Black Elk) is fairly controversial.

965. Cronon, William. *Changes in the land : Indians, colonists, and the ecology of New England.* NY: Hill & Wang, 1983. 241 p. 0809001586
Contrasts the two societies in terms of agriculture, relation to animals,notions of property, and varying degrees of respect for the land. Emphasizes the environmental impact of settlement by Europeans. HR.

966. Dauenhauer, Nora*, and Dauenhauer, Richard. eds. *Haa shukba, our ancestors : Tlingit oral narratives.* Seattle: Univ. of Washington, 1987. 514 p. : ill., map. 0295964952
Contains a wide variety of Tlingit legends gathered from tribal elders. Many detail the complex pattern of relationships among Tlingit clans and between them and the Earth.

967. Delage, Denys. *Bitter feast : Amerindians and Europeans in Northeastern North America, 1600-64.* Vancouver: UBC, 1993. 399 p. : ill., maps. 0774804513 (pbk) 0774804343
Contends that both the Amerindian and European civilizations had strong economic, religious and cultural traditions, and that each had things to learn from the other. Superior technology and greed fostered an unequal trading situation whose social and environmental impacts are still felt today.

968. Deloria, Vine*. *God is red : a native view of religion.* 2d ed. Golden, CO: North American, 1992, c1973. 313 p. 1555919049
AR: *The metaphysics of modern existence, We talk you listen.*
Contrasts Native American views on ecology and religion against the expansionistic beliefs that justified the conquest of people and the land. Contends that the pollution and plundering of the Earth will not cease until Christianity is replaced with a philosophy which reveres the interconnectedness of all living things. Controversial and HR.

969. Dooling, D. M., and Jordan-Smith, Paul. eds. *I become part of it : sacred dimensions in Native American life.* NY: Parabola, 1989. 291 p. : ill. 0930407075

These 30 tales from a variety of traditions demonstrate the importance of the sacred aspects of nature to cultural survival and how to live in a harmonious balance with the Earth.

970. Edmonds, Margot, and Clark, E. Elizabeth. eds. *Voices of the winds* : *Native American legends.* NY: Facts on File, 1989. 368 p. 0816020671
Contains over 130 legends, from a variety of tribal groups, about how and why Native Americans believe that spirit life animates all creatures.

971. Evans-Wentz, W. Y. *Cuchama and sacred mountains.* Ed. by Frank Waters and Charles L. Adams. Athens: Swallow, 1989, c1981. 227 p. 0804009082
A Tibetan Buddhist scholar compares the beliefs of Native Americans and various Eastern religions by focusing on sacred mountains. Editor Waters argues not just for a unification of traditional philosophies, but a synthesis of traditional and modern thought. HR.

972. Fire, John*, and Erdoes, Richard. *Lame Deer, seeker of visions.* NY: Washington Square, 1994, c1972. 317 p. : ill. 0671888021
A traditional Sioux relates much about the native way of life, especially spiritualism, religion, relationship to nature, ceremonies, problems and humor. Written during a revival of Sioux culture and pride, this book is similar to *Black Elk Speaks* and is HR.

973. Gedicks, Al. *The new resource wars* : *native and environmental struggles against multinational corporations.* Boston: South End, 1993. 270 p. 0896084639 0896084620 (pbk)
A study of native environmental resistance campaigns against hydroelectric, logging, oil drilling, and mining projects. Analyzes how racism has been used to spearhead the assault on native lands and how the defense of treaties can be used to protect both races. Essential source for multicultural environmental activism. HR.

974. Griffin, Arthur, and Griffin, Trenholme J. *Ah mo* : *Indian legends from the Northwest.* Surrey, BC: Hancock House, 1990. 64 p. 0888392443
Gathered in the 1880s by Arthur Griffin and retold by Trenholme Griffin for modern audiences, these legends "flow with the spirit of nature." #

975. Hausman, Gerald. *Tunkashila* : *from the birth of Turtle Island to the blood of Wounded Knee.* NY: St. Martin's, 1993. 264 p. : ill., map. 0312099282
Hausman retells over 80 stories in a "circle of highly anecdotal myths" about creation, love, loss, power, war, and eventual conquest. Weaves traditional native legends from across America into one common thread. HR AOT. #

976. Highwater, Jamake*. *The primal mind : vision and reality in Indian America.* NY: Harper & Row, 1981. 234 p. 0060148667
A highly philosophical and eclectic but very accessible consideration of the place of people in nature, the develoment of the mind, and the philosophical and social construction of reality. Calls on artists of all types to bridge the gap between ancient and modern modes of thought.

977. Hilbert, Vi. trans. and ed. *Haboo : native American stories from Puget Sound.* Seattle: Univ. of Washington, 1985. 204 p. : ill. 0295962704
A Skagit woman who was raised in a traditional home recounts 33 tales. Most illustrate points relating to human nature, the Indian social order, and the relationships between humans, other animals, natural forces, and the spirit world.

978. Horn, Gabriel*, and White Deer of Autumn*. *Native heart : an American Indian odyssey.* San Rafael, CA: New World Library, 1993. 293 p. 1880032077
Horn describes his life as a spiritual quest to preserve Indian culture, spirituality, and medicine; presents a proper relationship between people and all aspects of nature; and relates his work in the AIM survival schools. HR AMT. #

979. Hungry Wolf, Adolf*. *Teachings of nature.* Skookumchuck, BC: Good Medicine, 1989. 94 p. : ill. 0913990752
This handbook of outdoor knowledge from various tribes of North America lists wild plants and their uses for food and medicine, traditional gardening methods, etc.

980. Kerven, Rosalind. *Earth magic, sky magic : North American Indian stories.* NY: Cambridge Univ., 1991. 79 p. 0521362350
Includes folk tales and legends from twelve North American tribal groups that "convey a mystical atmosphere and bring out the Native Americans' affinity with and respect for the natural world."#

981. Lang, Julian. ed. and trans. *Ararapikva : creation stories of the people : traditional Karuk Indian literature from northwestern California.* Berkeley: Heyday, 1994. 110 p. : photos. 0930588657 (pbk) 093058869X
A bilingual book in English and Karuk featuring traditional creation stories emphasizing the unity of all things and respect for nature. Includes an introduction to Karuk history, culture and language.

982. Long, Haniel. *The marvelous adventure of Cabeza de Vaca.* Clearlake, CA: Dawn Horse, 1992. 63 p. : ill., map. 091880146X
A.k.a.: *The power within us* or *The interlinear of Cabeza de Vaca.*
A conquistador stranded in Florida walks to Western Mexico where he

learns the native ways and lives with natives as a healer, forsaking his earlier actions and beliefs for a life in close harmony with the Earth. HR.

983. Lopez, Barry Holstun. *Giving birth to Thunder, sleeping with his daughter* :
Coyote builds North America. NY: Avon, 1990, c1977. 169 p. 0380711117
Coyote's adventures with other animals illustrate creation myths, human
infallibilities, and the relationships of different species and forces in
nature. Incorporates 70 short legends from a wide variety of American Indian peoples. Somewhat risque, but HR AMT. #

984. Mander, Jerry. *In the absence of the sacred : the failure of technology and the*
survival of the Indian nations. SF: Sierra Club, 1991. 446 p. 0871567393
A powerful critique of how the inappropriate use of technology and lack of
respect for nature have combined to devastate both the land and the
indigenous people of North America. HR.

985. Margolin, Malcolm. *The Ohlone way : Indian life in the San Francisco-Mon-*
terey Bay Area. Berkeley: Heyday, 1978. 182 p. : ill. 0930588029 0930588010
(pbk)
Recreates an era 200 years ago when the SF-Monterrey area held one of the
largest indigenous populations in America. Describes the rich natural environment; a balanced (rather than exploitative) relationship with the environment; and an economic system based on sharing rather than
competition. HR.

986. ___. ed. *The Way we lived : California Indian stories, songs & reminiscences.*
Berkeley: Heyday, 1993. 248 p. : 92 photos. 093058855X
A collection of traditional, historic, and contemporary Native Californian
literature with commentaries by Margolin.

987. Martin, Calvin. *Keepers of the game : Indian-animal relationships and the fur*
trade. Berkeley: Univ. of Cal., 1982, c1978. 226 p. 0520046374
AR:*Indians, animals and the fur trade : a critique of Keepers of the game.*
Rejects the idea that Indians participated in the fur trade because their
lives were brutal and short so they were distracted by trinkets and alcohol.
Describes their lives as full and nonexploitive, and warns that since white
people will never understand native cosmology they had best look elsewhere for inspiration.

988. Mathews, John Joseph*. *The Osages : children of the middle waters.* Norman: Univ. of Oklahoma, 1982, c1961. 847 p. : ill. 0806117702
AR: *Wah'kon-tah, Talking to the moon.*
Describes traditional Osage society, first meetings with Europeans, conflict
and assimilation, the attempted renewal of Plains Indian society in the late

1900s, and their status in the mid-20th century. An excellent depiction of the changing relation between people and the land. HR.

989. Matthiessen, Peter. *Indian country.* NY: Penguin, 1992. 338 p. 0140130233
AR: *In the spirit of Crazy Horse.*
Argues that environmental desecration, alteration in the name of progress, and spiritual transgression are the same thing. Explores ten important instances where the white man's encroachments upon sacred grounds show the tragic effects of a confrontation which is bound to harm both sides.

990. McGaa, Ed*. *Mother Earth spirituality : native American paths to healing ourselves and our world.* SF: Harper & Row, 1990. 230 p. 0062505963
Teaches how to reconnect with and heal the Earth in this introduction to Native American philosophy, history, and rites. McGaa is a controversial figure who claims that he is simply continuing the work of the legendary Black Elk, but critics accuse him of cultural appropriation.

991. McGuire, Thomas R.; Lord, William B.; and Wallace, Mary G. eds. *Indian water in the new West.* Tucson: Univ. of Arizona, 1993. 241 p. 0816513929
Brings together the views of engineers, lawyers, ecologists, economists, and a Native American tribal leader to discuss how the legitimate claims of both Indians and non-Indians to scarce water in the West are being settled.

992. McLuhan, T. C. *Touch the Earth : a self-portrait of Indian existence.* Photos by Edward S. Curtis. NY: Promontory, 1991, c1971. 185 p. : photos. 083940000
Selections from speeches and writings of Indians living in all parts of the country between the 16th and 20th centuries. They speak with courtesy and respect of the land, of animals, of the objects that made up the territory in which they lived. HR AOT. #

993. Miller, Jay. ed. *Earthmaker : tribal stories from Native North America.* NY: Perigee, 1992. 175 p. 0399517790
These 23 legends with animal protagonists focus on creation and the unity of all living things. Contains four sections: Making the world, Adjusting the world, Shaping animals, and Awaiting humans.

994. Momaday, N. Scott*. *The way to Rainy Mountain.* Albuquerque: Univ. of New Mexico, 1976, c1969. 88 p. : ill. 0826304362
AR: *The ancient child.*
Combines Kiowa legends, personal experience, and spiritual insights to

depict how an indigenous people, the natural landscape, and a spiritual-artistic cosmology are being simultaneously destroyed. HR AOT. #

995. Mowat, Farley. *The desperate people.* rev. ed. NY: Bantam, 1984, c1959. 240 p. : ill., map. 0777042323X
Mowat revisited the Ihalimiuts only six years after writing *People of the deer* to discover that the extinction of their culture through deportation, commercial development, and slaughter of the caribou herd had accelerated.

996. ___. *People of the deer.* NY: Bantam, 1981, c1952. 304 p. : ill., map. 0553148176
Sequel: *The desperate people.*
Describes how the Arctic Ihalmiut culture (located west of Hudson's Bay) lives symbiotically with an apparently hostile landscape. Also covers threats posed by sport hunting, bounty hunting, and commercial development.

997. Nelson, Richard K. *Make prayers to the raven : a Koyukon view of the northern forest.* Chicago: Univ. of Chicago, 1986, c1983. 308 p. : ill. 0226571629
Combines historical research, folklore, extensive field interviews, and personal experience to demonstrate the inherent unity of Alaska's Koyukon Indians and their forest environment.

998. Pijoan, Teresa. *White wolf woman : Native American transformation myths.* Little Rock: August House, 1992. 167 p. 0874832012 0874832004 (pbk)
A collection of 37 transformation myths collected from the oral traditions of Native Americans showing the powers of certain animals as they move between human and nonhuman worlds. #

999. Robinson, Harry*. Ed. by Wendy Wickwire. *Nature power : stories from an Okanagan elder.* Seattle: Univ. of Washington, 1992. 272 p. 0295972238
AR Sequel: *Write it in your heart.*
Relates stories featuring the shoo-MISH, or nature helpers, that assist humans and sometimes provide them with special powers. #

1000. Standing Bear, Luther*. *Land of the spotted eagle.* Lincoln: Univ. of Nebraska, 1978, c1933. 259 p. : ill. 0803209649 0803258909
Depicts the life of the Sioux during the transition from traditional to reservation life, arguing for the survival of Indian cosmology in a time of rapid change. Includes information on the Sun Dance and other rituals in an attempt to preserve them.

1001. ___. *My Indian boyhood.* Lincoln: Univ. of Nebraska, 1988, c1931. 189 p. : ill. 0803291868 (pbk) 0803241933
Describes how he was taught Earth-centered values and acquired plant,

herb, hunting, and fishing skills. Intended to increase interracial understanding and compassion. HR for all ages. #

1002. ___. *My people, the Sioux.* Lincoln: Univ. of Nebraska, 1975, c1928. 308 p. : photos. 0803257937
A depiction of the life of the Sioux people during a period of transition from independence to conquest. Emphasizes the unity of all living things and the need to respect the Earth. HR. #

1003. Stone, Jana; Curtis, Mel; and Sharpe, Bonnie. *Every part of this Earth is sacred : Native American voices in praise of nature.* SF: Harper, 1993. 140 p. : photos. 0062508482
In photographs marked by a quiet beauty, and a text of Native American voices of praise, chants, celebrations, and prophecies of coming disaster, the authors honor the Native American way of traditional living in grace and harmony with nature.

1004. Summer Rain, Mary*. *Earthway.* NY: Pocket, 1992, c1990. 442 p. 0671706675
Provides a Native American-based, organic, naturopathic approach to healing the body, mind, and spirit, and a strong unification of human beings and the Earth. Includes practical advice on naturopathy, dream analysis, meditation, and related subjects.

1005. Sun Bear,* and Wabun Wind*. *Black dawn, bright day : Indian prophecies for the millenium that reveal the fate of the Earth.* NY: Simon & Schuster, 1992. 231 p. 0671759000
A Chippewa prophet describes the environmental future and tells where and how to survive the changes he has foreseen: climatic shifts, droughts, floods, acid rain and pollution, earthquakes, and volcanic eruptions. These severe changes will pave the way for a cleansing of the Earth and a new relationship between Earth and humanity.

1006. Sun Bear,* Wabun Wind*, and Shawnodese*. *Dreaming with the wheel : how to interpret and work with your dreams using the medicine wheel.* NY: Simon & Schuster, 1994. 318 p. : ill. 0671784161
The Medicine Wheel teaches that all things and beings are related and must be in harmony for the Earth to be in balance. By a better understanding of dreams, we can have a better understanding of our relationship to the Earth. Both philosophical and practical.

1007. Talayesva, Don C. *. *Sun chief; the autobiography of a Hopi Indian.* New Haven, CT: Yale, 1978, c1942. 480 p. 0300002270
The story of a man raised in traditional Hopi society, but educated in white

schools, who moved to California only to reject modern society and return to Hopiland. Offers a reconsideration and appreciation of the Hopi cosmology and lifestyle. Occasional sexual frankness or cultural criticism add spice to the stew.

1008. Tanner, John. *The falcon : a narrative of the captivity and adventures of John Tanner*. Intro. by Louise Erdrich. NY: Penguin, 1994, c1830. 280 p. 0140170227
The first-person account of a settler captured by Ojibwa Indians who accepted their way of life, was assimilated, and lived with them for 30 years.

1009. Tedlock, Dennis, and Tedlock, Barbara. compilers. *Teachings from the American Earth : Indian religion and philosophy*. NY: Liveright, 1992, c1975. 279 p. : ill. 0871401460
This anthology is a scholarly investigation into Native American cosmology focusing on their approaches to perception and the human presence in the natural world.

1010. Thompson, Lucy*. *To the American Indian : reminiscences of a Yurok woman*. Berkeley: Heyday, 1991, c1916. 292 p. : photos. 0930588479
Concerned that her culture was near extinction, a holy woman wrote a text in the style of oral legends to explain Yurok values, reveal their ceremonies and legends, and re-emphasize the importance of their connection with the Earth. She scorns white people who either destroy or appropriate Indian culture and Indians who have abandoned traditional values.

1011. Trimble, Stephen. *The village of blue stone*. NY: Macmillan, 1990. 58 p. : ill., map. 0027895017
Recreates the day-to-day life throughout a year in a Chaco Culture Anasazi pueblo where people, landscape, work, art, and religion are inseparable. HR. #

1012. Vecsey, Christopher, and Venables, Robert W. eds. *American Indian environments : ecological issues in Native American history*. NY: Books on Demand, 1994, c1972. 236 p. : ill. 0835731200
Essays by Vecsey, Venables, Calvin Martin, Wilbur Jacobs, William Hagan, Laurence Hauptman, Kai Erickson, Peter MacDonald, and Oren Lyons on the environment in native religions, Indians as ecologists, white vs. Indian land use practices, energy exploitation on reservations, sovereignty of Indian land, and inter-Indian resource disputes.

1013. Waters, Frank*. *Book of the Hopi*. Ill. by Oswald White Bear Fredericks. NY: Penguin, 1977, c1963. 393 p. : ill. 0140045279

According to Hopi prophecy the first three worlds were destroyed by self-ishness and materialism and the current fourth world is on the verge of destruction, followed by a new age of natural harmony. Waters presents Hopi ceremonialism and consciousness as a model for that harmony. HR.

1014. ___. *The Colorado*. Athens, OH: Swallow, 1985, c1946. 410 p. : ill 0804008647
In this consideration of the land and its inhabitants, Waters portrays white settlers as being exploitative and power oriented and Indians as intuitive and land-loving, but not as "noble savages" with all the answers. He suggests a synthesis between the two apparently opposing viewpoints. HR.

1015. ___. *The Earp brothers of Tombstone : the story of Mrs. Virgil Earp*. Lincoln: Univ. of Nebraska, 1976, c1960. 247 p. 0803208731 0803258380
Presents the Earp's violent, exploitative attitudes and actions as a symbol of the failure of white settlers to live harmoniously with the land as the Indians and Hispanic settlers had. Tension and misunderstanding led to a blind compulsion to dominate and destroy.

1016. ___. *Masked gods : Navaho and Pueblo ceremonialism*. 2d ed. Athens, OH: Swallow, 1984, c1950. 432 p. 0804006415
Demonstrates how these tribes recognize dichotomies within people and between them and nature: ceremonies to resolve dualities, create harmony, and regain a natural equilibrium.

1017. ___. *Pumpkin Seed Point : being within the Hopi*. Athens: Ohio Univ., 1981, c1969. 175 p. 0804006350
Waters attempts to find a viable fusion between the two worlds of the industrial-mechanical-rational and the organic-spiritual-intuitive. Suggests that Hopi myths and ceremonies recognize the depths of the unconscious mind that western society has repressed. HR.

1018. Wenzel, George W. *Animal rights, human rights : ecology, economy and ideology in the Canadian Arctic*. Toronto: Univ. of Toronto, 1991. 206 p. : photos. 0802059619 0802068901 (pbk)
Critiques attempts to ban seal hunting in the Arctic for failing to recognize that the native people have always lived in ecological balance with the land while relying on seals for furs and food. Animals and cultures both have a right to fair treatment and protection of extinction.

1019. Weyler, Rex. *Blood of the land : the government and corporate war against first nations*. Rev. ed. Philadelphia: New Society, 1992. 352 p. 1550921827 1550921835 (pbk)
Covers the government's retaliation against the AIM leaders, struggles

between "First Nations" and industrial development, and the history of the subjugation of Native Canadians and Americans. Persecuted activists include Anna May Aquash, Dennis Banks, Russell Means, Leonard Peltier, and Clyde and Vernon Bellecourt.

1020. Whaley, Rick, and Bresette, Walter. *Walleye warriors : an effective alliance against racism and for the Earth*. Fwd. by Winona LaDuke. Philadelphia: New Society, 1994. 272 p. : photos, maps. 0865712565 0865712573 (pbk)
How a multiracial alliance of Anishinabe, local residents, and activists defused protests against Indian spearfishing in Wisconsin. Includes information on AIM, Wisconsin Greens, related organizations, and the history of natural resource conflicts between Indians and American industry.

1021. Williamson, Ray A., and Farrer, Claire R. eds. *Earth & sky : visions of the cosmos in Native American folklore*. Albuquerque: Univ. of New Mexico, 1992. 299 p. : ill. 0826313175
Seventeen astronomers and folklorists consider many aspects of ethnoastronomy: basic description and history of the science; cosmogony and cosmology; the role of the sky in legends; and the unity of sky, Earth, and people.

Books for young adults:

1022. Benton-Banai, Edward. *The Mishomis book : the voice of the Ojibway*. Saint Paul: Red School House, 1988. 114 p. : ill.
Recounts the legends, customs, and history of the Ojibway Indians of Wisconsin. Emphasizes establishing a strong personal connection to the land and respect for all living things. HR.

1023. Bruchac, Joseph*. *Flying with the eagle, racing the great bear : stories from Native North America*. Mahwah, NJ: BridgeWater, 1993. 128 p. : ill. 0816730261 081673027X (pbk)
A collection of traditional tales which present the heritage of various Indian nations, including the Wampanoag, Cherokee, Osage, Lakota, and Tlingit. Many have natural or environmental themes. HR GR 5-9.

1024. Freedman, Russell. *Buffalo hunt*. NY: Holiday House, 1988. 52 p. : ill. 0823407020
Examines the importance of the buffalo in the lore and day-to-day life of the Indian tribes of the Great Plains. Describes hunting methods and the uses found for each part of the animal that could not be eaten.

1025. Grinnell, George Bird. *When buffalo ran*. Surrey, BC: Hancock House,

1993, c1920. 124 p. : ill. 0888392583
The tale of how a young Cheyenne boy in the late 18th century learned to hunt buffalo and deer, live in cooperation with his people, fight their enemies, and develop a philosophy in which people are an integral part of nature. #

1026. Hodge, Gene Meany. *Kachina tales from the Indian Pueblos*. Santa Fe, NM: Sunstone, 1993. 64 p. : ill. 0865341842
Stories about the spirits who manifest themselves sometimes as dolls, and sometimes as living players in Hopi ceremonies. They connect the Hopi and other Pueblo nations to the Earth and to the spiritual world. "Material in this book originally appeared in *The kachinas are coming* by Gene Meany Hodge . . . in 1936."#

1027. Jenness, Aylette, and Rivers, Alice. *In two worlds : a Yup'ik Eskimo family*. Boston: Houghton Mifflin, 1989. 84 p. : photos. 0395427975
Notable Children's Trade Book in Social Studies, 1989.
Documents the life of a Yup'ik Eskimo family, detailing the changes that have come about in the last 50 years from a seasonal life close to the land to a more modern existence with some tribal values and customs remaining. HR GR4-8.

1028. Mayo, Gretchen. *Earthmaker's tales : North American Indian stories about Earth happenings*. NY: Walker, 1989. 89 p. : ill. 0802768393 0802768407 (lib. bdg.)
Contains 16 North American Indian legends about the origins of thunder, tornados, and other weather phenomena. Illustrates the unity between traditional Indian culture and the natural environment. GR 4-8.

1029. Seattle, Chief* (Supposed author). *Brother eagle, sister sky : a message from Chief Seattle*. NY: Dial, 1991. 1 v. (unpaged) : col. ill. 0803709633
Supposedly this is a speech by a Suquamish Indian chief conveying his people's respect and love for the Earth and concern for its destruction. It was actually composed in the early 1970s. For all ages, but only if understood to be a modern white interpretation. #

6. Philosophical and Literary Nature Writing

The books in this chapter tend to emphasize the personal aspects of nature and related topics. Many consider cultural, scientific, and action-oriented issues, but the authors tend to draw inspiration more from humanistic than from scientific sources. This chapter provides a bridge between fairly rigorous nonfiction books and creative works of fiction. The Native American section also contains many philosophical and religious works. It should also be noted at the outset that given the reflective and sometimes challenging nature of the books in this chapter, fewer than 20% of them can be expected to be of interest to young adults.

Philosophy

Beliefs, attitudes, values, ethics, and philosophies shape actions. Philosopers have long pondered such questions as the meaning of life, humanity's place in the world, and the underlying principles of reality. Pliny the Elder (142) and *Ancient natural history* (116) provide a glimpse into classical thinking. The Enlightenment and the rise of rationalism featured dualistic thinking and had the effect of emphasizing separation over unity: man and woman, culture and nature, body and mind, etc. Belief in the underlying unity of all things, a rejection of rationalistic dualism, is increasingly popular today. Unity is emphasized in Taoism, Buddhism, Native American, Australian Aboriginal, and other Eastern or so-called primitive thinking. It was later adopted by Muir and other 19th century conservationists.

There are many aspects of ecophilosophy. Bateson and Roszak focus on the psychological ones, as does Estes (1150). Disch, Nash, Regan, Rolston, Stone, and Westra explore environmental ethics. The relationship between people and technology is considered by Brower, Eiseley, Hamilton, McKibben, Oates, Schneider, and Winner. Also see Berry (638-40), Thomas (153-54), Dubos (1084), and Fuller (876)

Holistic philosophies are proposed by Capra, Elgin, Fox, Kellert, Olson, Ornstein, Sheldrake, and Tucker. The Ecofeminism section contains related works by Gaard (1133), Gray (1155), Kraal (1159), Peterson (1169), Walker (1177), and Warren (1179). Books on Deep Ecology are located in the Green Movement chapter since Deep Ecology and has an inherent activist orientation.

Books for adults:

1030. Bateson, Gregory. *Mind and nature : a necessary unity.* NY: Bantam, 1988, c1979. 255 p. 0553345753
AR: A *sacred unity* .
This book develops some of the themes considered in *Steps to an ecology of mind* including the mechanization of self and the disembodiment of nature. Primarily for academics.

1031. ___. *Steps to an ecology of mind : collected essays in anthropology, psychiatry, evolution, and epistemology.* Northvale, NJ: Aronson, 1988, c1972. 545 p. 0876689500
Considers the relationship between the human mind, behavior, and nature. Contends that science has wrongly attempted to build a bridge betweeen form and substance, mind and nature. HR.

1032. Brock, Peter. *Variations on a planet.* Lawrencetown Beach, N.S.: Pottersfield, 1993. 126 p. : ill 0919001777
A former chemical engineer becomes a shepherd, teacher, cabin builder, potter, TV producer (*Land and Sea* for the CBC) and sailor. His career transition reflects a change in his basic beliefs about and attitudes toward nature and humanity's relation to it.

1033. Brower, Kenneth. *The starship and the canoe.* NY: Harper & Row, 1983, c1978. 270 p. : maps. 0060910305
While astrophysicist Freeman Dyson designs spaceships, his brilliant high school dropout son lives a nature-oriented life in a treehouse and builds a series of ever-larger kayaks made from high tech materials, but based on Northwest Indian designs. AOT. #

1034. Capra, Fritjof. *The turning point : science, society, and the rising culture.* NY: Bantam, 1987, c1982. 464 p. 0553343165
Rejects the Cartesian notion of the duality and separation of self and society, society and nature, etc. in favor of a more holistic systems view of life wherein everything is interconnected. HR.

1035. Corrington, Robert S. *Nature and spirit : an essay in ecstatic naturalism.* NY: Fordham Univ., 1992. 207 p. 0823213625 0823213633 (pbk)
AR: *Ecstatic naturalism.*
The focus of this inquiry shifts from a purely deconstructive reading of the world toward a metaphysical conception of nature and spirit which pro-

vides an ultimate meaning for the self. For sophisticated readers with a background in philosophy.

1036. Disch, Robert. ed. *The ecological conscience; values for survival*. Englewood Cliffs: Prentice-Hall, 1970. 206 p. 0132228289 0132228106
An eclectic collection of essays on the importance of incorporating environmentally positive values within all technological, economic, and political activities. Contribuors include Ginsberg, Commoner, McHarg, Merton, Leopold, Shepard, Fuller, and Snyder. This fine "time capsule" of environmental ethics is HR.

1037. Eiseley, Loren C. *The firmament of time*. NY: MacMillan, 1972, c1960. 182 p. 0689700687
AR: *The invisible pyramid*, the lost notebooks of Loren Eiseley
Contains six essays on evolution, nature, life, death, evolution, and humanity's place in nature. HR.

1038. ___. *The immense journey*. NY: Random House, 1990, c1957. 152 p. 0394701577
Traces evolution from its simple beginnings to the awesome complexity of the present and wonders about the future. Will technology create a fork in the road? This blending of scientific rigor and imagination is HR.

1039. ___. *The night country*. NY: MacMillan, 1987, c1971. 240 p. : ill. 0684198089
Eisely combines science and philosophy in this considerstion of how the past shapes attitudes and behavior in nature and people.

1040. ___. *The star thrower*. NY: Harcourt, Brace, Jovanovich, 1979, c1978. 319 p. 0156849097
AR: *The lost notebooks of Loren Eiseley*..
A collection of essays from books and journals selected by Eiseley focusing on his earlier pieces. The section "Nature and autobiography" concerns questions of our place in the natural world and appropriate conduct of life. "Science and humanism" attempts to bridge the gap in technological and humanistic thought and values.

1041. Elgin, Duane. *Awakening Earth : exploring the evolution of human culture and consciousness*. NY: Morrow, 1993. 382 p. 0688116213
AR: *Voluntary simplicity*.
Combines science and spirituality to analyze human evolution over 35,000 years, from hunting-gathering to possible future stages of planetary consciousness and integral awareness. This consideration of alternatives to ecological collapse and social anarchy is HR.

1042. Evernden, Lorne Leslie Neil. *The natural alien : humankind and environment.* 2d ed. Toronto: Univ. of Toronto, 1993, c1985. 172 p. 0802029620
Argues that humankind is now rootless and needs to rediscover its natural mindset and place by rediscovering the natural world.

1043. Fox, Michael W. *One Earth, one mind.* Malabar, FL: Krieger, 1984, c1980. 264 p. 089874752X
In an almost diarylike fashion Fox presents an agenda for the development of a new natural consciousness which will discard "tool-using" lifestyles to become a supremely thinking, doing, unifying being capable of forming a new evolutionary link with the Earth.

1044. Hamilton, Robert. *Earthdream : the marriage of reason and intuition.* NY: Green, 1990. 250 p. 1870098110
Describes a new cosmology arising from scientists becoming more intuitive and spiritual, and religious leaders becoming better informed and more politically involved. Aims to resolve the linked dilemmas of spiritual impoverishment and material waste.

1045. Kellert, Stephen R., and Wilson, Edward Osborne. eds. *The Biophilia hypothesis.* Washington: Island, 1993. 484 p. 1559631481
Contains 15 essays providing psychological, biological, cultural, symbolic, and aesthetic perspectives on biophilia (inherent love of nature) and its converse, biophobia. Authors include Kellert, Wilson, Nabhan, Shepard, Rolston, and Soule.

1046. Krishnamurti, J. *On nature and the environment.* SF: Harper, 1991. 115 p. 0062505343
Contains over 30 brief extracts from Krishnamurti's books, essays, speeches, and interviews on the nature of the individual, the nature of nature, and the place of the individual within nature. Although Krishnamurti rejected his status as a guru, this book will ironically probably appeal most to his followers and New Age devotees.

1047. McKibben, Bill. *The age of missing information.* NY: Random House, 1992. 261 p. 0394589335 0394576012
Contends that the mass media and popular culture may be technologically advanced, but that the depth of content of knowledge presented is superficial and trivial compared to the wealth of knowledge inherent in nature. HR.

1048. Meyers, Steven J. *On seeing nature.* Golden, CO: Fulcrum, 1987. 141 p. : photos. 1555910084
This extended essay is a journey into ourselves to analyze the preconceptions that cloud our experience of seeing nature.

1049. Myers, Norman, and Simon, Julian Lincoln. *Scarcity or abundance?* A *debate on the environment.* NY: Norton, 1994. 160 p. : ill. 0393035905
Simon maintains that the quality of human life improves and will continue to do so, while Myers argues that we stand on the brink of environmental catastrophe. Includes debate on population growth, preservation of biodiversity, global warming, the ozone layer, and air and water quality.

1050. Nash, Roderick. *The rights of nature : a history of environmental ethics.* Madison: Univ. of Wisconsin, 1989. 320 p. 0299118401
Nash sees the granting of respect and rights to nature as part of a natural progression that includes the abolition of slavery and women's suffrage. The industrial revolution subdued nature and granted more power to machines and mechanistic thought, but now rights are starting to be accorded to nature. A classic. HR.

1051. Oates, David. *Earth rising : ecological belief in an age of science.* Corvallis: Oregon State Univ., 1989. 255 p. 0870713574
Attempts to delineate a way of life that does not deny either hard science or our delight in the Earth as a unique and wonderful planet. Is there a legitimate role for people or are we simply "a planetary disease?" Considers why humans, particularly in the western tradition, find it hard to make peace with nature.

1052. Olsen, Marvin Elliott; Lodwick, Dora G.; and Dunlap, Riley E. *Viewing the world ecologically.* Boulder: Westview, 1992. 214 p. 081338298X
Examines a major paradigm shift occuring in contemporary American thought from a technological-social point of view to one which is ecologically based.

1053. Ornstein, Robert E., and Ehrlich, Paul R. *New world new mind : moving toward conscious evolution.* NY: Doubleday, 1989. 302 p. 0385239408
Maintains that technology and society have evolved much faster than human consciousness or wisdom. This has fueled the population explosion, warfare, and the environmental crisis. Recommends developing a whole "new mind" of thought processes, perceptions, impulses, and values.

1054. Regan, Tom. ed. *Earthbound : new introductory essays in environmental ethics.* NY: Random House, 1984. 371 p. 0394332687
Contains articles by William Aiken, Annette C. Baier, Alastair Gunn, Dale Jamieson, Edward Johnson, Tibor R. Machan, Tom Regan, Mark Sagoff, K. S. Shrader-Frechette, and Robert L. Simon on environmental ethics as applied to cities, pollution, political theory, energy, environmental law, ocean resources, agriculture, species preservation, moral theory, and related subjects.

1055. Rolston, Holmes. *Environmental ethics : duties to and values in the natural world.* Philadelphia: Temple Univ., 1988. 391 p. 087722501X
Considers the historical, economic, scientific, recreational, and religious value of nature. These values determine our duties to nature and form the basis of environmental ethics. HR.

1056. ___. *Philosophy gone wild : environmental ethics.* Buffalo: Prometheus, 1989, c1986. 269 p. 0879755563
Considers the nature and scope of the environmental ethic, criteria for policy development, application of the ethic to local areas, and understanding and implementing theory through personal encounters with nature.

1057. Roszak, Theodore. *The voice of the Earth.* NY: Simon & Schuster, 1992. 367 p. 0671867539
AR: *Bugs, Person/planet.*
Presents a form of ecophilosophy which seeks to bridge the centuries-old gap between the psychological and the ecological. Sees the needs of the planet and the needs of a person as a continuum.

1058. Schneider, Stephen Henry, and Morton, Lynne. *The primordial bond : exploring connections between man and nature through the humanities and sciences.* NY: Plenum, 1981. 324 p. 0306405199
Combines the humanities with the sciences to explore our place in nature, how humans have modifed it, and what actions are needed to reverse negative changes. An interesting study of the concept of natural cycles as patterns that connect people to nature and the sciences to the humanities.

1059. Sheldrake, Rupert. *The presence of the past : morphic resonance and the habits of nature.* NY: Vintage, 1989, c1988. 391 p. 0394759907
Challenges most of the fundamental assumptions of a mechanistic understanding of the universe. Theorizes that nature at every level has memory, and proposes a more organic and wholistic understanding of reality.

1060. Stone, Christopher D. *Earth and other ethics : the case for moral pluralism.* NY: Harper, 1988. 288 p. 0060914866
AR: *Do trees have standing?*
An environmental attorney attempts to build a legal ethic to prevent the destruction of the natural world based on both utilitarian resource considerations and the establishment of a legal "hierarchy of sentience." Although somewhat radical, this book falls short of a condemnation of "speciesism" per se. HR.

1061. Tucker, Mary Evelyn, and Grim, John A. eds. *Worldviews and ecology.* Lewisburg, PA: Bucknell Univ., 1993. 242 p. 0838752721

Considers a new global environmental ethic. A broadened context for a new ecological ethic is created by presenting various worldviews. Authors include J. Baird Callicott, Michael Tobias, Charlene Spretnak, George Sessions, Thomas Berry, and others.

1062. Westra, Laura. *An environmental proposal for ethics : the principle of integrity.* Lanham, MD: Rowman & Littlefield, 1994. 237 p. 0847678946 0847678954 (pbk)
Examines the Great Lakes Water Quality Agreement, whose purpose was to "restore integrity" to that ecosystem. Westra questions what environmental integrity means and whether there is a traditional moral doctrine that can support both the ideal and the obligations it creates.

1063. Winner, Langdon. *The whale and the reactor : a search for limits in an age of high technology.* Chicago: Univ. of Chicago, 1989. 200 p. 0226902110
Explains how technology goes beyond the scientific and industrial sphere to shape how we think and act. In an age of dwindling resources and overpopulation we need not overreact by rejecting technology, but do need to "re-tool" our thinking about how to use it.

1064. Worster, Donald. *Nature's economy : a history of ecological ideas.* New ed. NY: Cambridge Univ., 1985, c1977. 404 p. 0521267297
Correlates changes in ecological thinking to developments in science and consciousness from the 18th century to the present. He argues that the Romantics with their concepts of nature as a living entity and a "spirit of the woods" represent the lushest flowering of ecological thought. HR.

Religion and Spirituality

If philosophy is a study of underlying principles, one might consider spirituality to be the relation of the individual to those principles, and religion to be their institutionalization. The books in this section cover a very broad range of religious attitudes and beliefs regarding the environment.

Berry, Hull, McDaniel, McLuhan, Rockefeller, Rounder, and Willers provide general, comparative or ecumenical views. Specific religions covered here include Judaism (by Bernstein, Fisher, Rose, and Stein), Buddhism (by Badiner, Batchelor, the Dalai Lama, Halifax, Macy, and Matthiessen), Islam (by Khalid), Hinduism (by Prime), Taoism (by Watts), Pantheism (by Altman and Kaza), and New Age religions (by Foster, Kjos, LaChappelle, McClean, McFague, and Parrish-Harra.) Considerations of Goddess or Earth religions by Adams, Adler, Christ, Frymer-Krensky,

Gadon, and Starhawk are listed in the Ecofeminism section.

Most books in this section are on Christianity, and as the Rev. Washington Phillips observed so succinctly in his song *Denomination Blues*: "I can tell all you people it's a natural fact/not everybody understands the *Bible* alike." Some scholars consider the Biblical directive to multiply and subdue the Earth to be a cause of the environmental crisis: the environmental history and population scholars in Cultural Factors; Cronon, Delage, Deloria, and other authors in Native Americans; and the ecofeminist theologians named above. Breuilly acknowledges problems with conservative Christianity, but sees it as a potentially environmentally positive force. Dowd, Dubos, Epperly, Hallman, Meyer, and Oelschlager consider the best aspects of Christianity to be part of the solution to the environmental crisis.

Christian stewardship, which maintains that people are responsible for the care of God's earth, is explored by Bailey and Campolo. A simple, nonmaterialistic lifestyle known as voluntary simplicity is advocated by Crean and Vandenbroeck. For other related books see "Simplicity" in the subject index. Heidtke, Leax, Linzey, Pitcher, and Seed provide ideas for spiritually-based environmental action. Clarkson, LaChappelle, Roberts, Seed, and Simsic provide collections of prayers, meditations, or rituals. Dillard, Douglas, Easwaran, Foster, Milne, Norris, and Wetherall open their souls, offering personal experiences of God in nature.

Books for adults:

1065. Altman, Nat. *Sacred trees*. SF: Sierra Club, 1994. 244 p. : ill. 087156470X
Tracing history and myth from many of the world's cultures, Altman explores the special relationship people have felt with trees. Stresses that respect, reverence, and communion with the rest of nature are essential for the planetary healing.

1066. Badiner, Allan Hunt. ed. *Dharma Gaia : a harvest of essays in Buddhism and ecology*. Foreword by the Dalai Lama. Berkeley: Parallax, 1990. 265 p. : ill. 0938077309
Contains over 30 essays, poems, and meditations by Thich Nhat Hanh, Joanna Macy, Joan Halifax, Gary Snyder, John Seed and others on "green Buddhism," experiencing extended mind, Earth as a sentient being, and appropriate action through enlightened engagement with the material world. Includes a glossary, bibliography, and directory. HR.

1067. Bailey, Liberty Hyde. *The holy Earth*. Ithaca, NY: NY State College, 1980, c1915. 112 p. 0960531467
Argues for an understanding of nature by all people and for recognition that the Earth is humanity's essential resource, the very basis of our civi-

lization. Other themes relate to the agrarian tradition, individual responsibility, and rural life.

1068. Batchelor, Martine, and Brown, Kerry. eds. *Buddhism and ecology*. NY: Cassell, 1992. 128 p. : ill. 0304323756
Buddhists from Japan, Sri Lanka, Thailand, Vietnam, and Western cultures explore teachings on the environment from the various branches of Buddhist thought. Contains both traditional and modern essays, stories, pictures, and poems.

1069. Bernstein, Ellen, and Fink, Dan. *Let the Earth teach you Torah : a guide to teaching Jewish ecological wisdom*. Wyncote, PA: Shomrei Adamah, 1992. 184 p. 0963284819
Combines classic and contemporary Jewish sources with readings from nature writers. Designed for the high school curriculum, but also appropriate for families. #

1070. Berry, Thomas Mary. *The dream of the Earth*. SF: Sierra Club, 1990, c1988. 262 p. 0871566222
AR:*The universe story: an autobiography of planet Earth*.
Discusses the idea of the sacred in the context of aesthetic, imaginative, and creative experiences. Also covers sustainable development, bioregionalism, sexism, and American Indian ecology. HR.

1071. Berry, Thomas Mary, and Clarke, Thomas E. *Befriending the Earth : a theology of reconciliation between humans and the Earth*. Mystic, CN: Twenty-Third, 1991. 157 p. 0896224716
A dialogue on the crucial importance to both nature and religion of developing respect for nature and on appropriate conduct toward it based upon religious faith. They discuss environmentalism and human ecology in a Roman Catholic context.

1072. Bowman, Douglas C. *Beyond the modern mind : the spiritual and ethical challenge of the environmental crisis*. NY: Pilgrim, 1990. 139 p. 0829808477
A professor of religion and Presbyterian minister explores a fundamental shift in thinking that he believes is required for our children to inherit a healthy and livable Earth.

1073. Breuilly, Elizabeth, and Palmer, Martin. eds. *Christianity and ecology*. NY: Cassell, 1992. 118 p. : ill. 0304323748
An ecumenical group of liberal Christian activists identifies Biblical and other Christian teachings and attitudes which have helped create the current enviornmental crisis. They then suggest ways to bring about attitudinal shifts to improve global ecology from the same sources.

1074. Campolo, Anthony. *How to rescue the Earth without worshiping nature.* Nashville: T. Nelson, 1992. 213 p. 0840777728
Contends that conservative Christianity will lose many young adherents if it fails to address ecological concerns in a timely manner, as was the case with civil rights and Viet Nam. He offers Christian stewardship as an appropriate vehicle for merging ecological and spiritual concerns.

1075. Clarkson, Edith Margaret. *All nature sings.* Grand Rapids: Eerdmans, 1986. 138 p. 0802802257
A collection of short essays, poems, and prayers celebrating the spirit of God in nature, its natural blessings and positive transformational powers, and the possibility of understanding the creator best through creation itself.

1076. Dalai Lama XIV. *Worlds in harmony : dialogues on compassionate action.* Berkeley: Parallax, 1992. 159 p. : photos. 0938077775
The Dalai Lama, Daniel Goleman, Stephen Levine, Jean Shinodu Bolen, Daniel Brown, Jack Engler, Margaret Brenman-Gibson, and Joanna Macy discuss the relationship of Buddhism to a variety of world problems including war, inner-city violence, and the destruction of the natural environment.

1077. Deming, Alison Hawthorne. *Temporary homelands.* SF: Mercury House, 1994. 214 p. 1562790625
Contains twelve essays by a Walt Whitman Award-winning poet which consider her quest for a spiritual home in nature. Somewhat comparable to Annie Dillard's work.

1078. Dillard, Annie. *The Annie Dillard Reader.* NY: HarperCollins, 1994. 455 p. 0060171588
Contains excerpts from Pilgrim at Tinker Creek, Teaching a stone to talk, Holy the firm, and An American childhood; poems; the short story that The living was based on; and selected new work. HR AMT. #

1079. ___. *Holy the firm.* NY: Perennial Library, 1988, c1977. 76 p. 0060915439
Observations and meditations on the spiritual aspects of life, death, sacrifice, and nature. "The firm" refers to God's holy firmament of the Earth. AMT. #

1080. ___. *Pilgrim at Tinker Creek.* NY: Collins, 1988, c1974. 271 p. 0060915455
In almost rhapsodic prose Dillard considers how all living beings are united and utterly interdependent in a vast variety of natural processes and

relationships. She considers the spiritual aspects of ecology, with her vision ranging from the microscopic to the cosmic. This modern classic is HR AOT. #

1081. ___. *Teaching a stone to talk : expeditions and encounters.* NY: Collins, 1988, c1982. 175 p. 006915455
In this fascinating, highly personalized account of a variety of natural events and their meanings to human beings, Dillard encourages people to enrich their lives through an appreciation and understanding of nature and their place within it. AOT. #

1082. Douglas, David. *Wilderness sojourn : notes in the desert silence.* SF: Harper & Row, 1987. 112 p. 0060619910
Describes a one week solo trip into the desert which was both a physical and a spiritual journey. He comes to feel more humble, blessed, grateful, awed, and prayerful as he experiences a strong sense of the presence of God.

1083. Dowd, Michael. *Earthspirit : a handbook for nurturing an ecological Christianity.* Mystic, CN: Twenty-third, 1991. 117 p. : ill. 0896224791
A UCC minister combines quotes from theological and environmental leaders, anecdotes, and his own analysis in order to integrate modern scientific thought and the Bible as ecological spirituality. Includes exercises, prayers, meditations, a pledge of allegiance to the Earth, bibliography, and study questions. HR.

1084. Dubos, Rene J. *A God within : a positive philosophy for a more complete fulfillment of human potentials.* NY: Scribner, 1984, c1972. 325 p. 0684179792
AR: *The wooing of Earth, So human an animal, Only one Earth.*
Explores a "theology of the Earth" which harmonizes ecological concerns with spiritual, cultural, and humanistic traditions. Considers positive ecological thought and practice within the Judeo-Christian tradition.

1085. Easwaran, Eknath. *The compassionate universe.* Petaluma, CA: Nilgiri, 1989. 188 p. 0915132591 0915132583 (pbk)
Considers how to harness the restorative powers of nature by learning to recognize it and link it to the restorative powers within ourselves. As we restore ourselves we can learn to treat nature with greater respect.

1086. Ehrenfeld, David W. *The arrogance of humanism.* NY: Oxford, 1978. 286 p. 019502415X
Humanism (and materialism) as "the dominant religion of our time" (p.3) is anti-environmental, viewing nature in terms of economics or its utility to humans. Ehrenfeld pleads for a wiser fusion of humanistic and ecological thought.

1087. Epperly, Bruce Gordon. *At the edges of life : a holistic vision of the human adventure.* St. Louis: Chalice, 1992. 172 p. 082720020X
This synthesis of traditional Christianity, New Age spirituality and ecological awareness considers the spiritual aspects of death, holistic medicine, and process theology.

1088. Fisher, Adam. *Seder Tu Bishevat : the festival of trees.* NY: CCAR, 1989. 107 p. 0881230081
Celebrates this festival as a time of joy and thankfulness for the natural beauty of the Earth and a time to focus on human responsibility to care for God's world. Text in English and romanized Hebrew.

1089. Foster, Steven, and Little, Meredith. *The roaring of the sacred river : the wilderness quest for vision and self-healing.* NY: Prentice Hall, 1989. 225 p. 0137814453
A guide to rites of passage into the physical and metaphorical wilderness. Emphasizes spiritual aspects of nature from a New Age perspective

1090. Glacken, Clarence J. *Traces on the Rhodian shore; nature and culture in Western thought from ancient times to the end of the eighteenth century.* Berkeley: Univ. of Cal., 1967. 791 p. 0520023676
"Glacken relates social and natural phenomena to the dichotomy perennially occuring between man and nature. . . . an extraordinary synthesis of a vast literature of Western thought."-Davis. HR.

1091. Halifax, Joan. *The fruitful darkness : reconnecting with the body of the Earth.* SF: Harper, 1993. 271 p. : ill. 0062503693
Presents the Buddhist ways of silence, traditions, the mountain, language, story, nonduality, protectors, ancestors, and compassion to invoke spirituality, deep ecology, Christianity and shamanism. Particularly for deep ecologists, New Age advocates, and Buddhists.

1092. Hallman, David G. *A place in creation : ecological visions in science, religion, and economics.* Toronto: United Church, 1992. 233 p. 0919000800
AR: *Caring for creation.*
Explores the intrinisic connection of humans and the natural environment, God's will for harmony in all his creation, the restructuring of the economy for the sake of the environment, and political aspects of developing a more sustainable society.

1093. Heidtke, John. *Getting down to Earth : a call to environmental action.* NY: Paulist, 1993. 179 p. : ill. 0809195712
This workbook is designed to provide insights into the environmental crisis, develop ecological values and an action plan based upon them, under-

stand how the values of St. Francis, Muir, Thoreau, Schweitzer, Leopold, and Carson apply today, and show how to use a variety of resources such as books, films, music, and organizations. HR.

1094. Hull, Fritz. ed. *Earth & spirit : the spiritual dimension of the environmental crisis*. NY: Continuum, 1993. 224 p. 0826405754
An ecumenical group discusses appropriate religious perspectives and approaches to environmental problems through which humanity can move beyond anthropocentrism and materialism. Includes Thomas Berry, Joanna Macy, Stephnie Kaza, David Spangler, Brian Swimme, and 24 others.

1095. Kaza, Stephanie. *The attentive heart : conversations with trees*. NY: Fawcett Columbine, 1993. 258 p. : ill. 0449907791
A naturalist and meditator uses both scientific knowledge and Zen to better understand, appreciate, and conserve trees. By realizing our oneness with nature we will naturally preserve it and enrich our own spiritual lives. HR.

1096. Khalid, Fazlun M., and O'Brien, Joanne. *Islam and ecology*. NY: Cassell, 1992. 122 p. 0304323772
The *Qur'an* shows the central importance of concern for the world as part of the human responsibility to care for one another. Covers both historic beliefs and examples of current environmental action by Muslims including desert reclamation, animal husbandry, and changes in trade and scientific policies.

1097. Kjos, Berit. *Under the spell of Mother Earth*. Wheaton, IL: Victor, 1992. 204 p. 0896938506
This conservative Christian book warns against the pantheistic pitfalls of the environmental movement such as spiritism, Gaia, paganism, mysticism, witchcraft, global oneness, and the New Age movement.

1098. LaChance, Albert J. *Greenspirit : twelve steps in ecological spirituality, an individual, cultural, and planetary therapy*. Rockport, MS: Element, 1991. 216 p. 1852302631
A practical guide to withdrawing from addictive consumerism and living a profoundly down to Earth lifestyle which offers greater spiritual rewards. Combines contemporary cosmology and a 12-step recovery process. HR.

1099. LaChapelle, Dolores. *Earth wisdom*. Silverton, CO: Finn Hill Arts, 1984, c1978. 183 p. 0917270010
An extremely eclectic guide to liturgy and ritual designed to foster the spiritual reinhabitiation of the Earth by a more advanced human ecological consciousness. Depicts rituals, thoughts, and practices based upon healing the nature-culture duality.

1100. Leax, John. *Standing ground : a personal story of faith and environmentalism.* Grand Rapids: Zondervan, 1991. 127 p. 0310537916
This personal account of active opposition to the building of a radioactive waste dump near his home demonstrates how the author's resolve and concern for the Earth emerged from his religious faith and how his faith was deepened by the struggle.

1101. Maclean, Dorothy. *To honor the Earth : reflections on living in harmony with nature.* Photos by Kathleen Thormod Carr. SF: Harper, 1991. 122 p. : photos. 006250603X 0062505998
Describes her experiences communicating with the energy fields of plants and other natural objects. Presents "nature's viewpoints" in the forms of prayers, meditations, and observations she has received from these beings, each accompanied by a full-page color photo. For New Age readers.

1102. Macy, Joanna. *World as lover, world as self.* Berkeley: Parallax, 1991. 251 p. 0938077279
Describes dependent co-arising, a Buddhism concept which teaches humans to see and treat the world and all its creatures compassionately as an extension of the self. Provides examples of everyday applications of "the greening of the self" including meditations and the Council of all Beings. HR.

1103. Matthiessen, Peter. *Nine-headed dragon river : Zen journals,* 1969-1985. Boston: Shambhala, 1987, c1985. 288 p. 0394552512
Combines autobiography, a history of Buddhism in Japan and America, and lyrical descriptions of nature. Provides fresh perspectives on Buddhism and the relation between nature and culture in Japan. HR.

1104. McDaniel, Jay B. *Earth, sky, gods & mortals : developing an ecological spirituality.* Mystic, CN: Twenty-Third, 1990. 214 p. 0896224120
An eclectic synthesis of traditional and contemporary wisdom that points in the direction of an ecologically sound and nurturing faith.

1105. McFague, Sallie. *Models of God : theology for an ecological, nuclear age.* Philadelphia: Fortress, 1987. 224 p. 0800620518
Examines the metaphorical nature of all theology, focusing on 'the world as God's body' as the primary metaphor for God's relationship to creation. She presents both traditional and newer arguments for relating to God as mother, lover, and friend. HR.

1106. McLuhan, T. C. *The way of the Earth : encounters with nature in ancient and contemporary thought.* NY: Simon & Schuster, 1994. 576 p. : ill. 0671759396
Surveys how various cultures have struggled to make sense of their place in the universe and the human relation to nature and to God. Covers Greece,

Japan, aboriginal Australia, Africa, the Kogi of Columbia, and Native North Americans. HR.

1107. Meyer, Art, and Meyer, Jocele. *Earth-keepers : environmental perspectives on hunger, poverty, and injustice.* Scottdale, PA: Herald, 1991. 264 p. 083613544X
Examines the root causes of environmental degradation from a Mennonite perspective. Presents a "biblical theology on which to base Christian care for the Earth" and demonstrates links between the environment, the economy, conflict, energy, sustainability, and related subjects.

1108. Milne, Courtney. *The sacred Earth.* NY: Abrams, 1993, c1991. 246 p. : photos. 0810938316
An account of pilgrimages to over 160 sacred sites around the world including Stonehenge, the Outer Hebrides, Gethsemene, Ayres Rock, Easter Island, and natural meadows and glades which have served as sites for sacred ceremonies. Encourages enlightenment through reconnecting with the Earth.

1109. Norris, Kathleen. *Dakota : a spiritual geography.* NY: Ticknor & Fields, 1993. 224 p. 0395633206
Describes both the positive (low crime, close to nature) and negative aspects (gossip, insularity, missile sites and waste dumps) of life in a small South Dakota town. The land and her growing familiarity with the lives and rituals of Benedictine monks serve as twin fonts of her spiritual and personal growth. HR.

1110. Oelschlaeger, Max. *Caring for creation : an ecumenical approach to the environmental crisis.* New Haven: Yale, 1994. 273 p. 0300058179
Rejects the charge that the tenet of dominion over the Earth is the cause of environmental woes, arguing that religion is essential to solve them. Reviews church history in international affairs, explores ecological aspects of creation stories, and promotes an ecumenical approach to solving problems.

1111. Parrish-Harra, Carol E. *The book of rituals : personal and planetary transformation.* Santa Monica: IBS, 1990. 280 p. 1877880027
A New Age approach to personal spirituality couched in an ethic of environmental concern. Personal and planetary healing are presented as integrally linked, symbiotic processes.

1112. Pitcher, W. Alvin. *Listen to the crying of the Earth : cultivating creation communities.* Cleveland: Pilgrim, 1993. 157 p. 0829809619
Frustrated by the failure of social and political institutions to respond to the environmental crisis, Pitcher proposes grass-roots communities based

in neighborhoods and congregations which are commited to changing national priorities and personal lifestyles.

1113. Prime, Ranchor. *Hinduism and ecology.* NY: Cassell, 1992. 128 p. : ill. 030432373X
Human respect for the environment is shown to be at the heart of Hinduism in this collection that ranges from sacred texts to interviews with modern Hindus. The eternal essence of life, *Sanatan Dharma*, is found in all living beings and unites them to the Godhead.

1114. Roberts, Elizabeth J., and Amidon, Elias. eds. *Earth prayers : from around the world : 365 prayers, poems, and invocations for honoring the Earth.* Rev. ed. SF: Harper, 1993, c1991. 480 p. 0062508881
Includes prayers from a variety of established religions or by individual authors such as Walt Whitman, T. S. Eliot, Margaret Atwood, Robert Frost, May Sarton, Thich Nhat Hanh, Black Elk, Denise Levertov, and W.E.B. DuBois.

1115. Rockefeller, Steven C., and Elder, John. eds. *Spirit and nature : why the environment is a religious issue : an interfaith dialogue.* Boston: Beacon, 1992. 226 p. 0807077089 0807077097 (pbk)
Contains an ecumenical array of essays from a Symposium on Spirit and Nature by Audrey Shenandoah, Ismar Schorsch, Sallie McFague, J. Ronald Engel, Seyyed Hossein Nasr, Tenzin Gyatso, Robert Prescott-Allen, Steven Rockefeller, and John C. Elder. Appendix: United Nations World Charter for Nature.

1116. Rose, Aubrey. *Judaism and ecology.* NY: Cassell, 1992. 160 p. : ill. 0304323780
A study of Judaism's ceremonies, laws, and moral teachings toward the Creation. The contributions of modern Jews, particularly the current environmental policies of Israel, are covered in depth.

1117. Rouner, Leroy S. *On nature.* South Bend: Univ. of Notre Dame, 1984. 188 p. 026801499X
Contains essays on various religious and philosophical aspects of nature by W. V. Quine, Stephen Toulmin, Huston Smith, Dorion Sagan and Lynn Margulis, Carl Ruck, Robert Thurman, Tu Wei-Ming, Jurgen Moltmann, J.N. Findlay, Charles Hartshorne, and John C. Bennett.

1118. Seed, John; Macy, Joanna; Fleming, Pat; and Naess, Arne. *Thinking like a mountain : towards a Council of All Beings.* Philadelphia: New Society, 1988. 122 p. : ill. 0865711321
Contains spiritual and deep ecological readings, meditations, poems, ritu-

als, and workshop notes. Presents a group process which provides a context for deep ritual identification with the natural environment and "community therapy" in defense of the Earth. For personal or group use. HR.

1119. Shepard, Paul. *Man in the landscape : a historic view of the esthetics of nature*. 2d ed. College Station: Texas A&M, 1991, c1967. 290 p. : ill. 0890964211
AHR: *Nature and madness, The tender carnivore and the sacred game, Thinking animals*.
Considers the influence of Christianity on ideas of nature, the absence of an ethic of nature in modern philosophy, and the obsessive themes of dominance and control as elements of the modern mind. Identifies the transport of traditional imagery into new places as anachronistic cultural baggage. HR.

1120. Simsic, Wayne. *Earthsongs : praying with nature*. Winona, MN: Saint Mary's, 1992. 93 p. : ill. 0884892948
After considering ways in which nature can open people to spirituality and how prayer can intensify experience, Simsic offers 30 short prayer services. Each contains an opening, a psalm, a reading, a hymn, and a closing. This nondenominational Christian handbook is recommended for personal or group use.

1121. Stein, David E. A *Garden of choice fruit : 200 classic Jewish quotes on human beings and the environment*. 2d ed. Wyncote, PA: Shomrei Adamah, 1991. 103 p. 0963284800
Includes 200 sentence to paragraph length quotations and maxims from biblical, medieval, and modern Jewish sources.

1122. VandenBroeck, Goldian. *Less is more : the art of voluntary poverty*. NY: Inner Traditions, 1991, c1978. 336 p. 0892814314
An anthology of quotations and brief texts from scholarly, religious, and literary sources. How to live a life which is simpler in terms of material possessions but more personally and spiritually rewarding.

1123. Watts, Alan. *Nature, man and woman*. NY: Vintage, 1991, c1958. 209 p. 0394733118
AR: *Tao, the watercourse way*.
Contends that divisions between humanity and nature, man and woman, and the mundane and the sacred are due to Western dualistic thinking. Adopting a more inclusive Buddhist or Taoist philosophy and actively engaging in life according to more holistic principles can mend these rifts and the Earth.

1124. Wetherell, W. D. *Upland stream : notes on the fishing passion*. Boston: Little, Brown, 1991. 204 p. 0316931756
Reflections on the joy of fishing, nature, and reaching a deeper spiritual solace. Wetherell is a fisher of both fish and a higher consciousness. AOT. #

1125. Willers, W. B. ed. *Learning to listen to the land*. Washington: Island, 1991. 282 p. 1559631201 (pbk) 155963121X
Contains 24 essays focusing on the spiritual and ethical aspects of environmentalism. Authors include Lovelock, Wilson, Stegner, Shepard, Berry, Commoner, Abbey, Anne and Paul Ehrlich, Merchant, and Foreman.

Books for young adults:

1126. Appelman, Harlene Winnick, and Shapiro, Jane. *A Seder for Tu BÆShevat*. Rockville, MD: Kar-Ben Copies, 1984. 32 p. : ill. 0930494393
Describes the celebration of the Jewish arbor day, Tu bi-Shevat, and explains its history. The seder ritual discusses trees, reforestation, conservation, and recycling. For schools, synagogues, and families.

1127. De Jonge, Joanne E. *All nature sings*. Grand Rapids: Eerdmans, 1992, c1985. 141 p. : ill. 0802850650
Examines some of the less attractive wonders of nature, such as the slimy slug and hairy tarantula, and explains how all aspects of nature sing God's praise.

1128. Nehemias, Paulette. *A tree in Sprocket's pocket : stories about God's green Earth*. St. Louis: Concordia, 1993. 128 p. : ill. 0570047307
In this collection of Christian oriented short stories with accompanying activities, children confront environmental problems and learn to care for the Earth. GR 3-7.

1129. ___. *Wiggler's worms : stories about God's green Earth*. St. Louis: Concordia, 1993. 127 p. : ill. 0570047315
More stories with accompanying activities which raise children's environmental awareness and describe ways to help care for the Earth. GR 3-7.

Ecofeminism

Mother Earth, Mother Nature, The Goddess, Gaia: people have associated femininity with the Earth's beneficence.for millenia. Ecofeminism's deepest roots are in Goddess religions, but the two should not be equated with one another since the ecofeminist umbrella covers a broad range from ecstatic pantheists to atheists.

Ecofeminism also has more recent roots. Often excluded from the clergy, business, and politics, 19th century women were sometimes accepted as teachers, authors, illustrators, or naturalists. Rachel Carson's career can be viewed as the culmination of the earlier tradition of women in the natural sciences and the beginning of women taking leading roles in the transformation of society. In the early 1970s, segments of the women's and environmental movements began to coalesce as ecofeminism. Books by Bonta, Knowles, Merchant, and Norwood trace ecofeminist history. Historical precursors include Austin, Bird, Cox, Foote, Cather, and Gilman.

Gaard, Gray, Kraal, Patterson, Walker, and Warren provide general or philosophical overviews. Adams, Adler, Christ, Collard, Frymer-Kensky, Gadon, Reuther, and Starhawk explore spiritual aspects of the feminine, as does Watts in the Religion section (1123). Andrews, Dinnerstein, Estes, LaCapelle, and Williams consider psychology, sexuality, relationships, and the home. Native American or Australian Aboriginal feminism are examined by Allen, Bell, Niethammer, and Wallis. Activism is advocated and outlined by Biehl, Caldecott, Rodda, and Joni Seager. Bradford (1370) and Bookchin (1368) offer male perspectives on Ecofeminism and the Green Movement.

Books for adults:

1130. Adams, Carol J. ed. *Ecofeminism and the sacred.* NY: Continuum, 1993. 340 p. 082640586X
"Ecofeminism argues that the connections between the oppression of women and the rest of nature must be recognized to understand adequately both oppressions."-Adams, p. 1. Contains 20 essays on revisioning religion, envisioning ecofeminism, and embodying feminist spiritualities. Authors include Reuther, Kaza, Orenstein, Spretnak, and others. HR.

1131. ___. *The sexual politics of meat : a feminist-vegetarian critical theory.* NY: Continuum, 1992. 256 p. : ill. 0826405134
Examines the historical gender, race, and class implications of eating meat, and makes the links between the practice of butchering/eating animals and the maintenance of patriarchy.

1132. Adler, Margot. *Drawing down the moon : witches, Druids, Goddess-worshippers, and other pagans in America today.* Rev. ed. Boston: Beacon, 1987, c1986. 595 p. 0807022530
Refutes popular myths about paganism and presents a thorough analysis of both traditional and modern theory and practice. Describes these traditions as being based in ancient, animistic, Earth oriented religions and tells how

they have influenced both the environmental and feminist movements.

1133. Allen, Paula Gunn*. *The sacred hoop : recovering the feminine in American Indian traditions*. Boston: Beacon, 1992, c1986. 311 p. 0807046175
Documents the continuing utility of American Indian traditions and the crucial role of women in those traditions. Considers the relationship of people in general and Indian women in particular with nature and the connections between Native American thought and ecofeminism.

1134. Andrews, Valerie. *A passion for this Earth : exploring a new partnership of man, woman, & nature*. SF: Harper, 1992. 256 p. : ill. 0062500856
A compelling examination of how to restore our personal connection to the Earth and one another through a new understanding of nature, myth, human relationships, and society.

1135. Austin, Mary Hunter. *Earth horizon : autobiography*. Albuquerque: Univ. of New Mexico, 1991, c1932. 403 p. 0826313167
AR: *The flock*.
This autobiography provides an evocation of the the Southwestern landscape, an appreciation of the native American and Hispanic peoples, a mystical sense of union between land and people, and a foreshadowing of ecofeminist thought. HR.

1136. ___. *The land of journeys' ending*. Tucson: Univ. of Arizona, 1983, c1924. 459 p. 0186508070
A brilliant evocation of life, culture, and travel in the deserts of the Southwest. HR.

1137. ___. *The land of little rain*. NY: Penguin, 1988, c1903. 107 p. 0140170094
Austin's first published book focused on the desert and hill country between Death Valley and the Sierra. Her husband was a land and water hustler, but Austin took the position that the desert is precious as is and should not be developed. This classic is HR AMT. #

1138. Bell, Diane. *Daughters of the dreaming*. 2d ed. Minneapolis: Univ. of Minnesota, 1993, c1983. 342 p. : ill. 0816623988
Bell, a white Australian woman, was initiated into aboriginal rituals and discovered that Aboriginal women are not the subservient, second class citizens described by earlier anthropologists. The inseparable sustaining ideals of these women are land, love, and well being.

1139. Biehl, Janet. *Rethinking ecofeminist politics*. Boston: South End, 1991. 181 p. 0896083926 0896083918 (pbk)
Critiques ecofeminism, claiming that is has left the mainstream by rejecting

legacies of democracy and reason, and by presenting much of the natural world as part of a radical liberatory movement. Calls for a return to "social ecology" to reincorporate radical feminist theory into environmentalism.

1140. Bird, Isabella L. A *lady's life in the Rocky Mountains*. London: Virago, 1991, c1879. 318 p. 0860682676
"Collection of letters written by a Victorian woman who traveled from England to visit the Rocky Mountains in the fall and winter of 1873. Bird journeyed on horseback, seeking a religious, meditative experience of nature, and deploring the development by miners and entrepreneurs."-Anderson

1141. Boice, Judith. ed. *Mother Earth : through the eyes of women photographers and writers*. SF: Sierra Club, 1992. 142 p. : col. photos. 0871565560
Combines the work of 35 writers and 34 photographers in presenting the five realms of Earth: mineral, plant, animal, human, and oneness of all beings. AOT. #

1142. Bonta, Marcia. *Women in the field : America's pioneering women naturalists*. College Station: Texas A & M Univ., 1991. 299 p. : ill. 0890964890
Recounts the lives and accomplishments of 25 naturalists, botanists, entomologists, ornithologists, and ecologists who worked between the early 18th century and the 1960s. Particularly recommended for young women considering careers in these fields. #

1143. Christ, Carol P. *Laughter of Aphrodite : reflections on a journey to the goddess*. SF: Harper & Row, 1987. 238 p. 0062501461
The author journeys to Aphrodite's Temple where she learns that "the spirituality we need for our survival...encourages us to recognize limitation and mortality, a spirituality calling us to celebrate all that is finite...the indigenous traditions of Africa, Asia, America, and Europe."-Intro.

1144. Coleman, Jane Candia. *Shadows in my hands : a Southwestern odyssey*. Athens: Swallow, 1993. 117 p. 0804009724
Combines nature writing, history, and a poetic evocation of the land to describe a woman's gradual awakening to her own strength. Coleman is a poet, short story writer, and essayist who has won two Western Heritage Awards. AOT#

1145. Collard, Andree, and Contrucci, Joyce. *Rape of the wild : man's violence against animals and the Earth*. Bloomington: Indiana Univ., 1989, c1988. 204 p. 025331514X
Contends that problems such as the abuse of animals and pollution are not isolated, but are caused by a male-dominated, sexist, exploitative consciousness. Proposes a more wholistic ecofeminist approach. HR.

1146. Cox, Leona Dixon. *Single woman homesteader.* Dobbins, CA: Inkwell, 1991. 213 p. 0962768073
Recounts Cox's life as a hardscrabble homesteader and rancher who built her home for five dollars, raised or killed all her food, and did whatever was required to survive. Her accounts of nature, wildlife, hunting and fishing are quite vivid. Especially recommended for seniors.

1147. Diamond, Irene. *Fertile ground : women, Earth, and the limits of control.* Boston: Beacon, 1994. 202 p. 0807067725
Contends that the struggle for women's rights is inseparable from those for ecological sanity, the end of militarism, and opposition to industrialized nations' exploitation of poor ones. Outlines alternative practices such as organic farming, community-based child care, and land redistribution which are environmentally and socially desirable.

1148. Diamond, Irene, and Orenstein, Gloria Feman. eds. *Reweaving the world : the emergence of ecofeminism.* SF: Sierra Club, 1990. 320 p. 087156694X
Essays on reuniting all beings on Earth through a wholistic merger of feminist and ecological values. Authors include Charlene Spretnak, Paula Gunn Allen, Starhawk, Susan Griffin, Ynestra King, Carolyn Merchant, and 20 others.

1149. Dinnerstein, Dorothy. *The mermaid and the minotaur : sexual arrangements and human malaise.* NY: HarperCollins, 1991, c1976. 288 p. 0060905875
Contends that patriarchy and one-dimensional technological thought have warped sexuality and the raising of children and led to the destruction of the environment. Rejects dominion and hierarchy in favor of an feminist return to "the gifts of nature."

1150. Estes, Clarissa Pinkola. *Women who run with the wolves : myths and stories of the wild woman archetype.* NY: Ballantine, 1992. 520 p. 0345377443
A Jungian analyst explores women's spiritual lives through myth and archetypes, noting that both wolves and women are social, cooperative beings who have been misunderstood and oppressed. She urges reconnecting with the wildness in nature and the soul. HR.

1151. Foote, Mary Hallock. A *Victorian gentlewoman in the far West : the reminiscences of Mary Hallock Foote.* Rodman Paul, editor. San Marino, CA: Huntington Library, 1992. 416 p. : ill. 0873280571
AR: *The led-horse claim, The chosen valley, Cour d'Alene.*
Combines prose and drawings to vividly portray life in various rugged western mining and irrigation projects with her family. Natural history, art, literature, and cultural differences between east and west are considered. Foote was the model for Susan Burling Ward in Wallace Stegner's *Angle of respose.*

1152. Frymer-Kensky, Tikva Simone. *In the wake of the goddesses : women, culture, and the biblical transformation of pagan myth.* NY: Free, 1992. 292 p. 0029108004
Examines religious and cultural symbols to trace the development of the conceptions of nature, gender, and sexuality from the goddesses of ancient Mesopotamia to biblical monotheism. Coverage of religion and state is primary, with secondary consideration of nature.

1153. Gaard, Greta Claire. ed. *Ecofeminism : women, animals, nature.* Philadelphia: Temple Univ., 1993. 331 p. 0877229880 0877229899 (pbk)
Contains 12 essays by various academics and activists on interconnections between animals and nature, theory and practice, the UFW grape boycott, animal rights, ecology and Romanticism, wholistic ethics, cross-cultural critiques, and Native American cultures.

1154. Gadon, Elinor W. *The once and future goddess : a symbol for our time.* NY: Harper & Row, 1989. 405 p. : ill. 0062503464 0062503545 (pbk)
Studies the history of Goddess religions, their representation in art and mythology, and the contemporary re-emergence of Goddess worship and ecofeminism.

1155. Gray, Elizabeth Dodson. *Green paradise lost.* Wellesley, MS: Roundtable, 1981. 166 p. 093412027
Original title: *Why the green nigger?*
Calls on humanity to "re-myth" its place in the natural world. A truly ecofeminist future will both protect nature by rejecting sexist, overly exploitative, and hierarchical thought.

1156. Jensen, Joan M. *Promise to the land : essays on rural women.* Albuquerque: Univ. of New Mexico, 1991. 319 p. 0826312470
Combines her experiences on a farm commune, those of her immigrant grandmother, and research, to demonstrate that although laws prevented women from owning land, they played a crucial role in farming, settlement, and conservation. The patriarchal/exploitative attitudes of many settlers are contrasted with those of Native American women.

1157. Kelly, Petra Karin. *Thinking green! : essays on environmentalism, feminism, and nonviolence.* Foreword by Peter Matthiessen, intro. by Charlene Spretnak. Berkeley: Parallax, 1994. 175 p. 0938077627
This posthumous anthology of speeches and articles by a leader of the German Greens who "died mysteriously" in 1992, combines many aspects of Green visionary politics and grounds them in the practical concerns of everyday life. Links the common concerns of feminism, ecology, pacifism, and human rights. HR.

1158. Knowles, Karen. ed. *Celebrating the land : women's nature writings, 1850-1991.* Flagstaff: Northland, 1992. 136 p. 0873585453
Authors include Susan Cooper, Isabella Bird, Mabel Osgood Wright, Mary Austin, Marjorie Kinnan Rawlings, Rachel Carson, Maxine Kumin, Ursuala Le Guin, Ann Zwinger, Sue Hubbell, Hope Ryden, Annie Dillard, Gretel Ehrlich, and Terry Tempest Williams. This collection of both obscure and well known female naturalists is HR.

1159. Krall, Florence R. *Ecotone : wayfaring on the margins.* Albany: SUNY, 1994. 270 p. 0791419614 0791419622 (pbk)
This personal history of place is written from the perspective of a teacher, naturalist, and feminist. Uses the metphor of the biological ecotone as the boundary where inner and outer landscapes of the woman/environment continuum meet.

1160. LaBastille, Anne. *Women and wilderness.* SF: Sierra Club, 1984, c1980. 310 p. : ill. 0871568284
Explores the factors which have excluded women from the wild and offers profiles of 15 women who have overcome those factors to become naturalists, activists, artists, teachers, and journalists. Particularly recommended for female high school and college students. #

1161. ___. *Woodswoman.* NY: Penguin, 1991, c1976. 277 p. : photos. 0140153349
A young ecologist describes her primitive way of life among the mountains, forests, lakes, and meadows of the Adirondacks. #

1162. LaChapelle, Dolores. *Sacred land, sacred sex : rapture of the deep : concerning deep ecology and celebrating life.* Durango, CO: Kivaki, 1992, c1988. 384 p. 1882308115
An inquiry into the relationship betwen human sexuality and the natural world. Topics include animal-human relationships, seeing nature, sacred land, human-Earth bonding, and wilderness.

1163. Le Guin, Ursula K. *Dancing at the edge of the world : thoughts on words, women, places.* NY: Perennial, 1990. 306 p. 0060972890
A collection of talks, essays, and reviews written between 1976 and 1988. Le Guin excoriates opponents of abortion rights, destroyers of the environment, macho writers, inequalities that women still face, and the destruction of Native American cultures.

1164. Mellor, Mary. *Breaking the boundaries : towards a feminist green socialism.* London: Virago, 1992. 308 p. 1853812005
Argues that we face a stark choice betwen survivalism or a new socialism

based on feminist and green principles. Advocates bringing together elements of deep ecology, ecofeminism, spirituality, and revolutionary socialism in a synthesis to offer a political vision for the 21st century.

1165. Merchant, Carolyn. *The death of nature : women, ecology, and the scientific revolution*. NY: Harper, 1989, c1980. 372 p. 0062505955
Explores the historical and cultural connections between women's issues and ecology, science, and environmental degradation.

1166. ___. *Radical ecology : the search for a livable world*. NY: Routledge, 1992. 276 p. 0415906504
A lively discussion of the nature of radical ecology, the global ecological crisis, the benefits and limitations of the scientific worldview, environmental ethics in political conflict, deep ecology, spiritual ecology, ecofeminism, sustainable development, and the future of the radical ecology movement.

1167. Niethammer, Carolyn J. *Daughters of the Earth : the lives and legends of American Indian women*. NY: Collier, 1977. 281 p. : photos. 0025885804 0020961502 (pbk.)
Dismisses the stereotypical "squaw" in a chronicle of the lives of traditional native women. Presents their practical mastery, ceremonial importance, approach to sexuality, and role as guardians of the Earth as a model for modern women.

1168. Norwood, Vera. *Made from this Earth : American women and nature*. Chapel Hill: Univ. of North Carolina, 1993. 368 p. : ill. 0807820628
Traces contributions made by female naturalists as writers, illustrators, landscape designers, ornithologists, botanists, biologists, and conservationists from the early 19th century to the present. Analyzes challenges to the ideologies and values of male scientists and the rise of ecofeminism.

1169. Peterson, Brenda. *Nature and other mothers : reflections on the feminine in everyday life*. NY: HarperCollins, 1992. 216 p. 0060163135
The only way to save the natural world, Peterson argues, is to claim it as our own, not through aggressive male action such as "marking territory" or the arms race, but through more feminine and aboriginal means such as spells, magic, and worship.

1170. Plant, Judith. ed. *Healing the wounds : the promise of ecofeminism*. Philadelphia: New Society, 1989. 262 p. 0865711526 0865711534 (pbk)
Contains 26 essays by Susan Griffin, Ynestra King, Ursula Le Guin, Anne Cameron, Starhawk, Margo Adler, Joanna Macy, and others. Divided into four sections: The Meaning of Ecofeminism; Politics; Spirituality; and Community.

1171. Rodda, Annabel. *Women and the environment.* Atlantic Highlands, NJ: Zed, 1991. 180 p. 086232985X (pbk) 0862329841
Analyzes the impact of environmental problems and issues on women and the role women have played in development. Argues that environmental degradation is tied to unequal educational and employment opportunities, and that equality will improve environmental quality.

1172. Ruether, Rosemary Radford. *Gaia & God : an ecofeminist theology of Earth healing.* SF: Harper, 1992. 310 p. 0060670223
Despite their patriarchal orientation, Christian and classical thought offer examples of transformative, biophilic relationships which can be merged with ecofeminism and the Gaia perspective to create a new consciousness as a basis for eco-justice and personal transformation.

1173. Seager, Joni. *Earth follies : coming to feminist terms with the global environmental crisis.* NY: Routledge, 1993. 332 p. 0415907209
Offers an original ecofeminist analysis of the environmental crisis that focuses on the structures of power within institutions and the ways in which they are dominated by masculinist assumptions.

1174. Shiva, Vandana. *Staying alive : women, ecology, and survival in India.* North Highlands, NJ: Zed, 1989. 224 p. 0862328225 0862328233 (pbk)
Considers the development of technology and the notion of industrial progress, noting that contributions by women and humanists have been largely excluded. Analyzes deforestation, world hunger, and water pollution, and offers an ecofeminist alternative to the current situation. HR.

1175. Starhawk. *Dreaming the dark : magic, sex & politics.* New ed. Boston: Beacon, 1988. 272 p. 0807010251
A fascinating consideration of self-empowerment through feminism, witchcraft, goddess worship, paganism, and ecology. Includes rituals, chants, group exercises, and tips for political activism. This controversial ecofeminist classic is HR.

1176. Teal, Louise. *Breaking into the current : boatwomen of the Grand Canyon.* Tucson: Univ. of Arizona, 1994. 200 p. : ill. 0816514135 0816514291 (pbk)
Teal relates the story of eleven women, including herself, who overcame the sexism of the commercial rafting business to become successful guides in the Grand Canyon. Emphasizes their perserverance and love of nature. Especially for women interested in careers in recreation. AOT. #

1177. Walker, Alice. *Living by the word : selected writings, 1973-1987.* San Diego: Harcourt Brace Jovanovich, 1988. 196 p. 0151529000
Essays demonstrating a transcendent sense of connection with nature and

Earth's creatures. "Am I blue?" and "Everything is a human being" are particularly notable. HR AMT. #

1178. Wallis, Velma. *Two old women : an Alaska legend of betrayal, courage, and survival.* Fairbanks: Epicenter, 1993. 145 p. 0945397186
Western States Book Award for Creative Nonfiction, 1993.
A retelling of the Athabaskan legend wherein two old women are left behind by their tribe during a harsh winter. They make snowshoes and animal traps, and as their old ways return to them they thrive in the wilderness. AOT. #

1179. Warren, Karen J. ed. *Ecological feminism.* NY: Routledge, 1994. 176 p. 041508623X
Because of the historical linkage of patriarcy and the destruction of nature, feminism must have ecological roots and vice-versa. Deep ecology, animal rights, stewardship, and other movements are criticized from an ecofeminist perspective. Contains ten essays by various authors.

1180. Williams, Terry Tempest. *An unspoken hunger : stories from the field.* NY: Pantheon, 1994. 160 p. 0679432442
Advocates expanding human love to "an erotics of place" which can heal nature, our interpersonal relationships, and our own souls. Also covers predaor-prey relations, Georgia O'Keefe, Edward Abbey, feminism and bears, anti-nuclear demonstrations, and the tragedy of seven radiation–caused breast cancer deaths in her family. HR.

Nature Writing Anthologies

Given the great diversity of authors, styles, and subjects in the field of nature writing, anthologies serve as an excellent introduction for novices and a good refresher for the more seasoned. Such collections have proliferated in the last decade thanks to the work of Murray, Lyon, and other compilers.

Since some of these books contain both fiction and nonfiction and the same can be said about some of those listed in the literary anthologies chapter, readers are advised to check both places. Also see *Thinking Green* (2) *Green Perspectives* (17) *Being in the World* (24)*Words from the Land* (157) A *Thousand Leagues of Blue* (530) *The Desert Reader* (347) and A *Republic of Rivers* (360)

"New nature writing" is featured in *American Nature Writing, Nature's New Voices, On Nature's Terms,* and *Second Nature.* The rest contain both retrospective and contemporary selections.

Books for adults:

1181. Finch, Robert, and Elder, John. eds. *The Norton book of nature writing*. NY: Norton, 1990. 921 p. 0393027996
Contains 125 selections by British and American nature writers from the 18th century onward. Includes Coleridge, Irving, Audubon, Catlin, Emerson, Darwin, Thoreau, Melville, Ruskin, Powell, Twain, Muir, Henry James, Hopkins, Austin, Woolf, Dinesen, Lawrence, Beston, E.B. White, Dubos, Steinbeck, Orwell, Stegner, Hay, Mowat, Abbey, Zwinger, LeGuin, McPhee, Momaday, Hubbell, Dillard, Silko, T.T. Williams and many others. HR AOT. #

1182. Halpern, Daniel. ed. *On nature : nature, landscape, and natural history*. SF: North Point, 1987. 319 p. 086547284X
Contains over 20 selections from the journal *Anteus* . Authors include Annie Dillard, Gretel Ehrlich, John Hay, Italo Calvino, John Fowles, John Rodman, E. O. Wilson, Noel Perrin, and Leslie Marmon Silko. HR AOT. #

1183. Henricksson, John. ed. *North writers : a Strong Woods collection*. Minneapolis: Univ. of Minnesota, 1991. 292 p. 0816610306
Contains 34 historic and modern selections about the area north and west of Lake Superior's northern shore. Authors include William O. Douglas, Sigurd Olson, Henry Beston, and others.

1184. Lyon, Thomas J. ed. *This incomperable lande : a book of American nature writing*. NY: Penguin, 1991. 495 p. 0140144412
Contains 21 selections covering four centuries of some of the best American nature writing along with a very interesting introduction by Lyon. Authors include Abbey, Audubon, Carson, Dillard, Lopez, Muir, and Thoreau. HR AOT. #

1185. Lyon, Thomas J., and Stine, Peter. eds. *On nature's terms : contemporary voices*. College Station: Texas A&M Univ., 1992. 212 p. 0890965226
An excellent anthology of some of the best of the new nature writing. Includes work by Charles Bowden, William Kittredge, Marcia Bonta, Stephen Trimble, Rick Bass, Barry Lopez, Gary Nabhan, Gary Snyder, Edward Hoagland, and Terry Tempest Williams' brilliant "Undressing the bear." HR AOT. #

1186. Mills, Stephanie, and Carstensen, Jeanne. eds. *In praise of nature*. Washington: Island, 1990. 258 p. 1559630345 (pbk) 1559630353
Contains over 70 short pieces on Earth, air, fire, water, and spirit by a virtual who's who of environmentalists and nature writers past and present,

including Ed Abbey, Barry Lopez, Thoreau, Rachel Carson, Wendell Berry, John McPhee, Susan Griffin and many others. A short commentary accompanies each selection. HR AOT. #

1187. Murray, John A. compiler. *American nature writing*, 1994. SF: Sierra Club, 1994. 229 p. 0871564793
The inaugural volume of the best American nature writing of the past year which is a.k.a. *The best American nature writing*. Contains essays, journal entries, stories, excerpts from novels, and poems by established and new nature writers including Abbey, Bass, DeBlieu, Dillard, Hasselstrom, Hogan, Kittredge, Lopez, D. O'Brien, Schacochis, Wallace, and T.T. Williams. HR AOT. #

1188. ___. ed. *Nature's new voices*. Golden, CO: Fulcrum, 1992. 242 p. 155591117X
These essays by Jan DeBlieu, David Rains Wallace, Rick Bass, Dan O'Brien, Gretel Ehrlich, John Daniel, Terry Tempest Williams and emerging nature writers provide new perspectives not just on nature, but on nature writing itself. HR AMT, especially writers. #

1189. ___. compiler. *Wild Africa : three centuries of nature writing from Africa*. NY: Oxford, 1993. 318 p. 0195073770
Contains over 40 selections written 1770-1993. Includes tribal legends and works by Twain, Burton, Conrad, T. Roosevelt, Dinesen, Schaller, Matthiessen, Lopez, T.T. Williams, and others. AOT. #

Classic Nature Writing

This section contains general literary nature writing. Some of the authors listed here have other books in other chapters, e.g. Van Dyke is also listed in Deserts. Other well known authors have no books listed here, but can be found in other sections, e.g. Abbey in Deserts, Rivers, Wilderness, and Environmental Action Fiction or Zwinger in virtually every type of landscape known to woman. For this reason one should not assume that a given author is not included in *Earth Works* without checking the author index.

   "Classic" authors can be divided into three eras: Burroughs, Emerson, and Thoreau in the 19th century; Hay, Olson, Krutch, and Teale in the early-to-mid 20th; and Hall, Hoagland, Kumin, and McPhee from the mid-to-late 20th. Some of the most important classic nature writers such as Burroughs, Krutch, Olson, and Van Dyke were also significant naturalists, so their books tend to combine superb natural history writing with personal reactions and philosophical asides. Brooks' *Speaking for Nature* describes how such nature writers have influenced the envronmental movement.

    Thoreau's *Walden* is considered by many to be the most important work of nature writing, but his lesser-read *A Week on the Concord and Merrimac Rivers* and *The Maine Woods* are virtually as significant. They also contain vivid natural history writing and fascinating exchanges with his Native American guide. *Faith in a Seed*, a Thoreau manuscript which was not published until 1993, and *The Green Thoreau*, a quotation book, are also listed here. *Thoreau's World and Ours* and *Seeking Awareness in American Nature Writing* explore Thoreau's influence. Burleigh provides a biography of Thoreau for young adults.

Books for adults:

1190. Barney, William D. *Words from a wide land*. Denton: Univ. of North Texas, 1994. 194 p. 0929398645
Insights on people and places, natural phenomena, birds, animals, insects, etc. Takes the unusual and effective form of an annual journal compiled over many years, emphasizing long-term processes over short-term alterations.

1191. Beston, Henry. *The outermost house : a year of life on the great beach of Cape Cod*. NY: Holt, 1992, c1928. 218 p. 0805019669
This almost poetic evocation of a year spent on then-wild Cape Cod describes the natural history of the area, but more importantly establishes how living in the natural world can lead to a fuller understanding and appreciation of life to guide our actions. HR.

1192. Blanchet, M. Wylie. *The curve of time : the classic memoir of a woman and her children who explored the coastal waters of the Pacific Northwest*. Seattle: Seal, 1993, c1961. 170 p. 1878067273
Describes a family's expeditions along the B.C. coast in the 1920's, including encounters with bears, whales, cougars, fog, heavy seas, hunters and trappers, and Native Canadians. Good family reading. #

1193. Borland, Hal. *High, wide and lonesome*. Tuscon: Univ. of Arizona, 1990, c1956. 253 p. 0816511772
AR: *Hill country harvest, The history of wildlife in America, Hal Borland's book of days, Sundial of the seasons, Seasons, This hill, this valley, When the legends die*.
An autobiography of the noted *New York Times* nature writer's youth on the Colorado plains in the last days of the open range in the early 20th century. HR AOT. #

1194. Brooks, Paul. *Speaking for nature : how literary naturalists from Henry Thoreau to Rachel Carson have shaped America*. SF: Sierra Club, 1983, c1980. 304

p. 0395296102
AR: *The pursuit of wilderness.*
Considers how naturalists have used nature writing to raise public consciousness about the environment and stimulate conservation activity. Some of the major figures considered include Thoreau, Burroughs, Muir, Lanier, Powell, Beebe, Austin, Roosevelt, Leopold, DeVoto, and Carson.

1195. Burroughs, John. *Birch browsings : a John Burroughs reader.* Ed. by Bill McKibben. NY: Penguin, 1992. 231 p. 0140170162
AR: *John Burroughs' America.*
An anthology of essays, articles, and excerpts from the books of a pivotal figure who is equally notable as a pre-eminent naturalist of the northeast and as a literary critic who advanced the art of highly literate nature writing. AMT. #

1196. ___. *The birds of John Burroughs : a great naturalist's meditations and essays on bird watching.* Ed. by Jack Kligerman. Woodstock, NY: Overlook, 1989. 240 p. : ill. 0879513128
Contains selections about birds and birding from eleven of Burrough's books, many of which are out of print. Comments about nature in general, literature, and society are interspersed. These inspiring essays are most appropriate for birders and those who appreciate literate, highly personal commentary. HR.

1197. ___. *Deep woods.* Salt Lake City: Peregrine Smith, 1990. 218 p. 0879050306
This anthology of selections from seven books covers the Catskills, Alaska, the Rockies, and Maine, and offers an example of a highly sophisticated person developing a wholistic bond with the natural world. HR AOT. #

1198. ___. *John James Audubon.* Woodstock, NY: Overlook, 1987, c1902. 144 p. 0879512598
Audubon was such an inspiration for Burroughs that he wrote this biography covering the great ornithologist's life and work.

1199. ___. *Locusts and wild honey.* NY: Reprint Services, 1989, c1879. 255 p. 0781221811
Burroughs lively account his adventures hiking, observing and appreciating nature, and especially fishing. Essential for anglers.

1200. ___. *A sharp lookout : selected nature essays of John Burroughs.* Ed. by Frank Bergon. Washington: Smithsonian, 1987. 605 p. 0874742706 0874742714 (pbk)
An excellent anthology of Burroughs nature essays from a variety of books

and periodicals. May be the best Burroughs anthology. HR.

1201. ___. *Signs and seasons*. NY: Reprint Services, 1989, c1886. 289 p. 0781221846
Burroughs focuses on how to watch, experience, and become one with nature, thus nourishing the mind and soul.

1202. ___. *Ways of nature*. NY: Reprint Services, 1989, c1905. 279 p. 0781221927
Burroughs' critique of "the nature fakers" who passed off romanticized versions of nature and anthropomorphic animals as natural history. He argues that although animals have the "human" faculties of perception, sensation, memory, association of ideas, fear, jealousy, love, and other emotions, they lack deep reasoning or an innate sense of morality.

1203. ___. *Writings*. NY: Reprint Services, 1993, c1871. 26 v. 0781251419
Reprint of the complete works of Burroughs. HR.

1204. Dahl, Roald. *My year*. NY: Viking, 1994. 63 p. : col. ill. 0670853976
Dahl chronicles the passing of the year and reminisces on his life in the English countryside.

1205. Emerson, Ralph Waldo, and Thoreau, Henry David. *Nature by Ralph Waldo Emerson; Walking by Henry David Thoreau*. Boston: Beacon, 1991. 140 p. : ill. 0807015628
Nature c1836; *Walking* c1862.
Emerson laments that most people don't really see nature or realize they are part of it. His process for people to deeply experience nature has influenced countless writers. Thoreau considers the joys and benefits of walking as a mode of transportation which keeps people in touch with nature. HR AMT. #

1206. Flanner, Hildegarde. *At the gentle mercy of plants : essays and poems*. Santa Barbara: J. Daniel, 1986. 89 p. 093678413X
Lyrically describes California's natural landscapes and botanical richness and Flanner's lifelong dedication to the preservation of the wild.

1207. Fuller, Margaret. *Summer on the lakes in 1843*. Urbana: Univ. of Illinois, 1991, c1844. 156 p. : ill. 0252061640
A colleague of Emerson and Thoreau describes a trip to the Great Lakes, extolls the beauty of the mostly unspoiled country, and predicts the blight that would inevitably come with settlement.

1208. Hall, Donald. *Here at Eagle Pond*. NY: Ticknor & Fields, 1990. 141 p. :

ill. 089919978X
Contains 21 essays by the poet on country life, nature, poetry, and feeling
close to the land. "I see all around me emigrants from other places who
belong more preciously to this place than most old-timers do. But I see as
well...that overpopulation with its suburban density can disconnect us
from the land and its history."-Hall, p. xii.

1209. ___. *Seasons at Eagle Pond.* NY: Ticknor & Fields, 1987. 86 p. : ill.
0899195423
Sequel: *Here at Eagle Pond.*
The poet uses the cycle of the seasons to describe country life in New England, his relationship to nature, and how it effects his poetry. Reminiscent
of a Yankee version of Wendell Berry.

1210. Hay, John. *The bird of light.* NY: Norton, 1993. 158 p. 0393310019
Hay's eloquent personal observations on the natural history of the tern
serve as a backdrop for his themes of species and habitat preservation,
ecological thinking, and respect for nature. AOT. #

1211. ___. *The great beach.* 2d ed. NY: Norton, 1980, 1963. 131 p. : ill.
0393013677
AR: *Nature's year: the seasons of Cape Cod, In defense of nature, The undiscovered country, On nature.*
While exploring the outer bank of Cape Cod, Hay experiences a sense of
renewal and unity with nature.

1212. ___. *The immortal wilderness.* NY: Norton, 1987. 186 p. 0393310019
Hay uses the term wilderness not just for wild lands, but as "the great con-
tainer of life and death, the Earth's immortal genius. Nothing in wilderness
escapes the universal interdependence . . . Uncompromising but protec-
tive, it holds to its principles of renewal and diversity in all the facets of
nature." p. 14 & 16.

1213. Hoagland, Edward. *Balancing acts : essays.* NY: Simon & Schuster,
1994, c1992. 351 p. 0671892355
AR: *The Edward Hoagland reader, The moose on the wall.*
Fifteen essays about Thoreau, coral, Africa, cowboy poets, Ed Abbey,
Wyoming, the land, the Okefenokee Swamp, and various other topics. This
is the most nature oriented of Hoagland's many collections. HR.

1214. ___. *The courage of turtles : fifteen essays about compassion, pain, and love,.*
NY: Lyons & Burford, 1993, c1970. 256 p. 1558212159
Essays on nature, turtles, sports, rodeos and county fairs, bear hunting,
tugboats, writing, and miscellaneous subjects.

1215. ___. *Heart's desire, the best of* Edward Hoagland : *essays from twenty years.* NY: Summit, 1988. 429 p. 0318377446
Contains 35 pieces by the distinguished *New Yorker* essayist. About half the essays are on nature, with others on literature, society, and life in general. HR.

1216. ___. *Red wolves and black bears.* NY: Penguin, 1983. 273 p. 0140066861
Contains 19 essays about trapping wolves, "bearding" bears, city walks, Vermont communes, Texas ranchers, newts, and survival.

1217. ___. *The tugman's passage.* NY: Penguin, 1983. 208 p. : map. 0140066853
Contains 13 essays on such diverse topics as natural history, male-female relations, Africa, writing, outdoor adventure, and aging. Particularly notable is "A year as it turns," reprints of over 30 of his seasonal essays written for the *New Yorker.* HR.

1218. ___. *Walking the dead* Diamond *River : nineteen essays.* NY: Lyons & Burford, 1993, c1973. 340 p. 1558212167
Roughly half of the essays are about nature, aspects of rural life, and hiking in the wilderness. The rest are about the "joys and outrages of living in a large city."

1219. Jaques, Florence Page, and Jaques, Francis Lee. *Snowshoe country.* St. Paul: Minn. Hist. Soc., 1989, c1944. 110 p. : col. ill. 0873512367
John Burroughs Award for Nature Writing, 1946.
This journal carefully describes and depicts October through January in the border country of Minnesota and Canada. HR.

1220. Kanze, Edward. *The world of John Burroughs.* NY: Abrams, 1993. 160 p. : ill. 0810939703
Details the life of a pioneer conservationist and prolific author of both nature and literary studies. Describes his friendships with Whitman, Roosevelt, Muir, Edison and others. Includes historic photographs, quotes from Burroughs and his associates, and interviews with his granddaughter. #

1221. Kumin, Maxine. *In deep : country essays.* Boston: Beacon, 1988, c1987. 180 p. 0670814318
AR:*To make a prairie, Women, animals, and vegetables,* and her poetry.
A cycle of meditative essays following the seasons, depicting Kumin's rural lifestyle, and revealing how her soul and poetry are inspired by nature. Somewhat similar to Hubbell's A *country year.* HR AMT. #

1222. Lawrence, Louise de Kiriline. *The lovely and the wild.* Toronto: Natural

Heritage, 1987, c1968. 228 p. 0920474438
John Burroughs Medal for Nature Writing, 1968.
An account of the author's awakening sensitivity to birds and nature while living near a lake in southeast Ontario. HR.

1223. Lindbergh, Anne Morrow. *Gift from the sea.* NY: Pantheon, 1992, c1955. 132 p. 0679406832
"An enduring favorite, wise and gentle meditations on how women can carry the lessons found in a life close to nature (in this case, the author's seaside retreat) back into the crucible of everyday modern life."-Lorraine Anderson. HR.

1224. Luhan, Mabel Dodge. *Winter in Taos.* Taos, NM: Las Palomas de Taos, 1987, c1935. 237 p. : ill. 0911695508
"A lyrical celebration of Luhan's adopted land, structued around the seasonal cycle of death and rebirth, and richly integrating her emotional life with the physical landscape."-Lorraine Anderson

1225. McPhee, John. *The control of nature.* NY: Farrar, Straus, Giroux, 1989. 272 p. 0374522596
AR: *Outcroppings*, *Basin and range*, *Rising from the plains*.
Discusses the economic and environmental costs of efforts to control natural areas, focusing on the Mississippi Delta, Los Angeles, and Iceland. HR AMT. #

1226. ___. *The John McPhee reader.* NY: Farrar, Straus, Giroux, 1982, c1976. 409 p. 0374517193
Essays on basketball, the environment, tennis, oranges, the secret development of a visionary aircraft, poorly guarded plutonium, birch bark canoes, the Hebrides, and more. An excellent collection of his work through 1975, about half of which is about nature and the environment. HR AOT. #

1227. ___. *Pieces of the frame.* NY: Farrar, Straus, Giroux, 1975, c1963. 308 p. 0374514984
An eclectic collection of essays about travel and nature in Georgia, New Mexico, the Canadian Rockies, and Scotland; conservation; and sports. The title piece is a delightful tale about his family's search for the Loch Ness monster. AOT#

1228. Olson, Sigurd F. *Reflections from the North Country.* NY: Knopf, 1985, c1976. 172 p. : ill. 0394402650
"He sharpens our awareness of the beauty around us, gently warning us to leave behind our excess baggage of scientific sophistication and open ourselves to wonder. He reflects on our frontier heritage, ponders the meaning

of solitude...wholeness, cosmic rhythms, and the slow cycles of seasonal change."-Intro.

1229. Powys, Llewelyn. *Earth memories*. Bristol: Redcliffe, 1983, c1934. 147 p. 0317203088
Natural, autobiographical, and social essays set largely in pastoral England in the 1930s. Powys celebrated the emotional and spiritual power of nature, concluding that "We should grow less involved in society and more deeply involved in existence."

1230. Slovic, Scott. *Seeking awareness in American nature writing : Henry Thoreau, Annie Dillard, Edward Abbey, Wendell Berry, Barry Lopez*. Salt Lake City: Univ. of Utah, 1992. 203 p. 0874803624
Demonstrates how these contemporary nature writers follow Thoreaus' example of writing about the effect of nature on their consciousness and values as having primacy over writing about nature itself. HR.

1231. Teale, Edwin Way. *The American seasons*. NY: St. Martin's, 1990, c1956. 4 v. : photos. 0312044577
Pulitzer Prize, 1956.
AR: *The wilderness world of John Muir, Green treasury*.
Contains selected chapters from his books *North with the spring, Autumn across America, Journey into summer*, and *Wandering through winter*. Vividly portrays the passing of the seasons in America and how the seasons affect all of nature. HR.

1232. Thoreau, Henry David. *Faith in a seed : the dispersion of seeds and other late natural history writings*. Washington: Island, 1993. 300 p. : ill. 1559631813
Drawing on Darwin's theory of natural selection, Thoreau refutes the then widely accepted theory that some plants spring spontaneously to life. Remarkably, this is the first publication of this manuscript. Essential for Thoreau buffs. HR.

1233. ___. *The green Thoreau*. Comp. by Carol Spenard LaRusso. San Rafael, CA: New World, 1992. 65 p. 0931432963
A collection of brief quotations from several books by Thoreau divided into chapters on nature, technology, livelihood, living, possessions, time, and aspiration. Good for finding extracts for lectures and papers. HR AOT. #

1234. ___. *The Maine woods*. Intro. by Edward Hoagland. NY: Penguin, 1988, c1864. 442 p. 0140170138
Thoreau describes how his journeys into the woods, rivers, and mountains of Maine transformed his consciousness of nature and led him to believe that "in wildness is the preservation of the world." HR AOT. #

1235. ___. *Thoreau on birds : notes on* New England *birds from the Journals of* Henry David Thoreau. Boston: Beacon, 1993, c1910. 510 p. 0807085200
Original title: *Thoreau's bird-lore.*
Thoreau describes the bird life of New England and offers philosophical asides. Recommended for birders.

1236. ___. *Walden.* NY: Knopf, 1992, c1854. 298 p. 0679418962
Another ed. with an intro. by Ed Abbey pub. by Peregrine Smith, 1981. *The annotated* Walden (512 p.) pub. by Barnes and Noble, 1992. Children's version pub. by Philomel, 1990.
This account of Thoreau's simple lifestyle and nature study while living in a cabin near Walden Pond questions the values of industrial progress and explores the foundations of American transcendentalism. This extremely influential classic is HR AMT. #

1237. ___. *A week on the Concord and Merrimack rivers* ; *Walden, or, Life in the woods* ; *The Maine woods* ; *Cape Cod.* NY: Literary Classics of the U.S., 1985. 1114 p. 0940450275.
Thoreau's canoe journey with an Indian guide is an excursion through the wilds of nature and the soul. This collection of four of Thoreau's major works is HR AMT. #

1238. Van Dyke, John Charles. *The open spaces : incidents of nights and days under the blue sky.* Salt Lake City: Univ. of Utah, 1991, c1922. 297 p. 0874803764
AR:*Nature for its own sake.*
Van Dyke's autobiography. "What a strange feeling, sleeping under the wide sky, that you belong only to the universe. You are back to your habitat, to your original environment, to your native heritage."—p. 20.

Books for young adults:

1239. Burleigh, Robert. *A man named Thoreau.* NY: Atheneum, 1985. 31 p. : ill. 0689311222
ALA Notable Children's Book, 1985.
Presents the life and ideas of the renowned 19th-century American author. This introduction is HR GR 3-8.

1240. Thoreau, Henry David. *Walden.* NY: Philomel, 1990. 32 p. : ill. 0399221530
In this illustrated adaptation of Thoreau's famous work, a man retreats into the woods and discovers the joys of solitude and living simply with nature. HR GR 4-8, especially reluctant readers.

New Nature Writing

Thoreau's tendency to emphasive the influence of nature on the human soul over more straightforward natural history writing has influenced the contemporary genre known as New Nature Writing. Poetic, psychological, or philosophical asides, pleas for conservation or activism, and intimate personal revelations are common characteristics. These authors are primarily from the baby-boom generation, and many were strongly influenced by the anti-war, feminist, and environmental movements. I have included a few authors born slightly earlier such as Bonta, Finch, and Nichols, since their work fits the description above.

As noted in the introduction to Classic Nature writing, do not assume that such significant New Nature writers as Bowden, Lopez, or Williams are inexplicably excluded from *Earth Works* simply because they are not in this section. Please check the author index for additional works by the authors listed in this section and for those not listed here. Also see the anthologies *Nature's New Voices* (1188), *American Nature Writing* (1187), *and On Nature's Terms* (1185)

Books for adults:

1241. Adams, Noah. *Saint Croix notes*. Boston: Houghton Mifflin, 1992. 221 p. 0395597048
An NPR broadcaster's observations on life in the Saint Croix Valley in Wisconsin and Minnesota, much of it focusing on the land and the seasons that connect us to the natural world and the communities we live in.

1242. Arthur, Elizabeth. *Island sojourn*. St. Paul, MN: Graywolf, 1991. 220 p. 1555971490
AR: *Antarctic navigation* .
An account of her retreat from stress and confusion and subsequent journey of self-discovery in the B.C. wilderness. She and her husband learned simplicity, Indian lore, and natural ways. After a winter of isolation they realized they liked to visit the wilderness, but not stay.

1243. Bass, Rick. *Winter : notes from Montana*. Boston: Houghton Mifflin, 1992. 162 p. 0395611504
A daily journal of Bass' first winter in the last corner of America without electricity. Includes observations on nature, ecological patterns in winter, wilderness, and interdependence in country life.

1244. Baumgartner, Susan. *My Walden : tales from Dead Cow Gulch*. Freedom, CA: Crossing, 1992. 129 p. : ill. 0895945525
Environmental and personal observations on living in deeply rural areas

that are not true wilderness but possess natural and spiritual value.

1245. Bonta, Marcia. *Appalachian spring*. Pittsburgh: Univ. of Pittsburgh, 1991. 187 p. : maps. 0822936585
AR: *Appalachian autumn , Escape to the mountain*.
Through her observations on the natural world, Bonta aims to motivate others to "the third stage in humanity's relationship with nature, that of empathy with all nature for its own sake." AOT. #

1246. Daniel, John. *The trail home : essays*. NY: Pantheon, 1994. 223 p. 069754385
Daniel's aim is to apprehend the natural world more personally. He persuades us to conserve water, relate to nature in an individualistic manner, take political action to defend wildness, and take personal action to accept the wildness within us. AMT. #

1247. Dean, Barbara. *Wellspring, a story from the deep country*. Washington: Island, 1979. 208 p. 0933280017
A young woman moves to a square mile of wilderness land in northern California, developing a new relationship with nature by making major lifestyle and philosophical adjustments after 25 years of city life.

1248. Ehrlich, Gretel. *Islands, the universe, home*. NY: Viking, 1991. 196 p. 0670821616
Traveling between her Wyoming ranch, the wilds of Japan and the Channel Islands inspires Ehrlich to consider the seasons, natural patterns, forest fires, and our collective, universal "home." AOT. #

1249. ___. *The solace of open spaces*. NY: Viking, 1985. 131 p. 0140081135
AR: A *match to the heart*.
Autobiographical essays about ranching and living close to the land in Wyoming. HR.

1250. Finch, Robert. *Outlands : journeys to the outer edges of Cape Cod*. Boston: Godine, 1986. 175 p. 0879236191
Finch presents a lyrical consideration of the flora and fauna, tides and weather of Cape Cod. This fine companion to Beston's *The outermost house* is HR.

1251. ___. *The primal place*. NY: Norton, 1983. 243 p. 0393016234
This predecessor to *Outlands* is an even more meditative and reflective consideration of the natural world and the human mind in Massachusetts. HR.

1252. Fromm, Pete. *Indian Creek chronicles*. NY: Lyons & Burford, 1993. 184 p. : map. 1558212051

AR: *The tall uncut.*
When a naive college student accepts a wintertime job that entails living alone in the bitter cold Bitterroot wilderness, he is forced to deal with extreme conditions and with boredom which drives him to impetuous, perilous outings. Recommended AOT. #

1253. Gaines, Charles. A *Family place : a man returns to the center of his life.* NY: Atlantic Monthly, 1994. 195 p. 0871135604
A successful screen writer and his wife leave Hollywood for the wilds of Nova Scotia where they build a cabin, and in the process, rebuild their failing marriage and family life.

1254. Hubbell, Sue. *Broadsides from the other orders : a book of bugs.* NY: Random House, 1993. 296 p. : ill. 0679400621
AR: A *book of bees.*
This highly personalized account is less a scientific description of insects and their relation to humans as it is an appreciation of their variety, adaptability, beauty, and special abilities. For those who appreciate highly literate nature writing. HR AOT#

1255. ___. A *country year : living the questions.* NY: Perennial, 1987, c1985. 221 p. : ill. 0060970863
After a tough divorce Hubbell moves to the Ozarks to become a beekeeper, attaining self-sufficiency and inner peace. Presents a virtual running commentary of her interactions with and questions about nature through the course of the seasons. HR AOT. #

1256. ___. *On this hilltop.* NY: Ballantine, 1991. 195 p. 0345373065
Reconsiders some of the questions and issues she raised in A *country year* and raises some related ones. AOT. #

1257. Johnson, Cathy. *On becoming lost : a naturalist's search for meaning.* Salt Lake City: Peregrine Smith, 1990. 176 p. 0879053496
Johnson wanders a non-wilderness stretch of the Missouri River, finding nature among the litter and meaning in a single rock. Although "Awareness must be the most difficult state to achieve," you may find it if you wander with her. HR.

1258. Kappel-Smith, Diana. *Wintering.* NY: McGraw-Hill, 1986, c1984. 305 p. : ill. 0070341915
A personal journey into winter on a Vermont farm and out again, by a biologist and artist.

1259. Kinseth, Lance. *River eternal : the wonder of common and ashen days along-*

side a prairie river. NY: Viking, 1989. 115 p. 0670824410
Eloquently describes the powers of nature and its therapeutic value to the
human spirit. Builds an elegant bridge between nature and culture. HR
AMT. #

1260. LaBastille, Anne. *Beyond Black Bear Lake.* NY: Norton, 1988. 251 p. :
photos. 0393305392
La Bastille moves deeper into the woods, builds a cabin for $138.38, and
explores the wilderness. Chapters about nature and the wheel of the seasons alternate with passages about environmental concerns. AOT. #

1261. Lembke, Janet. *Looking for eagles : reflections of a classical naturalist.* NY:
Lyons & Burford, 1990. 181 p. 1558210776
A translator of Greek and Latin poetry discovers that much of what fasci-
nates her in nature today was also studied in classical times.

1262. ___. *Skinny-dipping : and other immersions in water, myth, and being human.*
NY: Lyons & Burford, 1994. 192 p. 1558212744
Lembke employs classical mythology, folk tales, and biblical allusions, to
comprehend and explain the natural world in a variety of settings, rural
and urban.

1263. Leschak, Peter M. *Seeing the raven : a narrative of renewal.* Minneapolis:
Univ. of Minnesota, 1994. 192 p. 0816624291
AR: *Letters from Side Lake.*
A collection of essays from the *New York Times, North Country Journal*, and
other periodicals about life and nature in Minnesota. Describes ice-cut-
ting, cold weather gardening, astronomy, the woods, wildlife, the cycle of
life and death, bereavement, and other subjects.

1264. Linehan, Don. *The mystery of things.* Lawrencetown Beach, N.S.: Pottersfield, 1989. 88 p. : ill. 0919001548
Explores the beauty and the mystery of the natural world of Nova Scotia.
Writing in the tradition of Thoreau and Annie Dillard, he meditates upon
the simple but provocative discoveries that nature offers up to the curious.

1265. Meyers, Steven J. *Lime Creek odyssey.* Golden, CO: Fulcrum, 1989. 116
p. 1555910378
While exploring the beauty of a small stream in the San Juan Mountains,
Meyers discovers universal paradoxes of humanity's relationship to the
natural world.

1266. Modzelewski, Michael. *Inside passage : living with killer whales, bald
eagles, and Kwakiutl Indians.* NY: HarperCollins, 1991. 184 p. : ill. 0060165332

Modzelewski learns the ways of nature in the northern wilderness, and in so doing develops greater self-reliance, respect for all life, and a deeper sense of himself.

1267. Nichols, John Treadwell. *A fragile beauty : John Nichols' Milagro country : text and photographs from his life and work.* Salt Lake City: Peregrine Smith, 1987. 146 p. : photos. 087905821
Nichols poetically describes his life in New Mexico from a boyhood journey in the 1950s, his moving there in 1969, and his emergence as an important author and activist. Discusses the land, the people, and the filming of his novel, *The Milagro beanfield war.* Includes excerpts from his novels. HR AOT. #

1268. ___. *If mountains die : a New Mexico memoir.* NY: Norton, 1994, c1979. 144 p. : photos. 0393311597
"This eloquent, moving, and often funny book is his account of how his life has been transformed by daily, intimate contact with this extraordinary landscape—at once hostile and nurturing—and by his growing sense of responsibility toward the land and the people who live there."-Intro. HR AOT. #

1269. ___. *On the mesa.* Salt Lake City: Peregrine Smith, 1986. 193 p. : photos. 0879052201
Nichols went to the Mesa for a vacation and retreat, but due to encroachments on the land and the fouling of the air and water, he re-engaged in struggles to preserve the area.

1270. ___. *The sky's the limit : a defense of the Earth.* NY: Norton, 1990. 1 v. (unpaged) : col. photos. 0393028658
Nichols combines his considerable photographic and writing skills to portray a vivid picture of the grandeur of northern New Mexico. He also offers advice on what we can do about today's major environmental crises. #

1271. Peterson, Brenda. *Living by water : essays on life, land & spirit.* Anchorage: Alaska Northwest, 1990. 134 p. 0882403583 0882404008 (pbk)
A series of lyrical, highly personal essays, most about the Puget Sound region, but encompassing Alaska, Florida, Brazil, and other areas. Peterson identifies strongly with other animals, especially dolphins.

1272. Sanders, Scott R. *The paradise of bombs.* Boston: Beacon, 1993, c1987. 155 p. 0807063436
Essays on hiking, nature, fathers and sons, and war fantasies based on Sander's childhood living in the woods surrounding a weapons testing ground. Considers the mysteries of human nature that propel us toward both life and destruction. #

1273. Stafford, Kim Robert. *Having everything right : essays of place*. NY: Penguin, 1987, c1986. 190 p. 014010254X
Western States Book Award, 1986.
"This book travels for place, custom, and story. As water is pilgrim, I know the urge to visit all the places I was healed. Water travels as a local inhabitant, as essence of tree, of capillary stone, of sunlight pillar in a meadow ablaze with grasshoppers. I want to learn place, custom, and story for my own home."-Stafford, p. 7-8. HR.

1274. Tisdale, Sallie. *Stepping westward : the long search for home in the Pacific Northwest*. NY: Harper, 1992, c1991. 284 p. : map. 0060975105
A poetic evocation of land, nature, and people in the Pacific Northwest. Describes the original natural wonders of the area, their destruction, what remains of nature, and how to live in balance with it. HR AMT. #

1275. Wheelwright, Jane Hollister, and Schmidt, Lynda Wheelwright. *The long shore : a psychological experience of the wilderness*. SF: Sierra Club, 1991. 202 p. 0871566257
A mother-daughter team of Jungian analysts discuss the benefits and joys experienced living on a large, wild ranch on the California coast, the sense of loss the sale of the ranch had upon them, and their realization of the importance of the direct experience of nature on human wholeness.

7. Environmental Action

As the books in previous chapters demonstrate, the Earth faces serious environmental challenges. This need not be a source of discouragement because there are many things individuals can do to make a positive impact both locally and globally. Conservation, perhaps the most traditional approach, is a strong and diverse movement which has spawned the theory and practice of Sustainable Development. Developing a Deep Ecological or at least a conservation-oriented philosophy can inspire people to be part of the growing Green Movement. A perusal of the books on Activism and Direct Action can help people decide where to focus their energies. Environmental Education is appropriate for people of all ages, providing a knowledge base for effective action. Ecotourism provides a combination of recreation and education. Greener Homes and Gardens can be safer, more economical, and more comfortable than traditional ones. A growing number of people in Business and Industry are employing ecological practices. The Vocational Guidance section can help people transform their concerns and knowledge into productive careers.

Conservation

Conservation is broadly defined here to include preservation, restoration, and land use. Dasmann, Orr, Owen, and Soule provide good introductions. Hornaday, Leopold, Marsh, Muir, Pinchot, Roosevelt, Schaefer, and the authors in *Voices from the Land* speak to us from the past. Huth, Strong, Stewart Wallach, and especially Nash provide historical and social analyses. Also see *Wilderness Tapestry* (2601). See the Wilderness and Mountain sections for the rest of Muir's books and the Social Factors section for related environmental history.

Allen, Brower, Cohen, Fox, Grove, and Pittman profile individual conservationists or organizations. Dunlap, Durrell, Houle, Hummel, Huxley, Livingston, MacNamee, Orr, and Reisner cover wildlife or plant protection. Related works can be found in the Endangered Species and Animal Rights sections.

Land use issues, land trusts, and greenways are described in books by Flink, the Institute for Community Development, Margolin, Morine,

Steiner, the World Wildlife Fund, and Wright. Other books on land use are located in Social Factors, Books by Baldwin, Goldsmith, Nilsen, and Yates go one step beyond conservation to natural restoration of damaged environments.

The books for young adults by Daniel, Douglas, the Fabers, Force, Harlan, Jezer, Kallen, and Tolan emphasize history and biography, often describing conservationists as role models. General conservation, land use, and wildlife preservation are covered by Bramwell, Patent, and Curtis. *Heaven Is Under Our Feet* is sort of a literary "We are the World" by various popular authors, celebrities, and rock stars.

Books for adults:

1276. Adams, John H., and Cahn, Robert. eds. *An Environmental agenda for the future.* Washington: Agenda, 1985. 155 p. 0961519304
Contains essays by the leaders of the Sierra Club, National Audubon Society, World Wildlife Fund, and the Conservation Foundation assessing key environmental issues in the past, present, and projected future. Proposes a conservation action plan.

1277. Allen, Thomas B. *Guardian of the wild : the story of the National Wildlife Federation, 1936-1986.* Bloomington: Indiana Univ., 1987. 212 p. : ill. 0253326052
Describes the NWF's formation as an effort to focus the work of many scattered conservation organizations on electoral politics, its political successes and failures, its educational work, and its role in preserving species and their habitats.

1278. Baldwin, A. Dwight; De Luce, Judith; and Pletsch, Carl. eds. *Beyond preservation : restoring and inventing landscapes.* Minneapolis: Univ. of Minnesota, 1994. 280 p. 0816623465 0816623473 (pbk)
Contains 20 essays by various authors on "the invented landscape," preservation vs. restoration, human sovereignty over nature, changing worldviews, native ecosystems, tropical forest restoration, restoring strip-mined areas, prairie restoration, and the political economy of restoration.

1279. Borrelli, Peter. ed. *Crossroads : environmental priorities for the future.* Washington: Island, 1988. 353 p. 0933280688 093328067X (pbk)
Contains 24 essays by Udall, Boyle, Commoner, Frome and others which assess the environmental victories and defeats of the past 20 years, describe the great diversity of environmental protection and conservation groups, and attempt to find effective paths to future environmental progress.

1280. Botkin, Daniel B. *Discordant harmonies : a new ecology for the twenty-first century.* NY: Oxford, 1992. 241 p. 01195674696
Presents several scenarios for the next century, demonstrating how economic and environmental decisions made today will shape the quality of life tomorrow. Argues that policymakers need a better understanding of the biosphere and of the effect of their policies upon it. HR.

1281. Brower, David. *For Earth's sake : the life and times of David Brower.* Salt Lake City: Peregrine Smith, 1990. 556 p. 0879050136
Sequel: *Work in progress.*
The first autobiographical volume by a founder of Friends of the Earth, the League of Conservation Voters, and Earth Island Institute, and long-time president of the Sierra Club, focuses on the origins of the modern conservation movement. HR.

1282. ___. *Work in progress.* Salt Lake City: Peregrine Smith, 1991. 348 p. 0879053747
In the proces of describing his life and struggles, Brower provides insights and recommendations on peace and security through a sustainable economy, conservation, militant environmentalism, alternative energy, and related subjects. HR.

1283. Cohen, Michael P. *The history of the Sierra Club, 1892-1970.* SF: Sierra Club, 1988. 567 p. 0871567326
Traces the history of the Club in its development from a hiking group to an international conservation organization.

1284. Commoner, Barry. *Making peace with the planet.* NY: New Press, 1992. 304 p. 1565840127
AR: *The closing circle, The poverty of power.*
A detailed and articulate examination of how a transition to alternative technologies and sustainable farming is both feasible and neccesary. HR.

1285. Dobson, Andrew. ed. *The Green reader : essays toward a sustainable society.* SF: Mercury House, 1991. 280 p. 1562790102 (pbk) 156279017X
An international collection of essays about the Green Movement and sustainable development by such notable authors as Rachel Carson, Aldous Huxley, E.F. Schumacher, Vandana Shiva, and others. HR.

1286. Dunlap, Thomas R. *Saving America's wildlife.* Princeton, NJ: Princeton Univ., 1991, c1988. 222 p. 069100613X
This history of wildlife preservation goes beyond the "good vs. bad" approach to show how and why Americans felt the way they did about wild animals, and how and why they have changed. Once viewed as an adversary

or resource, nature is now seen as a fragile system that must be protected.

1287. Durrell, Gerald Malcolm. *The ark's anniversary*. NY: Arcade, 1991. 179 p. : ill. 1559701404
A mixture of readable science with hilarious whimsy in this history of the Jersey Zoological Park, a world leader in conservation zoology, captive breeding, and public education. Four legged performers mix with two legged celebs such as David Niven, Princess Grace, Richard Adams, and Princess Anne. HR. #

1288. Engel, J. Ronald. *Sacred sands : the struggle for community in the Indiana Dunes*. Middletown, CN: Wesleyan Univ., 1986, c1983. 390 p. : ill. 0819561290
How a coalition of activists, artists, writers, and citizens successfully addressed a broad range of scientific, aesthetic, and philosophical concerns to save the Indiana Dunes. Includes ideas and strategies which other conservationists and environmentalists can appply elsewhere.

1289. Facts on File. *Masterworks of man and nature : preserving our world heritage*. NY: Facts on File, 1994. 496 p. : 600 col. photos. 0816031770
A photographic atlas depicting 360 sites with significant cultural or natural heritage. Includes essays on the preservation of both architectural and natural treasures. The motto here is "To save our heritage is to save the Earth." HR.

1290. Fisk, Erma J. *The peacocks of Baboquivari*. NY: Norton, 1987, c1983. 284 p. 0393304191
A 73 year old spends five months living alone in a cabin at the foot of the Boboquivari Mountains banding birds for the Nature Conservancy. For senior high school students through senior citizens. #

1291. Flink, Charles A., and Searns, Robert M. *Greenways : a guide to planning, design, and development*. Washington: Island, 1993. 351 p. : ill., maps. 1559631376
Greenways are protected open spaces that preserve nature in cities, suburbs, and rural areas. This how-to book covers the physical development of greenways, organizing and maximizing community resources and volunteers, forging partnerships, and principles of ecological design.

1292. Fox, Stephen R. *The American conservation movement : John Muir and his legacy*. Madison: Univ. of Wisconsin, 1985, c1981. 436 p. 0299106349
Original title: *John Muir and his legacy*.
Describes the origins of the conservation movement, focusing on the split between Muir's preservation philosophy and the so-called conservation for

use approach favored by Pinchot. Traces major issues and personalities in the movement from 1890 to 1975.

1293. Friedman, Mitch, and Lindholdt, Paul. eds. *Cascadia Wild : protecting an international ecosystem.* Bellingham, WA: Greater Ecosystem Alliance, 1993. 192 p. : ill., maps. 0939116359
A natural and cultural history of the North Cascades area covering both exploitation and conservation. Analyzes eco-politics in B.C.; threats from mining, logging, grazing, and pollution; the failure of single-species game management and loss of biodiversity; and the importance of conservation biology and alternative forestry practices.

1294. Frome, Michael. *Conscience of a conservationist : selected essays.* Knoxville: Univ. of Tennessee, 1989. 285 p. 0870496026
Relates conservation to the Southern Appalachians, forestry, ethics, pacifism, education, social justice, freedom of expression, and Henry David Thoreau. HR.

1295. Grove, Noel, and Krasemann, Stephen J. *Preserving Eden : the Nature Conservancy.* NY: Abrams, 1992. 176 p. : col. photos. 0810936631
For over 40 years the Nature Conservancy has purchased and managed nearly four million acres of wild and semi-wild land to protect natural habitats and foster biodiversity. Depicts the work of the Conservancy and the lands it protects. AOT. #

1296. Hornaday, William Temple. *Our vanishing wildlife: its extermination and preservation.* Salem, NH: Ayer, 1970, c1913. 411 p. : ill. 0405026749
AR:*Camp-fires on desert and lava, The American natural history.*
Hornaday demonstrated not just that wildlife was being eradicated, but that America's game laws, bounties, and other laws were direct contributors. Although primarily of historical value now, this book remains HR.

1297. Houle, Marcy Cottrell. *Wings for my flight : the peregrine falcons of Chimney Rock.* Reading, MS: Addison-Wesley, 1991. 187 p. 0201577062
Two young naturalists' efforts to save the endangered peregrine falcon initially arouse hostility from local people wishing to develop the area into a tourist site, but eventually the townspeople come to accept both them and the falcons. Particularly for activists facing similar resistance, and young people considering a career in the field. HR. #

1298. Hummel, Monte, and Pettigrew, Sherry. *Wild hunters : predators in peril.* Niwot, CO: Roberts Rinehart, 1992, c1991. 251 p. : ill., maps. 1879373270
AR: *Endangered spaces.*
Examines six large predators of Canada: grey wolf, polar bear, black bear,

grizzly bear, cougar, and wolverine. Considers their life histories, status, and threats to them, and develops an action plan for their survival. HR.

1299. Huth, Hans. *Nature and the American : three centuries of changing attitudes.* New ed. Lincoln: Univ. of Nebraska, 1991, c1957. 296 p. 0803272472
Traces American values from the utilitarian mode of settlers, to an emerging conservation consciousness arising from more free time and recreational activity in the late 19th century, to the present day. Shows how literature, science, and the arts contributed to the rise of conservationism. HR.

1300. Huxley, Anthony Julian. *Green inheritance : the World Wildlife Fund book of plants.* NY: Four Walls Eight Windows, 1992, c1985. 193 p. : col. photos. 0941423700
The goal of this book is to show, before it is too late, how rewarding our green inheritence is to mankind.

1301. Instit. for Community Economics. *The Community land trust handbook.* Emmaus, PA: Rodale, 1982. 230 p. : ill., maps. 0878574018 0878574395 (pbk)
A CLT is a non-profit corporation that can buy, sell, and hold land, generally for preservation and restoration. This practical book shows how to organize a CLT and how to recognize and avoid potential problems. Contains case studies. Despite its age, it is still relevant, valuable, and recommended.

1302. Kane, Hal, and Starke, Linda. *Time for change : a new approach to environment and development.* Washington: Island, 1992. 141 p. 1559631562 1559631554 (pbk)
By explaining the inextricable links among issues as seemingly unrelated as energy, population growth, transportation, women's issues, industry, housing, and health care, this book explores many aspects of sustainable development. Written for the 1992 Earth Summit, but still very useful.

1303. Katakis, Michael. ed. *Sacred trusts : essays on stewardship and responsibility.* SF: Mercury House, 1993. 281 p. : ill. 1562790560
Contains 30 essays by Wendell Berry, Gary Nabhan, Frederick Turner, Mary Catherine Bateson, Yvon Chouinard and others on stewardship, our moral and ethical responsibility to care for the Earth and preserve natural resources. Explores the fine line between interacting and interfering with nature. HR.

1304. King, Jonathan. *The Northwest greenbook : a regional guide to protecting and preserving our environment.* Seattle: Sasquatch, 1991. 209 p. : ill. 0912365412
Includes information on mitigating the negative effects of population growth, logging, toxics, and water and air pollution. Offers advice on the best methods of recycling and on evaluating and purchasing green products.

1305. Kreissman, Bern, and Lekisch, Barbara. *California, an environmental atlas & guide*. Davis, CA: Bear Klaw, 1991. 255 p. : maps. 0962748994
Details the geography, resources, ecology, natural history and environments of California. Given the complexity of the California environment and the plethora of various natural areas and bureaus governing them, this is very useful for conservationists, resource planners, activists, scientists, and travelers alike. HR.

1306. Leopold, Aldo. *Aldo Leopold's wilderness : selected early writings*. Harrisburg, PA: Stackpole, 1990. 250 p. : ill. 0811718646
Contains 26 selections, most written 1916-1925, divided into sections entitled, The road to restoration: game hogs, varmints, and refuges; Game management: a new science; The big picture: understanding the importance of land health; and Afterthoughts: reflections from afar.

1307. ___. *The river of the mother of God and other essays*. Madison: Univ. of Wisconsin, 1991. 384 p. 029912760X
Successfully integrates several themes: wilderness, predators, wildlife management, sustainable agriculture, forest conservation, grazing, and the land ethic. HR.

1308. ___. *A Sand County almanac, and sketches here and there*. NY: Oxford, 1987, c1949. 228 p. : ill. 0195053052
This classic introduces the "land ethic," considers humanity's duty to preserve the natural world, takes us through the passage of the seasons in Wisconsin, and provides philosophical asides. HR.

1309. Little, Charles E. *Hope for the land*. New Brunswick, NJ: Rutgers, 1992. 228 p. 0813518024
The need for a national land-use policy is advanced. We are urged to live by ethics and defend the integrity of the whole landscape, to get in touch with the spirit of the land. Describes how to perform land-use analysis.

1310. Livingston, John A. *The fallacy of wildlife conservation*. Toronto: McClelland & Stewart, 1981. 117 p. 0771053355
Why don't current government wildlife and resource conservation attempts work? Because the oxymorononic "conservation for use ethic" and political considerations almost inevitably favor immediate economic expansion.

1311. Marsh, George Perkins. *The Earth as modified by human action*. St. Clair Shores, Mich: Scholarly, 1976, c1874. 656 p. 0403001986
Original title: *Man in nature*, 1864.
Marsh was an early and highly influential advocate of the study of ecology, the practice of conservation, and stewardship. Some passages foreshadow

the Gaia concept. HR.

1312. Mathews, Jessica Tuchman. ed. *Preserving the global environment : the challenge of shared leadership.* NY: Norton, 1991. 362 p. 0393029115
Contains nine essays by various authors on overpopulation; deforestation; species loss; ozone depletion; preservation; energy and climate change; the world economy and international cooperation; making a transition to a preservation-oriented world; and needed changes in U.S. and international regulations.

1313. McNamee, Thomas. *Nature first : keeping our wild places and wild creatures wild.* Boulder, CO: Roberts Rinehart, 1987. 54 p. : ill. 0911797335
The greatest obstacle to nature conservation in the national parks and wilderness is the disparity between official boundaries and biological ones. He cites Yellowstone, with its astounding ten separate agencies, as an obvious case in point. #

1314. Morine, David E. *Good dirt : confessions of a conservationist.* NY: Ballantine, 1993, c1990. 195 p. 0345381475
Morine's Nature Conservancy oriented approach to conservation is "Stay focused, buy land!" He offers both practical advice and humorous anecdotes on fund raising techniques. Recommended for conservationists and land-trust activists.

1315. Muir, John. *John Muir, in his own words : a book of quotations.* Ed. by Peter Browning. Lafayette, CA: Great West, 1988. 98 p. 0944220029
A compendium of quotations from Muir's many books and articles.

1316. ___. *A thousand-mile walk to the gulf.* SF: Sierra Club, 1991, c1916. 141 p. 0871565919
Soon after the Civil War, the youngish Muir hiked from Kentucky to Florida's Gulf coast. His account focuses mainly on nature and conservation, but also includes political and social observations on the recently war-torn land and its people. AOT. #

1317. Nash, Roderick. *American environmentalism : readings in conservation history.* 3d ed. NY: Knopf, 1990. 364 p. 0670460590
Contains key writings on all aspects of conservation and environmentalism from the last 200 years by Black Elk, Thoreau, Marsh, Olmstead, Powell, Muir, Pinchot, Roosevelt, Leopold, Stegner, Commoner, Nader, Snyder, Abbey, Foreman and many others. HR.

1318. Nat'l Wildlife Federation. *Conservation directory : a list of organizations, agencies, and officials concerned with natural resource use and management.* Wash-

ington: The Federation, annual. ca. 450 p. 0945051549 (1993)
Describes nearly 2,000 government agencies, international, national, and
regional organizations and commissions, colleges and universities with
conservation programs, state environmental agencies and citizens' groups,
and Canadian government agencies and citizen's groups. HR.

1319. Nat'l Research Council. Water Science and Technology Board. *In situ
bioremediation : when does it work?* Washington: Nat'l Academy, 1993. 207 p.
0309048966
The use of microorganisms for on-site removal of contaminants is potentially cheaper and safer than existing methods, but remains controversial.
This book analyzes bioremediation in theory and practice. Although somewhat technical, it is accessible for many adults.

1320. Nilsen, Richard. ed. *Helping nature heal : an introduction to environmental
restoration.* Fwd. by Barry Lopez. Berkeley: Ten Speed, 1991. 152 p. : ill.
0898154251
Follows a catalog format to present short background articles, book and peri-
odical reviews, how to do it guides to tools and processes, lists of organizations, etc. Very useful for all libraries, organizations, and individuals. HR. #

1321. Oelschlaeger, Max. ed. *After Earth Day : continuing the conservation effort.*
Denton: Univ. of North Texas, 1992. 261 p. 0929398408
Proceedings of a conference on twenty years of conservation politics, envi-
ronmental science, economics and corporations, environmental philoso-
phy, and religion and conservation. Authors include Robert Paehlke,
George Sessions, Dolores LaChapelle, Elinor Gadon, and twelve others.

1322. Orr, Oliver H. *Saving American birds : T. Gilbert Pearson and the founding of
the Audubon movement.* Gainesville: Univ. of Florida, 1992. 296 p. : photos.
0813011299
This biography of a major conservationist relates how he and other members of the American Ornithologist's Union changed public attitudes, pro-
moted conservation laws, and revitalized bird conservation between 1891
and 1911.

1323. Owen, Oliver S., and Chiras, Daniel D. *Natural resource conservation : an
ecological approach.* 5th ed. NY: Macmillan, 1990. 538 p. : ill., maps.
002390111X
This college-level text covers principles of conservation and ecology,
human populations, soils, food production, water pollution, forest management, wildlife extinction and management, pesticides, solid wastes, air
pollution, mining, energy, and nuclear power. Recommended as both a text
and a reference book.

1324. Pinchot, Gifford. *Breaking new ground*. Washington: Island, 1987, c1947. 522 p. : ill. 0933280505
Both an autobiography and an account of the early years of the American conservation movement under T. Roosevelt. Pinchot was initially an ally of John Muir, but over time became one of the founders of the "multiple use" national forest concept.

1325. Pittman, Nancy P. *From The Land*. Washington: Island, 1988. 478 p. : ill. 0933280653 (pbk) 0933280661
A fascinating collection of essays from *The Land*, a magazine published by the Friends of the Earth from 1941-1954. Authors include John Muir, Gifford Pinchot, Jesse Stuart, Victor Hugo, Aldo Leopold, Rachel Carson, Alan Paton, and E. B. White.

1326. Roosevelt, Theodore. *Wilderness writings*. Salt Lake City: Peregrine Smith, 1986. 292 p. 0879052198
Contains selections from several of Roosevelt's books, focusing on his wilderness expeditions and on how they molded his conservation ethic.

1327. Schaefer, Paul. *Defending the wilderness : the Adirondack writings of Paul Schaefer*. Syracuse, NY: Syracuse Univ., 1989. 273 p. : photos. 0815602367
These pieces do more than chronicle 50 years of conservation work by capturing the essence of the country itself. Describes the beauty of the Adirondacks, its wild rivers and forests, its mountains and wildlife.

1328. ___. *Adirondack cabin country*. Syracuse, NY: Syracuse Univ., 1993. 195 p. : ill. 0815602758
A collection of essays, some dating back to the 1920s, on mountain men, hunters, nature, wilderness, and preservation.

1329. Soule, Michael E. ed. *Conservation biology : the science of scarcity and diversity*. Sunderland, MS: Sinauer Associates, 1986. 584 p. : photos. 0878937943
Expanded ed. of *Conservation biology : an evolutionary-ecological perspective*. An anthology of important research papers in four parts: Ecological Principles of Conservation, Consequences of Insularization, Captive Propagation and Conservation, and Exploitation and Preservation. HR.

1330. Strong, Douglas Hillman. *Dreamers & defenders : American conservationists*. Lincoln: Univ. of Nebraska, 1988. 295 p. : photos. 0803241615 0803291566 (pbk)
Enlarged ed. of *The conservationists*. Creates a history of the environmental movement through profiles of Thoreau, Olmstead, Marsh, Powell, Pinchot, Muir, Mather, Leopold, Ickes, Brower, Carson, and Commoner. HR.

1331. Troyer, James R. *Nature's champion* : B.W. *Wells, Tar Heel ecologist.* Chapel Hill: Univ. of North Carolina, 1993. 243 p. : ill. 0807820814
Describes Wells' work on the ecology of the Carolina coastal communities, bogs, savannahs, bays, and forests and his equally important contributions to public education and conservation.

1332. Wallach, Bret. *At odds with progress* : *Americans and conservation.* Tucson: Univ. of Arizona, 1991. 255 p. : maps. 0816509174
Ranging from the turn of the century to the 1980s and from Maine to California, Wallach demonstrates how the management of public and private lands has always been influenced by competing forces to develop and to preserve. This is clearly influenced by Carl Ortwin Sauer.

1333. Westman, Walter E. *Ecology impact assessment and environmental planning.* NY: Wiley, 1985. 532 p. : ill. 0471808954
Covers ecological impact assessment as a discipline; environmental law, public policy, and decision making; summarizing and evaluating impacts; and predicting impacts on the physical environment and the biota. This upper level college text is accessible to most adults with some scientific background.

1334. Worldwatch Inst. *State of the world* : *a Worldwatch Institute report on progress toward a sustainable society.* Ed. by Lester Brown and others. NY: Norton, annual. 039331171 (1994 ed.)
Examines options for making business a friend to the environment, for preserving cultural as well as biological diversity, and for beating swords into solar cells. Considers prospects for trade in a sustainable economy. HR.

1335. Wright, John B. *Rocky Mountain divide* : *selling and saving the West.* Austin: Univ. of Texas, 1993. 275 p. : maps. 0292790791
Colorado has over 25 land trusts with 42,000 acres of private land while Utah has only one trust of 220 acres. Wright contends that this is because Coloradans have embraced a responsible way to conduct real estate business while Mormon millenialism, high birth rates, and the belief that growth equals success have created a climate opposed to land trusts.

Books for young adults:

1336. Bramwell, Martyn. *The environment and conservation.* NY: Prentice Hall, 1992. 95 p. : col. ill. 013280090X
Provides an introduction to ecological processes and to the balance of nature. Demonstrates the negative effects of pollution on human and animal life. Explores conservation methods and efforts. GR 4-8.

1337. Curtis, Patricia. *Animals and the new zoos*. NY: Lodestar, 1991. 60 p. : photos. 0525673474
AR: *Animal rights.*
Describes the world's basic habitats and shows how some zoos try to exhibit the animals in environments like their natural ones. Emphasizes how zoos have played down the "freak show" role while establishing educational, preservation, and captive breeding programs.

1338. Douglas, William O. *Muir of the mountains*. SF: Sierra Club, 1994, c1961. 105 p. : ill. 0871565056
A biography of John Muir revealing the events and ideas that shaped America's pioneer conservationist and founder of the Sierra Club. HR.

1339. Faber, Doris, and Faber, Harold. *Nature and the environment*. NY: Scribner's, 1991. 296 p. : ill. 0684190478
Examines the life stories of 26 individuals from around the world who made notable contributions as naturalists, conservationists, or environmentalists. GR 5-9.

1340. Force, Eden. *John Muir*. Englewood Cliffs, NJ: Silver Burdett, 1990. 145 p. : photos. 0382099656
A biography of the naturalist who founded the Sierra Club and was influential in establishing the national park service. GR 6-10.

1341. Harlan, Judith. *Sounding the alarm : a biography of Rachel Carson*. Minneapolis: Dillon, 1989. 128 p. : ill. 0875184073
AR GR 3-6: Kathleen Kudlinski's *Rachel Carson, pioneer of ecology* .
Traces the life and achievements of the biologist who wrote about the sea and the dangers of pesticides. GR 5-8.

1342. Henley, Don, and Marsh, Dave. eds. *Heaven is under our feet*. NY: Berkley, 1992, c1991. 292 p. : ill. 0425135462
Authors, politicians, musicians, and actors offer short selections about nature conservation, with some commenting on Thoreau's *Walden*. Sort of an MTV homage to conservation. For teens and popular culture oriented adults. #

1343. Jezer, Marty. *Rachel Carson*. NY: Chelsea House, 1988. 111 p. : photos. 155546646X
Outstanding Science Trade Book for Children, 1988.
A biography of the marine biologist and author whose writings stressed the interrelation of all living things and the dependence of human welfare on natural processes. This may be the best young adult book about Carson. HR GR 5-9.

1344. Kallen, Stuart A. *Earth keepers*. Edina, MN: Abdo & Daughters, 1993. 46 p. : ill. 1562392115
Examines the lives of three pioneer naturalists and ecologists: John Muir, Rachel Carson, and Jacques Cousteau.

1345. Patent, Dorothy Hinshaw. *Habitats : saving wild places*. Hillside, NJ: Enslow, 1993. 112 p. : photos. 0894904019
Discusses the problems created by destruction of natural habitats and suggests ways that young people can help preserve them. GR 7-10.

1346. Pringle, Laurence P. *Saving our wildlife*. Hillside, NJ: Enslow, 1990. 64 p. : ill. 0894902040
Outstanding Science Trade Book for Children, 1990.
AR: *What shall we do with the land?*
Explains the importance of wildlife and the ways in which people are trying to save various species in North America. Encourages young people to get involved, providing directory information and other practical advice. HR.

1347. Tolan, Sally. *John Muir : naturalist, writer, and guardian of the North American wilderness*. Milwaukee: G. Stevens, 1990. 68 p. : ill. 0836800990
A biography of the naturalist who founded the Sierra Club and, as an early proponent of wilderness preservation, was influential in establishing the national park system. GR 4-8. Older teens should see Eden Force's *John Muir*.

Sustainable Development

The most important component of future conservation efforts may be Sustainable Development, a practice which refutes the argument that environmental protection is inconsistent with humanity's material needs. Conservation and protection are required now to insure sufficient resources in the future. Also see the Agriculture section for books about sustainable agriculture.

Books for adults:

1348. Browder, John O. ed. *Fragile lands of Latin America : strategies for sustainable development*. Boulder: Westview, 1989. 301 p. 0813377056
Contains essays primarily by geographers and anthropologists on nature-society relations and sustainable development as seen from a cultural ecology perspective.

1349. Brown, Lester Russell. *Building a sustainable society.* NY: Norton, 1981. 433 p. 0393300277
AR: *The twenty-ninth day. Renewable energy, the power to choose.*
Offers evidence for the feasibility of alternative ecological modes of social organization and production intended to direct our values toward more equitable, sustainable ways of life. Attempts to move society from conspicuous consumption to voluntary simplicity. HR.

1350. Brown, Lester Russell; Flavin, Christopher; and Postel, Sandra. *Saving the planet : how to shape an environmentally sustainable global economy.* NY: Norton, 1991. 224 p. 0393030709
A consideration of how to create a vibrant world economy which does not destroy the ecosystem upon which it is based. Refutes the argument that economic development and environmental quality are intrinsically contradictory.

1351. Chiras, Daniel D. *Lessons from nature : learning to live sustainably on the Earth.* Washington: Island, 1992. 289 p. 1559631066 (pbk) 1559631074
AR: *Beyond the fray : shaping America's environmental response.*
Defining the concept of sustainability in economic, biological, political, and ethical contexts, Chiras explores practical ways to apply the principles of sustainability to agriculture, industry, transportation, and other aspects of modern life.

1352. Cobb, Clifford W., and Cobb, John B. *The green national product : a proposed index of sustainable economic welfare.* Lanham, MD: Univ. of America, 1993. 343 p. : ill. 0819193224 0819193216 (pbk)
The green national product would tell us whether economic activity was making us better off or worse off by adding up the "goods" and subtracting the "bads." Proposes a specific "Index of Sustainable National Welfare" and encourages the development of alternative approaches.

1353. Daly, Herman E., Cobb, John B., and Cobb, Clifford W. *For the common good : redirecting the economy toward community, the environment, and a sustainable future.* 2d ed. Boston: Beacon, 1994, c1989. 544 p. 0807047058
AR: *Economics, ecology, ethics.*
Argues that existing economic thought is obsolete. Proposes a new set of economic theories regarding land use, production, taxation, etc. Recommends replacing the GNP with a measure which would take the distribution of goods and sustainability into account. HR.

1354. Dasmann, Raymond Fredric. *Environmental conservation.* 6th ed. NY: Wiley, 1986, c1959. 486 p. 047189141X
AR: *Last horizon , ecological principles for economic development, The conservation alternative.*
This classic text outlines basic ecological processes and terms, applying

them to conservation. Considers major issues such as population, preservation, energy, and natural resource conservation. Presents "ecodevelopment," a sort of sustainable development which promotes local self-reliance. HR.

1355. Engel, J. Ronald, and Engel, Joan Gibb. eds. *Ethics of environment and development : global challenge, international response.* Tucson: Univ. of Arizona, 1990. 264 p. : ill. 0816511837 0816512639 (pbk)
Contains 21 conference papers by authorities in environmental and developmental ethics. Develops the concept of sustainability as the ethical approach to reconciling the needs of environmental conservation with economic development.

1356. Holmberg, Johan. ed. *Making development sustainable : redefining institutions, policy, and economics.* Washington: Island, 1992. 362 p. 1559632135 1559632143 (pbk)
Contains eleven essays by various authors, primarily on the developing world. Includes discussions of environmental and development issues and presents a range of policy prescriptives for governments, development agencies, and others active in formulating the agenda for sustainable development.

1357. Khavari, Farid A. *Environomics : the economics of environmentally safe prosperity.* Westport, CT: Praeger, 1993. 189 p. 027594462X
Explores the ironic tragedy that although humanity's pursuit of economic goals inflicts irreparable devastation on the Earth, long-term economic progress is not acheived. Khavari analyzes this conundrum and recommends ways to channel economic activity in an ecologically sound manner.

1358. Lebel, Gregory G., and Kane, Hal. *Sustainable development : a guide to Our common future : the report of the World Commission on Environment and Development.* Washington: Global Tomorrow, 1989. 77 p. 019282080X
A valuable companion volume for the book *Our common future.* See entry 1364.

1359. Milbrath, Lester W. *Envisioning a sustainable society : learning our way out.* Albany: SUNY, 1989. 403 p. 0791401626 0791401634 (pbk)
Provides a penetrating analysis of how we have reached the point of unsustainability, why science and technology will fail to solve these problems, and how we as a society must change in order to avoid ecological catastrophe.

1360. Plant, Christopher, and Plant, Judith. *Turtle talk : voices for a sustainable future.* Philadelphia: New Society, 1990. 133 p. 0865711852 0865711860 (pbk)

Interviews with bioregionally oriented activists, visionaries, organizers and poets—all of whom share their passion for Earth, for human communities, and for creative change. HR.

1361. Repetto, Robert C. ed. *World enough and time : successful strategies for resource management Global Possible Conference.* New Haven: Yale, 1986. 147 p. 0140087680
Sequel: *The global possible.*
Considers successful cases of population stabilization, conservation, and sustainable development, and what strategies are effective in achieving sustainability.

1362. Thayer, Robert L. *Gray world, green heart : technology, nature, and the sustainable landscape.* NY: Wiley, 1994. 352 p. : ill. 047157273X
Considers the contemporary landscape as a conceptual battleground between our love of nature and our dependence on technology. Offers architects, environmental and city planners, geographers, and activists a new view of landscape in which technologies serve rather than dominate nature.

1363. Tisdell, C. A. *Environmental economics : policies for environmental management and sustainable development.* Brookfield, VT: Elgar, 1993. 259 p. 1852786396
Covers microeconomic policies to control pollution, methods for including environmental risk in project evaluation, cost-benefit analysis for sustainable development, and strategies for global conservation. For policymakers, teachers, and students with some background in microeconomics.

1364. World Commission on Environment and Development. *Our common future.* NY: Oxford, 1987. 383 p. 019282080X
Considers "the marriage of ecology and economics," sustainable development, the world economy, population, food security, species and ecosystems, energy, industry, cities, and the interlinked issues of peace, security, development, and the environment. For a valuable guide see entry 1358. HR.

Green Movement and Deep Ecology

The principles of the interconnectedness and inherent value of all living things fueled the conservation movement. They are central to a school of thought known as Deep Ecology which has in turn inspired a broad and sometimes radical Green Movement. Some of the political failures and compromises of the 1970s led frustrated theoreticians and activists to "go deeper." This phrase is often credited to Norwegian philosopher Arne Naess, whose articles and lectures were very influential from the early 1980s to the present. His *Ecology, Community, and Lifestyle* may still be the most significant book on this subject.

Snyder's sparkling poetry and essays and Lopez' nature writing and fiction helped provide a creative spark, while Bookchin, Devall, and Sessions helped Naess lay the foundations of Deep Ecology. Contemporary views are provided by Bishop, Coleman, Dobson, McLaughlin, Mills, Tobias, and Young. Its critics include Bradford, Dickens, Kaufmann, and Rubin in this section, and Warren in Ecofeminism (1179), A more positive ecofeminist view of Deep Ecology is expressed by LaChappelle (1162) and implied by Walker (1177).

The Green Movement provides a vehicle to move Deep Ecological principles into action. Bramwell, Dunlap, Gottlieb, Manes, Marshall, Mowrey, Sale, Scheffer, Shabecoff, and Wall describe this movement and link it to earlier developments. Books by Foreman, Norton, Porritt, Rifkin, Snow, and Tokar provide a strong link to the many forms of Activism covered in the next section.

Books for adults:

1365. Bishop, Peter. *The greening of psychology : the vegetable world in myth, dream, and healing.* Dallas: Spring, 1991. 237 p. : ill. 088214345X
Presents a deep ecological approach to vegetable consciousness and its relation to the human mind through myth, folktales, dreams, and archetypal psychology. Attempts to radically change typical notions about consciousness and identity.

1366. Bookchin, Murray. *The ecology of freedom : the emergence and dissolution of hierarchy.* Rev. ed. NY: Black Rose, 1991. 446 p. 092168973X
AR: *Renewing the Earth.*
Traces the destruction of nature to the rise of patriarchy in the late Neolithic era. Only by establishing patterns of domination in society—eventually leading to the class system, sexism, and racism—could civilization subjugate nature.

1367. ___. *Remaking society : pathways to a green future.* Boston: South End, 1990. 222 p. 089608373X
Argues that environmental degradation is part and parcel with racism, sexism, classism, and nationalism. Explores how to find the roots of these interconnected crises and remedy them.

1368. ___. *Toward an ecological society.* Buffalo: Black Rose, 1991, c1980. 315 p. 0919618995

Rejects both capitalism and Marxism as inimical to the natural world. Foreshadows deep ecology by rejecting "shallow" environmentalism and calling for a deeper examination of our beliefs and practices. Like Schumacher and Sale he argues in favor of "human scale" institutions.

1369. Bookchin, Murray, and Foreman, Dave. *Defending the Earth : a dialogue between Murray Bookchin and Dave Foreman*. Boston: South End, 1991. 147 p. 0896083837
Two noted environmental philosophers and activists discuss diversity and alliance building, racism in the Green movement, the influence of institutions on the environment, and related issues. HR.

1370. Bradford, George. *How deep is deep ecology? With an essay-review on woman's freedom*. Ojai, CA: Times Change, 1989. 84 p. : ill. 0878100350
Bradford finds much good in deep ecology, but criticizes the movement—especially Earth First!—for antipathy toward people, and failure to develop a social critique. His proposed solution is ecofeminism and an agrarian social transformation.

1371. Bramwell, Anna. *Ecology in the 20th century : a history*. New Haven, CT: Yale, 1989. 292 p. 030043430
Contends that the Green movement is not a new phenomenon, but is based on the work of scientist-activists in the U.S., England, and Germany. A.k.a. *Ecology and history : the greening of the west*.

1372. Coleman, Daniel A. *Ecopolitics : building a green society*. New Brunswick, NJ: Rutgers, 1994. 236 p. 081352055X (pbk) 0813520541
Argues that although people have been wasteful, "by blaming ourselves as individuals, we let governments and corporations off the hook." Critiques the causes of the environmental crisis. Presents political strategies based on ecological responsibility, participatory democracy, environmental justice, and community action.

1373. Devall, Bill. *Simple in means, rich in ends : practicing deep ecology*. Salt Lake City: Peregrine Smith, 1988. 224 p. 0879052473
A consideration of how to live in a way which is enriching to human lives without impoverishing the Earth's resources. Demonstrates how to "explore the ecological self," engage in effective political action, and employ deep ecology principles at work and at home. HR.

1374. Devall, Bill, and Sessions, George. *Deep ecology : living as if nature mattered*. Salt Lake City: Peregrine Smith, 1985. 266 p. 0879052473
Deep ecology rejects reform, Marxist, conservationist, and other philosophies in order to create a new consciousness of nature and culture. HR

1375. Dickens, Peter. *Society and nature : towards a green social theory.* Philadelphia: Temple Univ., 1992. 203 p. 0877229686 0877229694 (pbk) Contends that Marxism is the best foundation for social theory linking humans and nature. Dismisses deep ecology, ecofeminism, the Green Movement, and the Gaia hypothesis. His critique of other movements is of greater interest than his hackneyed Marxist rhetoric.

1376. Dobson, Andrew. *Green political thought : an introduction.* NY: Routledge, 1992, c1990. 224 p. 0415090792 Considers the underlying principles of green political thought; the difference between green politics and the existing ideologies of socialism, liberalism, and conservatism; and the difference between "light green" and more radical "dark green" thinking.

1377. Dunlap, Riley E., and Mertig, Angela G. eds. *American environmentalism : the U.S. environmental movement, 1970-1990.* Philadelphia: Taylor & Francis, 1992. 121 p. 0844817309 Contains eight essays exploring changes in the movement over the last 20 years from several perspectives: national, grassroots, African-Americans, deep ecology and radical environmentalists, global, and public opinion. HR.

1378. Foreman, Dave. *Confessions of an eco-warrior.* Rev. ed. NY: Crown, 1993, c1991. 224 p. : ill. 051788058X The cofounder of Earth First! describes the rise and methods of radical environmental activism and why certain controversial tactics are not only justifiable, but essential. Describes the shifts of thinking and action required to forestall further environmental devastation. HR AMT. #

1379. Gottlieb, Robert. *Forcing the spring : the transformation of the American environmental movement.* Washington: Island, 1993. 413 p. 1559631236 Broadens the definition of the environmental movement as a quest for wilderness preservation to a discussion of the social movements that arose in response to urban and industrial forces. Combines the work of the traditional conservation movement with that of social and human health activists.

1380. Kauffman, Wallace. *No turning back : dismantling the fantasies of environmental thinking.* NY: Basic, 1994. 256 p. 0465051189 A former president of two environmental groups argues that the environmental movement has developed an irrational view of the world due to internal politics. An otherwise valuable critique is weakened by hyperbole and overgeneralization.

1381. Killingsworth, M. Jimmie, and Palmer, Jacqueline S. *Ecospeak : rhetoric and environmental politics in America*. Carbondale: Southern Illinois Univ., 1992. 312 p. 08093175098
Critiques the ways environmentalists have used rhetoric to try to influence human consciousness and behavior, rejecting dualistic "us versus them" ecospeak. Argues that the concept of sustainability may be a dialectic capable of overcoming the division of environmentalism and development.

1382. Lopez, Barry Holstun. *Crossing open ground*. NY: Vintage, 1989. 208 p. 0679721835
In 14 essays written 1978-1988, Lopez urges people to go deep within themselves for a new sense of place and self to create a "re-enchantment of the world" which is essential to our survival. Similar to Gary Snyder's work. HR AOT. #

1383. Manes, Christopher. *Green rage : radical environmentalism and the unmaking of civilization*. Boston: Little, Brown, 1991. 291 p. 0316545325
A tightly argued, philosophically sophisticated defense of radical environmentalism. Traces the rise of Deep Ecology to environmental-political failures in the 1970s. HR.

1384. Marshall, Peter H. *Nature's web : rethinking our place on Earth*. NY: Paragon House, 1994. 528 p. 1557786526
Studies the role of religion, philosophy, and science in the development of ecological thought over time. He examines deep ecology, social ecology, and ecofeminism in presenting his vision of "libertarian ecology."

1385. McLaughlin, Andrew. *Regarding nature : industrialism and deep ecology*. Albany: SUNY, 1993. 280 p. 0791413837 0791413845 (pbk)
Contends that industrialism lies at the heart of our current ecocrisis, and that since both capitalism and socialism are both based on industrialism, a new "radical ecocentrism" must be based on deep ecology and radical economics.

1386. Meyer-Abich, Klaus Michael. *Revolution for nature : from the environment to the connatural world*. Denton: Univ. of North Texas, 1993. 145 p. 0929398696
Traces current problems to the "incomplete Enlightenment," which viewed the environment primarily as a resource for humans. Recommends a deep ecological philosophical shift. Includes recommendations for practical lifestyle changes.

1387. Mills, Stephanie. *Whatever happened to ecology?* SF: Sierra Club, 1989. 253 p. 0871566583

The autobiography of a woman who worked in the vanguard of the ecology movement in the 1970s and her attempts to apply bioregional ideals to her current life. HR.

1388. Mowrey, Marc, and Redmond, Tim. *Not in our back yard : the people and events that shaped America's modern environmental movement.* NY: Morrow, 1993. 496 p. 0688106447
Relates 99 stories about how the Green movement is a grassroots effort, with individuals or groups of neighbors confronting local environmental hazards and then networking with others. Portrays both victories and losses due to compromise. Critiques the failure of "nimby" activists to address global population and energy problems.

1389. Naess, Arne. *Ecology, community, and lifestyle : outline of an ecosophy.* NY: Cambridge Univ., 1989. 223 p. : ill. 0521344069
Naess coined the term "deep ecology" to distinguish between environmental reforms which maintain an affluent elite and the more radical fundamental equality of all species. Introduces "ecosophy," which maintains the uniqueness of humans while still recognizing the unity of all living things. HR.

1390. Norton, Bryan G. *Toward unity among environmentalists.* NY: Oxford, 1994, c1991. 287 p. 0195093976
Argues in favor of environmentalists of "the moralist and pragmatic stripes" working together in order to change not only industrial practices, but the way we think about the Earth.

1391. Pepper, David; Perkins, John; and Youngs, Martyn. *The roots of modern environmentalism.* London: Routledge, 1990, c1986. 246 p. 0685303365
Offers a provocative critique of environmental degradation in capitalist states and a less convincing argument that the former Soviet Union's poor environmental record is not a refutation of Marxist ecology since it was never a truly communist state.

1392. Porritt, Jonathon. *Seeing green : the politics of ecology explained.* NY: Blackwell, 1985. 252 p. 0631138935
This compelling introduction to the Green Movement provides a good overview of environmental politics along with philosophical and practical suggestions for change in our dominant political idealogies. HR.

1393. Rifkin, Jeremy. *Biosphere politics : a cultural odyssey from the middle ages to the new age.* SF: Harper, 1992. 388 p. 0062506951
Sets a new course for science, economics, and politics grounded in humanity's newfound responsibilites to a living planet. Asks people to journey beyond suburban culture, profligate consumption, the global shopping

center, and escalating resource wars into a new biosphere world.

1394. Rubin, Charles T. *The green crusade : rethinking the roots of environmental-ism.* NY: Free, 1994. 320 p. 0029275253
Attributes the rise in ecological consciousness almost entirely to a small group of writers and activists, denying that the general public may have noted environmental problems on its own. Critiques leaders for "careless use of science" and totalitarian solutions. Makes some thoughtful points among many banal generalizations.

1395. Sale, Kirkpatrick, and Foner, Eric. *The green revolution : the American environmental movement, 1962-1992.* NY: Hill & Wang, 1993. 124 p. 0809052180
Tracing its impetus to the publication of *Silent Spring*, Sale divides the movement into four periods: sixties seedtime, doomsday decade, Reagan reaction, and endangered Earth. Profiles leading organizations and individuals, and traces successes, failures, and new challenges.

1396. Scheffer, Victor B. *The shaping of environmentalism in America.* Seattle: Univ. of Washington, 1991. 249 p. : ill. 029597060X
AR: *A voice for wildlife, The year of the whale, The year of the seal, The seeing eye.* Covers the philosophical and historic roots of environmentalism, the rise of grass-roots organizations, trends in law and legislation, and setbacks of the Reagan years.

1397. Shabecoff, Philip. *A fierce green fire : the American environmental movement.* NY: Hill & Wang, 1993. 352 p. : photos. 0809084597
Traces the rise of the American environmental movement from colonial times through Thoreau, Parkman, and Marsh in the 19th century, the revitalization provided by Rachel Carson, the advances of the seventies, and the setbacks of the Reagan-Bush era.

1398. Snow, Donald. *Inside the environmental movement : meeting the leadership challenge.* Washington: Island, 1992. 295 p. 1559630272 1559630264 (pbk)
Presents the findings of a two-year study by the Conservation Leadership Project analyzing over 10,000 groups. Recommends a renewed focus on grassroots activity, greater minority participation, and merging the spirit of volunteerism with the benefits of professionally managed organizations.

1399. ___. ed. *Voices from the environmental movement : perspectives for a new era.* Washington: Island, 1992. 237 p. 1559631333 1559631325 (pbk)
Addresses leadership concerns including politics, ethics, science, academia, women, minorities, and the international community. Calls for a transformation of leadership. Includes essays by C. Jon Roush, Nathaniel

Reed, Joanna Underwood, Charles Jordan and Donald Snow, Sally Ranney, Aldemaro Romero, Daniel Simberloff, James Crowfoot, and Jack Lorenz.

1400. Snyder, Gary. *The old ways : six essays.* SF: City Lights, 1977. 96 p. 0872860914
Resurrects the wisdom and skill of those who studied the universe first hand for millenia, both inside and outside of themselves: the Old Ways. Envisions a future planet based on ancient wisdom and a "reinhabitation" of the land. HR AMT. #

1401. ___. *The practice of the wild : essays.* SF: North Point, 1990. 190 p. 0865474540
AR: Snyder's poetry, esp. the Pulitzer winning *Turtle Island*, *Axe handles*, *Earth household*, and *No nature.*
In these powerful and lyrical essays Snyder describes how to reattain unity with nature through understanding, loving, respecting, and living on the land. By doing so we can restore the ecological balance of the planet and experience a higher level of peace and consciousness. HR AMT. #

1402. Snyder, Gary, and McLean, William Scott. *The real work : interviews & talks,* 1964-1979. NY: New Directions, 1980. 189 p. 0811207609
A fascinating anthology of interviews and lectures on a wide range of related topics: nature, philosophy, Buddhism, Christianity, poetry, ecology, the power of legend and myth, American Indian thought, living on the land, activism, etc. HR AMT. #

1403. Tobias, Michael. *Deep ecology.* San Diego: Avant, 1988. 296 p. 0932238130
AR: *After Eden.*
An intentionally interdisciplinary anthology of deep ecology writings by Gary Snyder, Arne Naess, Paul Shepard, Murray Bookchin, Roderick Nash, Dolores LaChapelle, and others. Davis recommends this as a companion to Devall and Session's *Deep ecology*, entry 1374. HR.

1404. Tokar, Brian. *The green alternative : creating an ecological future.* 2d ed. San Pedro, CA: Miles, 1992. 183 p. 0936810238
Focuses on the development of the Green movement in North America. Explores ecophilosophical principles, grassroots activism, political organizing, bioregionalism, and related social justice issues.

1405. Wall, Derek. ed. *Green history : a reader in environmental literature, philosophy, and politics.* NY: Routledge, 1994. 273 p. 0415079241 041507925X (pbk)
Traces ecological writing through history via excerpts from the works of Blake, Thoreau, Percy and Mary Shelley, Muir, Orwell, Huxley, Emma Gold-

man, Alice Walker, Lewis Mumford, and many others. Early accounts of environmental and ecological problems are offered. HR.

1406. Young, John. *Sustaining the Earth : the past, present & future of the green revolution.* Cambridge: Harvard, 1991. 235 p. 0674858204
Original title: *Post environmentalism.*
"Post environmentalism" is a synthesis of political, economic, and moral idealogies meant to maintain a healthy planet with sustainable resources without diminishing democratic values and institutions.

Activism and Direct Action

It's hard to imagine how there could be a rebel without a cause these days given so many good causes. Directories to environmental groups include *The Harbinger File, Your Resource Guide to Environmental Organizations, People of Color Environmental Groups, The California Environmental Directory,* and *The Activist's Almanac.* Davis, Hunter, Scarce, Zakin, and Zisk offer analyses or personal accounts of some of these organizations. Biographies or interviews are provided by Carty, List, Loeffler, Mowat, and Wallace. Buzzworm's *Earth Journal* and Wild's *Earth Care Annual* review the year's work in the field and also provide directory information.

Caldicott, Canby, Corson, Dashefsky, Day, Johnson, Naar, Porritt, Rothkrug, Steger, and Wann describe a wide spectrum of actions. Bregman, Erickson, Fogelman, Gorczynski, Jorgensen, and Tanner demystify working through government regulatory channels. For those who have concluded that working through the government is either only a partial solution or a complete waste of time, Foreman and Haywood's controversial but practical *Ecodefense: A Field Guide to Monkeywrenching* suggests more direct approaches.

Other authors describe environmental action in specific areas: Freudenburg, Goldman, and Haun on health issues and/or toxic wastes; Caldicott and Gyorgy on nuclear power and weapons; Gorman and Thorne-Miller on oceans; and Echeverria on rivers and hydroelectric power.

Racial minorities are often confronted with a cruel double bind: poverty and high unemployment rates combined with more toxic waste sites and polluting industries in their neighborhoods than in more affluent ones. The emerging environmental equity movement to correct this injustice is profiled by Bullard, Hochrichter and Szasz. Also see the Native American chapter.

Dehr, Earthworks Press, Elkington, Goodman, Lewis, McVey, Miles, Pringle, and Wheeler provide action overviews for young adults. Anderson, Brown, Hamilton, Gay, and Landua describe various groups young people can join, while Preusch and Sailer profile young environmentalists. Newton focuses on pollution abatement, Hirschi on oceans and forests, and Aylesworth on government.

Books for adults:

1407. Baldwin, J. ed. *Whole Earth ecolog : the best of environmental tools & ideas.* NY: Harmony, 1990. 128 p. : ill. 0517576589
Contains reviews of products, services, software, books, and periodicals. Practically applies ecological principles to everyday life through simplicity, appropriate technology, community activism, and bioregionalism. This offshot of the *Whole Earth catalog* is updated quarterly by the *Whole Earth review.* HR. #

1408. Bregman, Jacob I., and Mackenthun, Kenneth Marsh. *Environmental impact statements.* Boca Raton, FL: Lewis, 1992. 279 p. 0873714938
Describes the purposes, legal basis, process, and public participation aspects of EISs. Covers their application to resources, climatology and floodplains, biology, surface water, groundwater, air, noise, hazards and nuisances, historic and cultural resources, transportation, and socioeconomics.

1409. Bullard, Robert D. ed. *Confronting environmental racism : voices from the grassroots.* Boston: South End, 1993. 259 p. : maps. 0896084477 0896084469 (pbk)
Conference papers exploring the disproportionate effects of industrial dumping, toxic landfills, uranium mining, waste incinerators, etc. on poor people and people of color. Attempts to intensify the development of environmentalism among minority groups and to shift policies away from primarily protecting the affluent.

1410. ___. *Dumping in Dixie : race, class, and environmental quality.* 2d ed. Boulder: Westview, 1994, c1990. 195 p. : photos, maps. 0813319633
Equity, fairness, and the struggle for social justice are among the topics covered.

1411. ___. *People of color environmental groups.* Atlanta: Environmental Justice Resource Center, biannual. 200 p. : ill.
Includes descriptions of organizations with a strong focus on people of color in the environmental movement and environmental equity. A great networking tool.

1412. ___. ed. *Unequal protection : environmental justice and communities of color.* SF: Sierra Club, 1994. 400 p. 0871564505
Contains essays by lawyers, academics, activists, and journalists demonstrating that minority communities bear the negative economic, health,

and quality of life impacts of environmental problems to a much higher degree than white communities. Environmental equity is proposed as a solution.

1413. Buzzworm Magazine. *Earth journal : environmental almanac and resource directory*. Boulder, CO: Buzzworm, annual. 447 p. : ill. 0960372253 (1993 ed.) This state-of-the-planet report contains the year in review, Earth issues, regional reports, arts and entertainment, the ecohome, ecotravel, ecoconnecting, and the *Whole Earth digest*. HR.

1414. Caldicott, Helen. *If you love this planet : a plan to heal the Earth*. NY: Norton, 1992. 231 p. 0393030458
AR: *Missile envy*.
Discusses such hazards as ozone depletion and the greenhouse effect, atmospheric degradation, deforestation, toxic pollution, species extinction, overpopulation, first world wealth vs. third world debt, and public apathy. Demonstrates how to help remedy these problems through lifestyle changes and activism. AOT. #

1415. ___. *Nuclear madness : what you can do!* Rev. ed. NY: Norton, 1994, c1982. 240 p. 0393036030
Provides a concise overview of the dangers of nuclear energy with important information on radiation, the connection between nuclear energy and weapons, nuclear waste, plutonium, mutual assured destruction, and anti-nuclear organizing tactics. HR for activists.

1416. Canby, Thomas Y. *Our changing Earth*. Washington: Nat'l. Geographic, 1994. 199 p. : col. photos. 0870449109
Explores problems and solutions, emphasizing the role of ordinary citizens as well as scientists in reversing environmental degradation, restoring ravaged areas, and protecting the Earth. International in scope.

1417. Caplan, Ruth. *Our Earth, ourselves : the action-oriented guide to help you protect and preserve our planet*. NY: Bantam, 1990. 340 p. : ill. 0553348574
Includes background information and "what we can do" about global warming, the ozone layer, air pollution, source reduction and recycling, the nuclear threat, oceans, and soil. Includes an extensive section on becoming an environmental activist. #

1418. Carty, Winthrop P., and Lee, Elizabeth. *The rhino man and other uncommon environmentalists*. Washington: Seven Locks, 1992. 177 p. : photos. 0929765109
The inspiring stories of some of the women and men who comprise the UN Global 500 Roll of Honor. Includes an appendix listing all winners and a

nomination form. Recommended for all ages, particularly young people considering an environmental career. #

1419. Corson, Walter Harris. *The Global ecology handbook : what you can do about the environmental crisis.* Boston: Beacon, 1990. 500 p. : ill. 0867085013
Companion to the PBS series, *Race to save the planet.*
Demonstrates how individuals can help solve the problems of overpopulation, overdevelopment, toxic foods, loss of biological and habitat diversity, overcutting rain forests, toxic wastes, etc. Outlines a plan for a sustainable future.

1420. Dashefsky, H. Steve. *Environmental literacy : everything you need to know about saving our planet.* NY: Random House, 1993. 298 p. : ill. 0679412808
Includes 1,100 one to five paragraph entries on a wide range of subjects including environmental activism, government agencies, law, conservation strategies, hazards, and natural processes. Explains opposing or varied points of view on many issues. HR.

1421. Davis, John. ed. *The Earth first! reader : ten years of radical environmentalism.* Salt Lake City: Peregrine Smith, 1991. 272 p. 0879053879
Contains 40 articles from *Earth First! journal* on protest and direct action, green politics, wilderness preservation, land use, biodiversity, deep ecology, ecospiritualism, and related subjects. Authors include Abbey, Snyder, Mills, Deval, Nabhan, and others. HR.

1422. Day, David. *The environmental wars : reports from the front lines.* NY: St. Martin's, 1990. 310 p. 0312044186
Original title: *The Eco wars : true tales of environmental madness* .
Details environmental abuses and activist responses encompassing every position from terrorism to pacifism. Covers murders of conservationists, endangered species traffic, animal welfare, interspecies communication, use of animals in war, anti-extinction campaigns, biological weapons, chemical pollution, deforestation, desertification, and the antinuclear movement.

1423. Deal, Carl. ed. *The Greenpeace guide to anti-environmental organizations.* Berkeley: Odonian, 1993. 110 p. 1878825054
An expose of over 50 industrial groups which establish "environmental" organizations to drain money and time from the public, sponsor misleading legislation and initiatives to scuttle real environmental legislation, and spread misinformation. Find out who they are and who funds them. HR AOT. #

1424. Echeverria, John D.; Barrow, Pope; and Roos-Collins, Richard. *Rivers at risk : the concerned citizen's guide to hydropower.* Washington: Island, 1989. 217

p. : ill. 0933280831 0933280823 (pbk)
Describes how small hydroelectric projects threaten wild rivers and land with impoundment, diversion, road access, and other destruction. Provides background on this threat and offers tools for fighting small scale hydropower. HR.

1425. Erickson, Paul A. *A practical guide to environmental impact assessment.* San Diego: Academic, 1994. 268 p. : ill. 0122415558
A practical guide to the technical and scientific concepts addressed in any comprehensive assessment of project-mediated impacts on the environment. Doesn't assume that the reader is professionally trained in any particular natural or social sciences discipline.

1426. Fogleman, Valerie M. *Guide to the* National Environmental Policy Act : *interpretations, applications, and compliance.* NY: Quorum, 1990. 309 p. 0899304869
This book is intended to help agency personnel, contractors and consultants, public interest groups, and environmental activists interpret the NEPA. Some legal background helpful, but not required.

1427. Foreman, Dave, and Haywood, Bill. *Ecodefense : a field guide to monkey-wrenching.* Exp. 3d ed. Tucson: Abzug, dist. by Ned Ludd, 1993. 350 p. : ill. 0963775103
Includes practical, field tested information on disabling heavy equipment, removing billboards, tree spiking, stake pulling, cutting or scaling fences, propaganda, security, etc. A good book for bloodless scholars to debate, but better use can be made by rebels with a good cause. HR for radical direct action.

1428. Freudenberg, Nicholas. *Not in our backyards! : community action for health and the environment.* NY: Monthly Review, 1984. 304 p. 0853456534 0853456542 (pbk)
Contains case studies describing how people in many communities have effectively organized to oppose local environmental health hazards. Includes advice on how to take advantage of opportunities and how to avoid common problems and pitfalls.

1429. Goldman, Benjamin A. *The truth about where you live : an atlas for action on toxins and mortality.* NY: Random House, 1991. 416 p. : ill., maps. 0812918983
Includes extensive maps, tables, and text about health risks such as toxic wastes, diseases, and other health hazards in your own community or others. Provides directions on how to find government documents and other information on these and related subjects.

1430. Gorczynski, Dale M. *Insider's guide to environmental negotiation.* Chelsea, MI: Lewis, 1991. 242 p. 0873715098
Provides insight into the negotiation process, both formal and informal, private and public. Offers valuable tips on techniques such as using the media to your advantage and developing effective strategies. For industry, environmental groups, government officials, lobbyists, and others involved in negotiation.

1431. Gyorgy, Anna. *No nukes : everyone's guide to nuclear power.* Boston: South End, 1979. 478 p. : ill., maps. 0896080072 0896080064 (pbk)
A definitive survey of the nuclear issue, ranging from descriptions of the inner workings of nuclear power plants, to the rise of the movement against them, to a survey of alternatives. Somewhat dated, but still valuable, particularly to anti-nuclear activists.

1432. *The Harbinger file : a directory of citizen groups, government agencies and environmental programs concerned with California Environmental issues.* Santa Cruz: Harbinger, biannual. 365 p.
The subtitle describes this perfectly. Lists more non-governmental organizations and groups than the *California environmental directory*, entry 1452. The two books complement each other well. HR.

1433. Haun, J. William. *Guide to the management of hazardous waste : a handbook for the businessman and the concerned citizen.* Golden, CO: Fulcrum, 1991. 212 p. 1555910653
Information on hazardous waste management written in clear, nontechnical language for the businessperson who needs to know the law and for the citizen who needs general information or a guide to action. HR.

1434. Hofrichter, Richard, and Gibbs, Lois Marie. eds. *Toxic struggles : the theory and practice of environmental justice.* Philadelphia: New Society, 1993. 260 p. 0865712700 1550922130
Documents the environmental justice movement led by the very people who suffer most from corporate ecological devastation: people of color, women, and low-income, working-class populations. Contributors include Robert Bullard, Winona LaDuke, Ynestra King, Cesar Chavez, and others. Essential for environmental equity activists.

1435. Hunter, Robert. *Warriors of the rainbow : a chronicle of the Greenpeace movement.* NY: Holt, Rinehart & Winston, 1979. 454 p. 0030437369 0030437415
A former Greenpeace president's story on its founding and development covering campaigns against whaling, nuclear weapons, and fur seal hunting as well as fiscal-organizational struggles and renewal. Due to a tenden-

cy to stereotype, this book is more valuable as an organizational history than as a consideration of the issues.

1436. Jessup, Deborah Hitchcock. ed. *Guide to state environmental programs.* 3d ed. Washington: Bureau of Nat'l. Affairs, 1994. 578 p. : maps. 0871798476
Designed to lead people easily through legal and bureaucratic environmental hoops. Each state-specific chapter contains practical information on state regulatory programs. Essential for those engaged in legal or regulatory work. HR.

1437. John, DeWitt. *Civic environmentalism : alternatives to regulation in states and communities.* Washington: CQ, 1994. 347 p. 0871879549 0871879484 (pbk)
Civic environmentalism focuses on decentralized, bottom-up initiatives using new tools to address newly recognized environmental problems. Presents three cases: reduction of agricultural chemicals in Iowa, restoration of the Everglades, and energy conservation in Colorado.

1438. Johnson, Lorraine. *Green future : how to make a world of difference.* Markham, Ont: Penguin, 1990. 231 p. : ill. 0140123016
Provides some background on environmental problems, tips for a greener lifestyle, advice on environmental groups and lobbying, and a guide to homes and appliances. The organizational and political info is primarily Canadian-oriented, but the superb section on home building and buying is also of interest to Americans. HR.

1439. Jorgensen, Eric P. *The Poisoned well : new strategies for groundwater protection.* Washington: Island, 1989. 415 p. 0933280564
Sponsored by the Sierra Club Legal Defense Fund. Describes groundwater contamination; how individuals and groups can effect policy; federal laws, rules, and programs; and case studies of local and state projects. For policymakers and activists. HR.

1440. List, Peter C. ed. *Radical environmentalism : philosophy and tactics.* Belmont, CA: Wadsworth, 1993. 276 p. 0534177905
Dave Foreman, Ed Abbey, Murray Bookchin, and members of Earth!First, Sea Shepard, Animal Liberation and similar organizations provide first-person accounts of radical environmental theory and practice. HR.

1441. Loeffler, Jack. *Headed upstream : interviews with iconoclasts.* Tucson: Harbinger House, 1989. 194 p. 0943173213
A collection of lively interviews with individuals who do not accept the myth of progress as practiced in modern America and have a strong, ecologically-based view of life. Interviewees include Abbey, Foreman, Hardin,

Nichols, Snyder, Udall and others.

1442. Mowat, Farley. *Rescue the Earth! : conversations with the green crusaders.* Toronto: McClelland & Stewart, 1990. 282 p. 0771066848
Interviews a wide range of Canadians including grass-roots activists, organizational executives, environmental educators and attorneys, philosophers, and sustainable developers. The activists tend to come off as more committed, while the organizational types seem rather compromised.

1443. Naar, Jon. *Design for a livable planet : how you can help clean up the environment.* NY: Perennial, 1990. 338 p. 0060551658
Considers garbage, toxics, water pollution, air pollution, acid rain, deforestation, global warming, radiation, renewable energy, law, direct action including civil disobedience, and personal lifestyles. Contains a directory and other practical information. HR.

1444. Porritt, Jonathon. *Save the Earth.* Atlanta: Turner, 1991. 208 p. 0563208473
A dramatic statement of the damage already done to the Earth, and an appeal to citizens of every nation to change their ways as voters, citizens, parents, and consumers.

1445. ___. *Where on Earth are we going?* London: BBC, 1992. 232 p. : ill. 0563208473
Companion volume to the BBC series. Covers finding more environmentally appropriate ways of meeting food and energy needs, revamping education and health services, organizing an effective Green movement, and providing fulfilling work in environmentally friendly industries.

1446. Rothkrug, Paul, and Olson, Robert L. *Mending the Earth : a world for our grandchildren.* Berkeley: North Atlantic, 1991. 219 p. 1556430914
Explores how to reconceive and reform consumerism, productivity, housing, food production, global alliances, disposal of toxics, defense, and energy use.

1447. Scarce, Rik. *Eco-warriors : understanding the radical environmental movement.* Chicago: Noble, 1990. 291 p. 096226833X
Introduces the politics and practices of radical environmental organizations that advocate nonviolent resistance and/or militant action against corporate and government abuse of the environment. Groups include Earth First!, Sea Shepards, Greenpeace, and the Animal Liberation Front. AOT. #

1448. Seredich, John. *Your resource guide to environmental organizations.* Irvine, CA: Smiling Dolphins, 1991. 514 p. : ill. 1879072009
Includes the purposes, programs, accomplishments, volunteer opportuni-

ties, publications, and membership benefits of 150 environmental organizations. Thorough, well organized, and HR.

1449. Steger, Will, and Bowermaster, Jon. *Saving the Earth : a citizen's guide to environmental action.* NY: Knopf, 1990. 306 p. : ill. 0394584317
An informative overview of the interaction of atmosphere, land, water, and people. Describes means of individual action, avenues of governmental action, and organizational resources.

1450. Szasz, Andrew. *Ecopopulism : toxic waste and the movement for environmental justice.* Minneapolis: Univ. of Minnesota, 1994. 216 p. 0816621748
0816621756 (pbk)
Ecopopulism is a grassroots effort to link struggles against racism, sexism, and other social problems with environmentalism. Traces the rise of ecpolulism to public concern about toxic waste disposal. Analyzes the development of the movement in its efforts to reconcile nature and human activity.

1451. Thorne-Miller, Boyce, and Catena, John. *The living ocean : understanding and protecting marine biodiversity.* Washington: Island, 1991. 180 p.
1559630647
Represents the first effort to define marine ecosystems, investigate threats to them, and concentrate on what must be done to halt further destruction. Covers all types of marine ecosystems.

1452. Trzyna, Thaddeus C., Caughman, Jennifer Trzyna, and Childers, Roberta. eds. *California environmental directory : a guide to organizations and resources.* 5th ed. Sacramento: Cal. Inst. for Public Affairs, 1993. 128 p. : maps. 0912102985
A "User's guide to who does what" related to specific subject areas. Describes U.S. government agencies in California, state, interstate, regional, and local agencies, associations and independent centers, and university research and public service programs. A good complement to entry 1432.

1453. Wallace, Aubrey. *Eco-heroes : twelve tales of environmental victory.* SF: Mercury House, 1993. 232 p. : ill. 1562790331
AR: *The new environmental handbook.*
Profiles the winners of the Golman Environmental Awards who have overcome great adversity: Wangari Maathai, Kenya; Christine Jean, France; Samuel LaBudde, U.S.; Catherine Wallace, New Zealand; Jeton Anjain, Marshall Islands; Eha Kern and Roland Tiensuu, Sweden; Michael Werikhe, Kenya; Colleen McCrory, Canada; Wadja Egnankou, Ivory Coast; Lois Gibbs, U.S.; Medha Patkar, India; Beto Ricardo, Brazil. #

1454. Walls, David. *The activist's almanac : the concerned citizen's guide to the lead-*

ing advocacy organizations in America. NY: Simon & Schuster, 1993. 431 p. : ill.
0671746340
American Libraries Outstanding Reference Source, 1994.
Provides the name, address, phone number, e-mail address, history, funding, tax status, publications, political action committees, foundations,
accomplishments, and goals of 105 organizations. Includes conservative as
well as liberal organizations. HR.

1455. Wann, David. *Biologic : environmental protection by design*. Boulder: Johnson, 1990. 284 p. : ill. 1555660487
An EPA analyst presents a plan to design our economy like a mature
ecosystem instead of an immature one, arguing that new ideas from many
sources are essential.

1456. Wild, Russell. ed. *The Earth care annual*. Emmaus, PA: Rodale, annual.
Sponsored by the National Wildlife Foundation. Covers the year's work in
environmental action relating to coral reefs, endangered species, garbage
and sewage, the greenhouse effect, ozone depletion, pesticides, tropical
forests, urban ecology, waterways, and wilderness. HR.

1457. Zakin, Susan. *Coyotes and town dogs : Earth! First and the environmental
movement*. NY: Viking, 1993. 483 p. 0670836184
Demonstrates how Earth! First arose due to the failure of lobbying and
other legal action. Describes such actions as putting the "crack" on Glen
Canyon Dam, the fight over Black Mesa, and tree spiking. Analyzes the
modern environmental movement and the relations between Earth! First
and more moderate groups such as the Sierra Club. HR.

1458. Zisk, Betty H. *The politics of transformation : local activism in the peace and
environmental movements*. Westport, CT: Praeger, 1992. 256 p. 0275940578
Surveys seven "pacesetter" areas (LA, SF, Boston, Portland, Or., western
Massachusetts, and coastal Maine) to determine the reasons for their success in citizen activism, concensus building, cooperation, strategies, goal
setting, and accomplishments.

Books for young adults:

1459. Anderson, Joan. *Earth keepers*. San Diego: Harcourt Brace, 1993. 1 v.
(unpaged) : ill. 0152421998
Discusses the work of three environmental protection groups who are striving to help save the Earth from destruction. GR 3-7.

1460. Aylesworth, Thomas G. *Government and the environment : tracking the*

record. Hillside, NJ: Enslow, 1993. 104 p. : photos. 0894904000
Describes the many aspects of environmental pollution, the growing
awareness of the problem, and the role of the federal government in formulating a policy to protect the environment. GR 7-10.

1461. Brown, Michael Harold, and May, John. *The Greenpeace story*. NY: Dorling Kindersley, 1991. 192 p. : col. photos. 1879431025
An overview of 20 years of effective, nonviolent, but sometimes confrontational activism. Covers campaigns to save whales, dolphins, fur seals, and
stop the dumping of toxic and radioactive waste in the ocean. #

1462. Dee, Catherine. *Kid heroes of the environment*. Berkley: Earth Works,
1991. 96 p. : ill. 1879682125
Profiles 25 young people, ages 7-17, who have worked at home, school, in
the community, and nationally to make positive contributions to the environment. Details what they did, how they did it, and what others can do.
Provides directory information for 15 groups children and young adults can
join and 8 honoring "kid heroes." HR for GR 4-12, teachers, and parents. #

1463. Dehr, Roma, and Bazar, Ronald M. *Good planets are hard to find! : an
environmental information guide, dictionary, and action book for kids and adults*. NY:
Firefly, 1990. 39 p. : col. ill. 0919597092
Includes short definitions and action ideas on environmental terms and
such related concepts as cooperation, love, joy, and war. GR 4-10 (particularly reluctant readers) parents, and teachers. #

1464. Earth Works Group. *50 simple things kids can do to save the Earth*. Kansas
City: Andrews & McMeel, 1990. 156 p. : ill. 0836223012
Explains how specific things in a child's environment are connected to the
rest of the world, how using them affects the planet, and how the individ-
ual can develop habits and projects that are environmentally sound.
Loaded with practical tips. HR.

1465. Elkington, John. *Going green : a kid's handbook to saving the planet*. NY:
Puffin, 1990. 111 p. : ill. 0140345973
Notable Children's Trade Book in Social Studies, 1990.
A guide to saving the environment, including explanations of ecological
issues and projects. Shows how to do a "green audit" of your home and
community. HR GR 3-8.

1466. Gay, Kathlyn. *Caretakers of the Earth*. Hillside, NJ: Enslow, 1993. 104 p.
: photos. 0894903977
Describes various ways both individuals and groups can get involved in
helping protect the environment both locally and globally. GR 6-10.

1467. Goodman, Billy. *A kid's guide to how to save the planet*. NY: Avon, 1990. 137 p. : ill. 038076041X
Provides more background information on natural history, ecological process, and environmental problems than most books of its type and covers successful environmental actions by young people. Each chapter concludes with "What you can do." HR GR 4-8.

1468. Gutnik, Martin J. *Recycling : learning the four R's, reduce, reuse, recycle, recover*. Hillside, NJ: Enslow, 1993. 104 p. : ill. 0894903993
Describes different types of trash, where they come from, their detrimental impact on the environment, and ways recycling can help solve the problem. GR 6-10.

1469. Hamilton, John. *ECO-groups : joining together to protect the environment*. Edina, MN: Abdo & Daughters, 1993. 31 p. : ill. 1562392107
Provides information about the history and purpose of such environmental groups as the Sierra Club, the Nature Conservancy, Greenpeace, and Kids for Saving Earth. GR 3-8.

1470. Hirschi, Ron. *Save our oceans and coasts*. NY: Delacorte, 1993. 72 p. : photos. 0385310773 0385311265 (pbk)
Discusses the characteristics, animal life, and importance of oceans and coastal areas and ways to protect these habitats. GR 3-8.

1471. ___. *Save our forests*. NY: Delacorte, 1993. 72 p. : photos. 0385310765 0553372394 (pbk)
Discusses the characteristics and importance of different kinds of forests and the life they support. Stresses the importance of preserving forests. GR 3-8.

1472. Landau, Elaine. *Environmental groups : the Earth savers*. Hillside, NJ: Enslow, 1993. 112 p. : photos. 0894903969
Describes various organizations which actively work to protect the environment, including the National Audubon Society, Greenpeace, Rainforest Alliance, and Sierra Club. GR 7-10.

1473. Lewis, Barbara A., and Espeland, Pamela. *The kid's guide to social action : how to solve the social problems you choose, and turn creative thinking into positive action*. Minneapolis: Free Spirit, 1991. 185 p. : ill. 0915793296
Resource guide for children for learning political action skills that can help them make a difference in solving social problems at the community, state, and national levels. Offers a strong, ten-step process covering lobbying, letter writing, public relations, etc. HR GR 5-10.

1474. McVey, Vicki. *The Sierra Club kid's guide to planet care and repair.* Ill. by Martha Weston. SF: Sierra Club, 1993. 84 p. : ill. 0871565676
Explains how human activities are destroying the balance of nature and suggests ways to prevent further damage. Includes profiles of young people who have made a difference. HR GR 4-8.

1475. Miles, Betty. *Save the Earth : an action handbook for kids.* Rev. ed. NY: Knopf, 1991. 118 p. : ill. 0678917314
An overview of the environmental problems of land, atmosphere, water, energy, plants, animals, and people. Includes projects and advice on becoming an environmental activist. HR GR5-9.

1476. Newton, David E. *Taking a stand against environmental pollution.* NY: Watts, 1990. 157 p. : photos. 0531109232
Examines current environmental issues and suggests ways to become involved in solving problems created by pollution and related causes. Includes profiles of classic and contemporary conservationists, emphasizing the influence that ordinary teenagers can have. GR 7-10.

1477. Preusch, Deb; Barry, Tom; and Wood, Beth. *Red ribbons for Emma.* Stanford, CA: New Seed, 1981. 47 p. : photos. 0938678078
The inspiring story of traditional Navajo shepherd Emma Yazzie's struggle against the destruction of rural land by power companies and the corrupt tribal government. HR GR 3-8.

1478. Pringle, Laurence P. *Restoring our Earth.* Hillside, NJ: Enslow, 1987. 64 p. : photos. 0894901435
Outstanding Science Trade Book for Children, 1987.
Discusses the ecological restoration of prairies, marshes, forests, rivers, and other damaged environments of North America. Offers suggestions for people wishing to volunteer for restoration projects. HR GR 4-8.

1479. Sailer, John. *A Vogt for the environment.* Summertown, TN: Book Co., 1993. 95 p. 0913990345
Relates how teenage environmentalist Tanja Vogt convinced McDonald's and other businesses and schools in her area to stop using styrofoam. A good example of how one committed young adult can make a difference. HR for teens, parents, and teachers. #

1480. Wheeler, Jill C. *Earth kids.* Edina, MN: Abdo & Daughters, 1993. 39 p. : ill. 1562391992
Highlights the activities of several young environmentalists who are working to save our planet and suggests ways in which the reader can get involved. GR 3-7.

Environmental Education

Perhaps even less productive than the rebel without a cause is the rebel without a clue. Environmental and Outdoor Education is a prerequisite to effective action and a lifelong activity. This section contains books for teachers and other environmental educators, parents, children, and young adults. Because of the wide span of appropriate age levels and since some activities require adult supervision, there is not a separate section for young adults, but detailed notes on age levels and uses are provided.

Some material will be primarily of interest to teachers, including catalogs and sourcebooks by Acorn Naturalists, Makower, and Facts on File. Educational philosophy or critiques of environmental education are provided by Bowers, Orr, Reardon, and Van Matre. Allman, the California Dept. of Education, Charles, Hocking, and Waage cover curriculum and teaching methods. Two books from the early 20th century by Comstock and Seton still have wide appeal despite their age.

Books for parents and teachers on effective child-rearing and nurturing ecological values are provided by Carson, Cornell, Liedloff, Nabhan, and Tilsworth. Foreman correctly suggests that Robertson's *Signs Along the River* is an excellent book for parents to use with very young children or for middle readers to read on their own, preferably outdoors.

Bowden, Cook, Leslie, and O'Brien employ a seasonal approach to nature while Arnosky and Johnson demonstrate how sketching can provide both education and pleasure. MacDonald links peace and environmental education. Arthur, Bruchac, Padilla, and Wilkins provide Native American perspectives. Bruchac's books are particularly recommended.

Most children and some adults enjoy nature projects. The books by Beller, Bonnet, Bruchac, Burnie, Cohen, Cornell, Dashefsky, Herman, Rains, Rybolt, and Schwartz and the seasonal and art books mentioned above contain literally thousands of projects.

Books for various levels:

1481. Acorn Naturalists. *Science and environmental education catalog : celebrating the spirit of discovery.* Tustin, CA: Acorn Naturalists, annual. 70 p. : ill.
Lists and evaluates over 2,000 books, magazines, sound recordings, kits, games, models, tools, puppets, and instruments useful to environmental educators. For a free copy call 1-800-422-8886. For educators and group leaders at all levels, preschool to adult. HR.

1482. Allison, Linda. *The wild inside : Sierra Club's guide to the great indoors.* SF: Sierra Club, 1988. 144 p. : ill. 0316034347
Introduces the basic principles of physics, geology, weather, electricity, ecol-

ogy, and natural history through everyday activities in a house. HR GR 3-8.

1483. Allman, S. Audean; Kopp, O. W.; and Zufelt, David L. *Environmental education : a promise for the future : curriculum guidelines and activities for classroom teachers.* Boston: American, 1982. 196 p. 0896410854
Includes background information and student projects on over 120 topics and extensive resource lists.

1484. Arnosky, Jim. *Drawing from nature.* NY: Lothrop, Lee & Shepard, 1987, c1982. 64 p. : ill. 0688070752
Companion volume to the PBS series. ALA Notable Children's Book, 1982. Teaches you how to study and observe nature and provides practical advice and philosophical insights on drawing water, land, plants, and animals. HR GR 5-adult. #

1485. ___. *Sketching outdoors in all seasons.* Woodstock, VT: Countryman, 1993. 180 p. : ill. 0688062903
ALA Notable Children's Book Award.
Includes drawings of plants, animals, landscapes, and other aspects of nature in each season, with comments from the artist on how and why he drew them. Also available separately as *Sketching outdoors in spring* , *Sketching outdoors in summer*, etc. HR GR 5-adult. #

1486. Arthur, Claudeen; Bingham, Sam; and Bingham, Janet. *Between sacred mountains : Navajo stories and lessons from the land.* Tucson: Univ. of Arizona, 1984, c1982. 287 p. : ill. 0816508550 0816508569 (pbk)
Covers the history and the economic, legal, and cultural questions touching Navajo land today. Written as a cultural history and environmental reader for Navajo students, it is recommended for all high school students. #

1487. Beller, Joel. *Experimenting with plants : projects for home, garden, and classroom.* NY: Prentice Hall, 1987, c1985. 154 p. : ill. 0668059893 0668059915 (pbk)
These experiments cover basic horticulture, environmental factors of plant growth, organic gardening, hydroponics, etc. Useful for classroom or home use. For GR 5 (with assistance) to adults. #

1488. Bonnet, Robert L., and Keen, Dan. *Environmental science : 49 science fair projects.* Blue Ridge Summit, PA: Tab, 1990. 124 p. : ill. 0830673695 0830633693 (pbk)
Describes projects in environmental science suitable for the classroom or a science fair. GR 5-9, although adult assistance is required on some projects.

1489. Bonney, Bruce F., and Drury, Jack K. *The backcountry classroom : lesson plans for teaching in the wilderness.* Merrillville, IN: ICS, 1992. 272 p. : ill.

0934802181
AR: *The Wilderness educator: the Wilderness Education Association's curriculum guide.*
Covers environmental ethics and backcountry conservation practices, bathing and washing, campsite selection, expedition behavior, fires, latrines, leadership, packing, hygiene, trail technique, water safety, food preparation, navigation, and safety. Recommended for outdoor educators, recreational leaders, and recreationists.

1490. Bowden, Marcia. *Nature for the very young : a handbook of indoor and outdoor activities.* NY: Wiley, 1989. 232 p. : ill. 047162084X (pbk) 0471509752
Provides a variety of activities on a season-by-season basis, including preparatory and follow-up activities. Includes stories, songs, and pictures. Designed for nature centers or lower elementary-level teachers, this book is also recommended for parents of younger children.

1491. Bowers, C. A. *Education, cultural myths, and the ecological crisis : toward deep changes.* Albany: SUNY, 1993. 232 p. 0791412555 0791412563 (pbk)
An examination of how the educational process perpetuates cultural myths contributing to the ecological crisis. For teachers and other environmental educators. HR.

1492. Brown, Vinson. *The amateur naturalist's handbook.* NY: Prentice Hall, 1992, c1980. 448 p. : ill. 0130237396
AR: *Investigating nature through outdoor projects.*
A comprehensive field guide covering concepts of nature study; observing, studying, classifying, and collecting plants, animals, and minerals; climate; ecology; animal behavior; exploring; and nature in the city. HR for novice to intermediate naturalists. #

1493. Bruchac, Joseph*. *Native American stories.* Golden, CO: Fulcrum, 1991. 145 p. : ill. 1555910947
ALA Notable Book for Young Adults nominee, 1991.
A collection of Native American tales and myths excerpted from *Keepers of the Earth*, focusing on the relationship between humanity and nature. HR GR 4-8, for home and school use. #

1494. ___. *Native American animal stories.* Golden, CO: Fulcrum, 1992. 135 p. : ill. 1555911277
American Bookseller Pick of the Lists, 1992.
Contains 24 stories excerpted from *Keepers of the animals*. HR GR 4-8, for home and school use. #

1495. Bruchac, Joseph*, and Caduto, Michael J. *Keepers of the Earth : native American stories and environmental activities for children.* Golden, CO: Fulcrum,

1988. 234 p. : ill. 1555910270
NY State Outdoor Education Assn. Art and Literary Award, 1990.
A selection of traditional tales from various Indian peoples each accompanied by instructions for related activities dealing with aspects of the environment. HR GR 4 to adult, for home and classroom use. #

1496. ___. Keepers of the animals : Native American stories and wildlife activities for children. Golden, CO: Fulcrum, 1991. 2866 p. : ill. 1555910882
American Bookseller Pick of the Lists, 1991, Association of Children's Booksellers Choice Award, 1992.
A collection of animal stories intended to foster a greater respect, understanding, and care for animals and nature. HR GR 4 to adult, for home and classroom use. #

1497. ___. Keepers of life : discovering plants through Native American stories and Earth activities for children. Golden, CO.: Fulcrum, 1994. 288 p. : ill. 1555911862
A collection of Native American stories and activities providing lessons in botany, ecology, and human ecology. HR GR 4 to adult, school or home use. #

1498. ___. Keepers of the night : Native American stories and nocturnal activities for children. Golden, CO: Fulcrum, 1994. 160 p. : ill. 1555911773
Includes eight carefully selected Native North American stories with hands-on activities that promote responsible stewardship of the Earth and human beings through learning to understand and care for nature at night. HR GR 4 to adult, home or school use. #

1499. Burnie, David. How nature works. Pleasantville, NY: Reader's Digest, 1991. 192 p. : col. ill. 0895773910
Includes 100 projects on various natural history, biology, and ecology topics. Most can be done by teens on their own, and those requiring adult assistance are so identified. For school or home use. #

1500. Cajete, Gregory A. Look to the mountain : an ecology of indigenous education. Durango, CO: Kivaki, 1994. 243 p. : ill. 1882308654
Provides natives and non-natives alike a thoroughly researched account of modern American education from a tribal perspective. Considers the ecological orientation of indigenous education as well as its spiritual and mythic roots.

1501. Cal. Dept. of Education, and Cal. Energy Commission. Compendium for human communities : environmental education. Sacramento: The Dept., 1994. 179 p. : ill.
Evaluates 84 curriculum kits and guides at grade levels K-3, 4-6, 7-9, and

10-12. For each item it provides an overall rating; bibliographic information; report card and discipline emphasis guide; comments on general content, presentation, pedagogy, and teacher usability; and facsimilies of two pages. HR for all teachers.

1502. Carson, Rachel, and Pratt, Charles. *The sense of wonder*. NY: Harper, 1987, c1965. 89 p. 0060914505
How parents can instill a sense of wonder about nature "so indestructible that it |will| last throughout life, as an unfailing antidote against boredom...the sterile occupation with things that are artificial, the alienation from the sources of our strength."-Intro. HR for parents and teachers.

1503. Charles, Cheryl L., and Lackey, Joanna. *Project* WILD : *secondary activity guide*. Rev. ed. Boulder, CO: Project WILD, 1986. 288 p. : ill.
AR: *Project learning tree*.
An interdisciplinary, supplementary environmental and conservation education program for K-12 teachers. This guide presents a variety of nature study and environmental activities for secondary students. HR.

1504. Cohen, Joy, and Pranis, Eve. *GrowLab* : *activities for growing minds*. Burlington, VT: Nat'l. Gardening Assn., 1990. 307 p. : ill. 091587332X
Provides a variety of activities and experiments based upon a four by eight foot indoor garden. Includes sections titled plants alive, generation to generation, diversity of life, and sharing the global garden. For elementary school teachers and environmental educators.

1505. Cohen, Michael J. A *field guide to connecting with nature*. Eugene, OR: World Peace Univ., 1989. 85 p. 0939170061
Outlines 110 self-paced activities to help people use all their senses, rationality, and personal resources to understand and relate to natural phenomena and processes. Sections titled: experiencing, sensing, thinking, embracing, joining, discovering, unifying, avoiding, and spacemaking. For individual or group-use. #

1506. ___. *How nature works* : *regenerating kinship with Planet Earth*. Walpole, NH: Stillpoint, 1988. 265 p. : ill. 0913299456
Explains our profound connection with nature, how we harm ourselves when we oppose the natural order, how our separation from nature underlies many social ills and prevents us from experiencing Earth as a living organism. Includes study guides. For individuals or high school classes. AOT. #

1507. Comstock, Anna Botsford. *Handbook of nature study, with a new foreword*. Ithaca: Comstock, 1986, c1911. 887 p. : ill 0801419131

This classic guide to nature study became known as *The nature Bible* because of the sensitive advice it gave on such topics as children's attitudes toward death from predation and returning creatures to their natural habitat after study. HR. #

1508. Cook, Amber. *Nature crafts for all the seasons*. NY: Sterling, 1993. 128 p. : ill. 0806986026
Lists 52 crafts projects incorporating natural materials, including complete instructions, lists of materials, tool lists, and photographs for each. Unfortunately, it does not cover how to minimize the impact of collecting natural materials. For GR 4-adult. #

1509. Cornell, Joseph Bharat. *Sharing the joy of nature : nature activities for all ages*. Nevada City, CA: Dawn, 1989. 167 p. : ill. 0916124525
AR: *Sharing nature with children, Listening to nature.*
Contains nature games and activities designed to awaken enthusiam, focus attention, provide direct experience, and share inspiration. HR for all environmental educators.

1510. Cornell, Joseph Bharat, and Deranja, Michael. *Journey to the heart of nature : a guided exploration*. Nevada City, CA: Dawn, 1994. 128 p. : ill. 1883220068
Provides an in-depth exploration of a personally selected part of nature. Combines stories from the lives of John Muir, Jim Corbett, J. Allen Boone, and others with reader activities such as sketches, questions, maps, interviews, poetry, etc. For personal and group use. #

1511. Dashefsky, H. Steve. *Environmental science : high-school science fair experiments*. Blue Ridge Summit, PA: TAB, 1993. 177 p. : ill. 083064587X 0830645861 (pbk)
Contains 24 experiments and projects exploring the soil, aquatic ecosystems, applied ecology, global warming, deforestation, indoor pollution, alternative energy sources, and related subjects. Includes materials lists, procedures, data analysis, and suggested further research. HR GR 9-adult. #

1512. Diagram Group. *Environment on file*. NY: Facts on File, 1991. 288 p. : ill., maps. 0816026955
Contains over 280 loose-leaf, reproducible instructional charts, graphs, and images offering clear depictions and comparisons. Each chapter includes an historic introduction, background information, and important issues. HR GR 4-adult. #

1513. ___. *Nature projects on file*. NY: Facts on File, 1992. 288 p. : ill. 0816027056

Contians 70 experiments, demonstrations and projects issued in loose-leaf, easily reproducible form. Topics include the Earth, animals, ecology, energy, weather, plants, pollution, and environmental quality. HR GR 4-adult. #

1514. Duensing, Edward, and Millmoss, A. B. *Backyard and beyond : a guide for discovering the outdoors.* Golden, CO: Fulcrum, 1992. 248. : Ill. 1555910718
Explains how to observe and appreciate nature in your own local environment. Includes information on tracking and trailing, observing wildlife, and collecting plant specimens. HR. #

1515. Durrell, Gerald Malcolm, and Durrell, Lee. *The amateur naturalist.* NY: Dorling Kindersley, 1993. 320 p. : ill. 0863188702
AR: *How to shoot an amateur naturalist.*
The Durrells offer background information and practical advice on how to become an amateur naturalist and, more important, how to think like a naturalist. Describes projects and experiments performed in various environments from the city to wilderness. #

1516. Gardner, Robert. *Experimenting with energy conservation.* NY: Franklin Watts, 1992. 128 p. : ill. 0531125386
Discusses energy and its conservation and provides experiments to investigate the topic. Includes plans for six large projects such as solar cookers and a model solar home.

1517. George, Jean Craighead. *Acorn pancakes, dandelion salad & other wild dishes.* NY: HarperCollins, 1994. 182 p. : ill. 006021550X
Earlier ed. entitled: *The wild, wild cookbook.*
A field guide for finding, harvesting, and cooking wild plants, arranged by season. GR 5-9.

1518. Herman, Marina Lachecki; Passineau, Joseph F.; Schimpf, Ann L.; and Treuer, Paul. *Teaching kids to love the Earth.* Duluth, MN: Pfeifer-Hamilton, 1991. 175 p. : ill. 0938586424
Includes 186 outdoor activities meant to instill and encourage curiosity, exploration, discovery, sharing, and passion for the Earth. Each section includes stories, activities, "Did you know?" and resources. HR for parents and teachers.

1519. Hocking, Colin. *Global warming & the greenhouse effect : teacher's guide.* Berkeley: Univ. of Cal., 1992. 174 p. : ill. 0912511753
Provides information on these potentially discouraging subjects in a way that encourages student empowerment and activism. Contains guides to 48 50–minute classroom activity sessions, simulations, experiments, and

discussions. GR 6-10.

1520. Hocking, Colin; Coonrod, Janice A.; and Barber, Jacqueline. *Acid rain : teacher's guide.* Berkeley: Univ. of Cal., 1992, c1990. 163 p. : ill. 0912511745
Shows the teacher how to present the subject of acid rain in a way that encourages them to feel engaged and empowered, rather than discouraged and overwhelmed. Contains guides to 48 50–minute classroom activity sessions, simulations, experiments, and discussions. GR 6-10.

1521. Johnson, Cathy. *The local wilderness : observing neighborhood nature through an artist's eye.* NY: Prentice Hall, 1987. 175 p. : ill. 0136101712
Covers how to observe nature and develop a personal perspective toward it along with practical guidance on drawing. For artistic teens and adults. #

1522. ___. *The Sierra Club guide to sketching in nature.* SF: Sierra Club, 1990. 228 p. : ill. 0871565544
A gifted illustrator invites you to deepen your understanding and appreciation of nature through sketching and shows you how to do it. #

1523. Leslie, Clare Walker. *Nature all year long.* NY: Greenwillow, 1991. 56 p. : col. ill. 0688091830 (trade) 0688091849 (lib.)
Describes the different plants, animals, and landscapes that can be seen outdoors each month of the year. Includes experiments, crafts, and other related activities.

1524. Liedloff, Jean. *The continuum concept : allowing human nature to work successfully.* Reading, MS: Addison-Wesley, 1985, c1977. 172 p. 0201050714
Describes how to rear non-violent children by role modeling an endangered South American tribe that lives in a symbiotic relationship with the land. HR for parents and teachers.

1525. Lingelbach, Jenepher. ed. *Hands-on nature : information and activities for exploring the environment with children.* Woodstock, VT: Vermont Inst. of Natural Science, 1986. 233 p. : ill. 0961762705
Intended to enable novice leaders to teach nature study successfully and to offer creative approaches for experienced educators. Contains sections on adaptations, habitats, cycles, and designs, each with eight sets of activities. Though designed for elementary school use, it can be adapted for older students. HR for environmental educators and parents.

1526. MacDonald, Margaret Read. *Peace tales : world folktales to talk about.* Hamden, CT: Linnet, 1992. 116 p. : ill. 0208023291 (pbk) 0208023283
A collection of 34 folktales from cultures around the world, reflecting different aspects of war and peace, including peace between different nations,

peoples of different races, and a more peaceful human relationship to the Earth. Stresses cooperation, patience, and understanding. Intended particularly for GR 3-7, but HR for all levels. #

1527. Makower, Joel; Tienvieri, Nancy; and Poff, Cathryn. eds. *The Nature catalog*. NY: Vintage, 1991. 327 p. : ill. 0679733000
Covers studying nature in the wild or in the classroom through listings of sites, organizations, publications, activities, computer software and other resources. Subjects include aquariums, arboreta, astronomy, birding, caves, coasts and wetlands, deserts, earthquakes and volcanoes, ecotourism, endangered species, ecology, forests, marine mammals, oceans, weather, zoos, and much more. HR. #

1528. McVey, Vicki. *The Sierra Club wayfinding book*. SF: Sierra Club, 1989. 88 p. : ill. 0316563404
ALA Notable Children's Book Award, 1989.
Describes how humans have developed systems using their senses, landmarks, maps, navigation, and signs in the natural world to find their way around. Includes activities, games, and experiments illustrating the principles of wayfinding. HR GR 3-adult. #

1529. Nabhan, Gary Paul, and Trimble, Stephen. *The geography of childhood : why children need wild places*. Boston: Beacon, 1994. 216 p. : photos. 0807085243
Describes the psychological importance of children understanding and identifying with nature and "the wildness within us." Despite being criticized for anti-intellectual prejudices it is recommended for parents and teachers.

1530. O'Brien, Margaret, and Shepherd, Ursula L. *Nature notes : a notebook companion to the seasons*. Golden, CO: Fulcrum, 1990. 135 p. : ill. 1555910564
A functional field notebook, featuring line drawings and text focusing on plant life, wildlife and natural history. Includes periodic blank spaces to record your observations and drawings. AOT. #

1531. Orr, David W. *Ecological literacy : education and the transition to a postmodern world*. Albany: SUNY, 1992. 210 p. 0791408736 0791408744 (pbk)
Argues that ecological literacy is the core of all practical learning. Focuses primarily on sustainabilty and raising ecological awareness. Contains a syllabus for ecological literacy.

1532. Padilla, Stan. *A Natural education : Native American ideas and thoughts*. Rev. ed. Summertown, TN: Book Co., 1994. 40 p. : ill. 0913990140
A collection of quotations from traditional Native Americans on the impor-

tance of educating young people in the natural way, learning how to listen and observe directly from nature.

1533. Rainis, Kenneth G. *Nature projects for young scientists.* NY: Watts, 1989. 142 p. : ill. 0531107892
A collection of nature projects and experiments exploring five kingdoms of life ranging from bacteria to plants and animals. Contains good background information on the nature of science and on experimentation as a process of inquiry.

1534. Reardon, Betty, and Nordland, Eva. eds. *Learning peace : the promise of ecological and cooperative education.* Albany: SUNY, 1994. 234 p. : ill. 0791417565 (pbk) 0791417557
Contains ten essays by American and Russian authors on the question "What are the consequences for education of our knowledge about the problems of the planet at the end of the 20th century?" Sponsored by the Project on Ecological and Cooperative Education.

1535. Robertson, Kayo. *Signs along the river : learning to read the natural landscape.* Boulder, CO: Roberts Rinehart, 1986. 58 p. : ill. 091179722X
An excellent introduction to wildlife watching, natural history, and ecology. Teaches children how to "read" the natural environment. HR for parents, teachers, and children. #

1536. Rockwell, Robert E.; Sherwood, Elizabeth A.; and Williams, Robert A. *Hug a tree : and other things to do outdoors with young children.* Mt. Rainier, MD: Gryphon House, 1983. 106 p. : ill. 0876591055
Covers how children learn, preparing children to learn, basic skills such as observing and classifying, and activities for the classroom or field. HR for parents and teachers of preschoolers and early elementary students.

1537. Rybolt, Thomas R., and Mebane, Robert C. *Environmental experiments about land.* Hillside, NJ: Enslow, 1993. 96 p. : ill. 0894904116
Presents experiments that explore the properties and erosion of soil, recycling, and organic waste. GR 5-9.

1538. ___. *Environmental experiments about life.* Hillside, NJ: Enslow, 1993. 96 p. : ill. 0894904124
Uses experiments to explain ecosystems, life cycles, and interactions of life and environment, including pollution and conservation. GR 5-9.

1539. ___. *Environmental experiments about air.* Hillside, NJ: Enslow, 1993. 96 p. : ill. 0894904094
Uses text and experiments to provide information about the air around us

and about pollution and other problems related to our atmosphere. GR 5-9.

1540. ___. *Environmental experiments about water.* Hillside, NJ: Enslow, 1993.
96 p. : ill. 0894904108
Presents experiments that focus on the properties of water, its cycle in
nature, pollution problems, and methods of purification. GR 5-9.

1541. Schwartz, Linda. *Earth book for kids : activities to help heal the environment.*
Santa Barbara: Learning Works, 1990. 184 p. : ill. 0881601950
Creative ideas with easy-to-follow instructions show kids how to make their
own paper, compare phosphate levels in detergents, test the effects of oil
pollution, conduct a recycling survey, create a trash sculpture, redesign a
package, chart a flush, measure acidity, etc. Unfortunately, the basic activities for GR 3-6 are mixed with those for teens, who may find the tone
demeaning.

1542. Seton, Ernest Thompson. *Book of woodcraft and Indian lore.* NY:
Stevens, 1994, c1912. 590 p. : ill. 1885529112
A classic outdoor guide covering woodcraft, camping, scouting, signaling,
outdoor games, health and medicine, natural history, forestry, songs and
stories, and Native American beliefs, ceremonies, and lore. This Boy Scout
book from 1912 is rather quaint, but is still a useful, comprehensive handbook. #

1543. Shaffer, Carolyn, and Fielder, Erica. *City safaris : a Sierra Club explorer's
guide to urban adventures for grownups and kids.* SF: Sierra Club, 1987. 185 p. :
ill. 087156713X 0871567202 (pbk)
Views the city as an ecosystem, and stresses the bond between nature and
civilization. Provides ways to explore nature through a variety of activities
and suggests types of appropriate environmental action. HR for teens, parents, and teachers. #

1544. Sisson, Edith A. *Nature with children of all ages : activities & adventures for
exploring, learning & enjoying the world around us.* NY: Prentice Hall, 1987,
c1982. 195 p. : ill. 0136104444 0136104363 (pbk)
Includes background information and field activities about trees, plants,
seeds, invertebrates, amphibians, birds, mammals, seashores, ponds, wetlands, winter, ecology, and living in tune with the environment. For parents
and teachers.

1545. Swain, Roger B. *Saving graces : sojourns of a backyard biologist.* Boston:
Little, Brown, 1991. 138 p. : ill. 0316824712
Nature and wildlife aren't just in the wilderness, but as close as your own
back yard. This book explains how to look for animals, study, and savor

them. Covers the rewards and satisfaction of a personal experience and deeper understanding of other animals. AOT. #

1546. Tilsworth, Debbie J. *Raising an Earth friendly child : the keys to your child's happy, healthy future.* Fairbanks: Raven, 1991. 194 p. : ill. 0962744670
Encourages parents to instill environmental consciousness in their children (through age 12) through a variety of projects and experiments.

1547. Van Matre, Steve. *Earth education : a new beginning.* Warrenville, IL: Inst. for Earth Education, 1990. 334 p. 0917011023
AR: *Earthkeepers* (for ages 10-12), *Sunship Earth* (10-12), and *Conceptual encounters I and II* (10-12 and 13-14).
Contends that environmental education has been led astray through trivialization, dilution with other agendas, and cooptation by the industries and agencies that created the environmental crisis. Proposes an alternative path to strengthen knowledge of and connection with the Earth and develop less wasteful lifestyles. HR for all environmental educators.

1548. Waage, Frederick O. ed. *Teaching environmental literature : materials, methods, resources.* NY: MLA, 1985. 191 p. 0873523083 0873523091 (pbk)
Provides models and ideas for courses in environmental literature and, more broadly, courses that combine humanities and environmental studies disciplines. Contains 20 contributions by various authors. HR for college faculty.

1549. Wilkins, Marne. *The long ago lake : a book of nature lore and crafts.* SF: Chronicle, 1989. 160 p. : ill. 0877016321
The author recollects her childhood summers in the Wisconsin Lakes district and the nature lessons her family learned from their Indian friends. Nature crafts projects are included. GR 5-9.

Ecotourism and Recreation

Travel and outdoor sports can also be highly educational. Some forms of recreation such as motor sports almost always create pollution and damage the landscape, but even the most well-meaning hiker can also be unintentionally harmful. The books in this chapter tend to include or even emphasize minimizing the impact of the activity and learning from nature.
Paddle down a river with Abbey, Haig-Brown, Jerome, Kane, Krakel, Rennicke, Starkel, Stensaas, or Wood. Head for the hills with Baron, Douglas, Ferguson, Graydon, or Heacox. Go camping with Jacobson and

Richards and catch a few fish with Leeson or the old master angler Izaak Walton. (Hunting is covered in the Human-Animal Relations section.) Explore the waters off Hawaii with Farber or Sutherland.

Take a hike with Franzine, Halsey, the Jenkins, Moser, Newman, or Ross. (Also see the Wilderness and Mountain sections.) Try Strickland's anthology *Shank's Mare* if you'd enjoy hiking with the likes of Abbey, Colin Fletcher, Gary Snyder, John Muir, Loren Eiseley, Henry David Thoreau, or other famous authors. Also see the books by LaBastille (1160-61) or Teal (1176) for books on women in river sports and other wilderness activities.

A broad range of activities is described by Rice, White, or Wilber, and in the anthology *Out of the Noosphere*. If you're traveling with kids bring Hodson along, too, and pay close attention as Meyers demonstrates the art and science of *How to Shit in the Woods*.

Although conservation and hiking groups have championed low-impact hiking and camping for decades, the business of Ecotourism is a more recent development. Collections or conference proceedings edited by Boo, Kusler, Nelson, or Whelan explore various aspects of ecotourism. They are intended primarily for academics and tour operators. Guides provided by *Buzzworm* magazine, Edelman, Foehr, Geffen, and Holing are more useful for travelers.

Weidensaul's *Seasonal Guide* books cover large regions and emphasize natural history and ecotourism. Local guidebooks and field guides to plants, animals, and birds are too numerous to include. The best sources for local guides and advice are often local (not chain) bookstores, libraries, sporting goods stores, and recreational or conservation groups.

Books for adults:

1550. Abbey, Edward. *Down the river*. NY: Plume, 1991, c1983. 242 p. 0452265630
Abbey summarizes 23 years experience rafting and canoeing on southwestern rivers, offering naturalistic, environmental, historic, literary, philosophical, economic, and personal commentary. A great book to take on a trip down a river or into your own mind. HR AOT. #

1551. ___. *The hidden canyon : a river journey*. Photos by John Blaustein, intro. by Martin Litton. NY: Penguin, 1977. 135 p. : col. photos. 014004678X
Abbey, Blaustein, and Litton recreate Powell's 1869 dory trip through the Grand Canyon. HR AOT. #

1552. Barron, T. A. *To walk in wilderness : a Rocky Mountain journal*. Englewood, CO: Westcliffe, 1993. 168 p. : col. photos. 1565790391
A personal account of the author's experiences in the Rocky Mountains, primarily Colorado, accompanied by 132 lavish color photos by John Fielder. #

1553. Boo, Elizabeth. *Ecotourism : the potentials and pitfalls*. Washington: World Wildlife Fund, 1990. 2 v 0942635140 (v.1) 0942635159 (v.2)
Explores the link between tourism and protected natural areas; impacts of nature tourism; and comparison of the status of nature tourism in Belize, Costa Rica, Dominica, Equador, and Mexico.

1554. Brown, Tom, and Watkins, William Jon. *The tracker : the story of Tom Brown, Jr., as told to William Jon Watkins*. NY: Berkley, 1984, c1978. 190 p. 0425101339
Describes the unique adventures of Tom Brown, Jr. in the New Jersey Pine Barrens using skills learned from an old Apache tracker. Conveys a legendary American experience.

1555. Buzzworm Magazine. *The Buzzworm magazine guide to ecotravel : 100 unforgettable adventures throughout the world, where the new ecotraveler will find the rewarding delights of travel with the Earth in mind*. Boulder, CO: Buzzworm, 1993. 239 p. : ill. 09603722288
The subtitle describes it perfectly. Includes a complete ecotravel outfitters directory. HR for adult travelers.

1556. Douglas, William O. *Of men and mountains*. SF: Chronicle, 1990, c1950. 338 p. 0877017123
The noted Supreme Court Justice describes his youthful adventures living and hiking in the Cascade and Wallowa Mountains. He tells how these experiences made him both an ardent conservationist and a stronger person. HR AOT. #

1557. Edelman, Jack R. *Field courses, seminars, safaris, and eco-tours in the natural sciences : a directory for teachers and students*. Golden, CO: North American, 1994. 250 p. 1555919235
Lists nearly 200 institutions worldwide which offer courses, eco-tours, and workshops. Includes information on sample courses, cost estimates, equipment required, prerequisites, college credits, financial aid, living arrangements, etc. HR for teachers, students, and ecotourists. #

1558. Farber, Thomas. *On water*. Hopewell, NJ: Ecco, 1994. 183 p. 0880013583
Reflections on the Pacific Ocean (primarily off Hawaii, South Sea islands, and California) and water in general. This "Sadhu with surfboard" (p. 113) rides and writes waves of myth, natural history, literature, philosophy, and sport.

1559. Ferguson, Gary. *Walking down the wild : a journey through the Yellowstone Rockies*. NY: Simon & Schuster, 1993. 204 p. 0671768514
An inspiring account of a 500 mile solo backpack in the Rockies portraying

a range of emotional experiences from fear to transcendence. Describes the history and natural history of the area, threats posed by development, and the contradiction between our spiritual desire for wilderness and economic demands which destroy it.

1560. Foehr, Stephen. *Eco-journeys : the world guide to ecologically aware travel and adventure.* Chicago: Noble, 1993. 351 p. 1879360217
Classifies journeys as "Safe Thrills, Comfortable Wilderness, Family Trips, Not Ordinary Trips, Special Trips, and Raw Adventure." Most destinations are areas where ecotourism is used for preservation and economic development. Lists environmentaly sensitive tour operators. HR.

1561. Frazine, Richard Keith. *The barefoot hiker : a book about bare feet.* Berkeley: Ten Speed, 1993. 98 p. 0898155258
Frazine advocates hiking barefoot as a sensuous and pleasurable experience, contending that most foot injuries are shoe-related. Also offers a step-by-step guide to acclimating your feet to the changing seasons.

1562. Geffen, Alice M., and Berglie, Carole. *Ecotours and nature getaways : a guide to environmental vacations around the world.* NY: C. Potter, 1993. 324 p. : map. 0517880687
Arranged by global regions, this book identifies top tour operators; provides details on cost, duration, and mode of travel; and provides advice on selecting a tour, preparing for tours, health and safety, and touring with children.

1563. Graydon, Don. ed. *Mountaineering : the freedom of the hills.* 5th ed Seattle: Mountaineers, 1992. 447 p. : ill. 0898863090 (pbk) 0898862019
Covers clothing and equipment, camping and food, navigation, modes of travel, climbing (including rock, snow, and ice climbing, and alpine rescue) the cyle of snow, and lightning. Emphasizes "clean climbing" and low-impact wilderness use.

1564. Haig-Brown, Roderick Langmere. *A river never sleeps.* NY: Lyons & Burford, 1991. 352 p. : ill. 1558210938
A personal account of and observations on fishing, riparian ecology, the seasons, and country life in British Columbia.

1565. Halsey, David, and Landau, Diana. *Magnetic north : a trek across Canada.* SF: Sierra Club, 1990. 252 p. : col. photos, maps. 0871567466
A young man starts the first modern trek across Canada by foot, dogsled, and canoe, is abandoned by his companions, but is later joined by photographer Peter Souchuk and a coyote dog who complete the 4700 mile trek with him. Describes their hardships, accomplishments, friendships, and search for peace and wholeness in the wilderness. #

1566. Heacox, Kim. *In Denali : a photographic essay of Denali National Park & Preserve, Alaska*. Santa Barbara: Companion, 1992. 96 p. : 105 col. photos. 0944197191 0944197183 (pbk)
Primarily photos of North America's highest mountain, accompanied by a text which emphasizes conservation. Includes tables charting the human history of the area, mammals, birds, and wildflowers.

1567. Hodgson, Michael. *Wilderness with children : a parent's guide to fun family outings*. Harrisburg, PA: Stackpole, 1992. 133 p. : ill. 0811724166
Describes how to prepare for and enjoy hiking, backpacking, cross-country skiing, canoeing, and biking trips. Includes practical tips on almost all aspects of such travel including first aid and making a minimal impact. Helps adults to see the world through the eyes of a child. HR.

1568. Holing, Dwight. *Earthtrips : a guide to nature travel on a fragile planet*. LA: Living Planet, 1991. 209 p. : ill. 1879326051
Emphasizes involvement in conservation and preservation activities as part of traveling. Includes directories of travel-study and volunteer vacation programs, ecotravel companies, expedition outfitters, and scientific research programs

1569. Jacobson, Cliff. *Campsite memories : true tales from wild places*. Merrilville, IN: ICS, 1994. 151 p. : ill. 0934802882
A professional canoe guide relates his adventures in Minnesota and Canada with occasional dollops of natural history and reverence for nature, and ironic asides such as power boating Indians and Eskimos who have never paddled a canoe.

1570. Jenkins, Peter. *A walk across America*. NY: Crest, 1983, c1979. 367 p. : photos. 0449204553
Sequel: *The walk west*.
A young man who felt disillusioned of the American dream backpacked nearly 5,000 miles with his dog in a journey toward understanding and wholeness. Describes the perils and rewards of his journey through the wilds and among the people. #

1571. Jenkins, Peter, and Jenkins, Barbara. *The walk west : A walk across America 2*. NY: Morrow, 1992, c1981. 368 p. : photos. 0688112714
Jenkins and his new wife hit the road again for "a two year honeymoon," more adventures and a deeper understanding of nature, their country, and themselves. #

1572. Jerome, Christine. *An Adirondack passage : the cruise of the canoe Sairy Gamp*. NY: HarperCollins, 1994. 224 p. : ill. 0060164352

The author and her husband retrace a month-long canoe trip made in 1883 by George Washington Sears. She combines her journal and that of Sears to imagine what the area was like a century ago and to trace the natural and cultural changes which have taken place. HR.

1573. Kane, Joe. *Running the Amazon*. NY: Vintage, 1990, c1989. 313 p. : photos, maps. 067972902X
The account of an incredible voyage by raft, kayak, and foot the entire 4,200 mile length of the Amazon, from 18,000 feet in the Andes to the ocean. This thrilling adventure story which includes perspectives on natural history and conservation is HR.

1574. Krakel, Dean. *Downriver : a Yellowstone journey*. SF: Sierra Club, 1988. 250 p. 0871567857
Krakel spent ten years kayaking, canoeing, and hiking in this area. Chronicles his adventures, the natural environment, the people who live and work there, tourism, and the struggles between preservationists and developers. #

1575. Leeson, Ted. *The habit of rivers : reflections on trout streams and fly fishing*. NY: Lyons & Burford, 1994. 240 p. 1558213007
English Prof. Leeson bags a full creel of fish, natural history, and philosophy. His description of opening one's entire self to nature is literate and moving. Reminiscent of Lopez and MacLean. HR.

1576. Mele, Audre. *Polluting for pleasure*. NY: Norton, 1993. 224 p. 0393035107
A boat designer outlines the environmental costs of pleasure boating including annual pollution equivalent to 40 *Exxon Valdez* spills, smog, emphysema, cancer, and mass extinctions. Proposes replacing the wasteful two stroke outboard motor, first with four stroke and then with electric motors. Recommends regulations.

1577. Meyer, Kathleen. *How to shit in the woods : an environmentally sound approach to a lost art*. Rev. ed. Berkeley: Ten Speed, 1994. 96 p. : ill. 0898156270
This is both a practical and an amusing book on an oft-ignored aspect of outdoor hygiene and low-impact recreation. Chapter titles include: Anatomy of a crap, Digging the hole, When you can't dig a hole, Trekker's trots, For women only: how not to pee in your boots, and What? Not pee?#

1578. Moser, David S., and Schad, Jerry. eds. *Wilderness basics : the complete handbook for hikers & backpackers*. 2d ed. Seattle: Mountaineers, 1993. 224 p. : ill. 0898863481
Covers wilderness ethics, conditioning, preparation, outfitting, food, navi-

gation, weather, traveling with children, coastal travel, mountain travel, desert travel, winter camping, first aid, and search and rescue. Continually emphasizes low impact camping and wilderness preservation. HR.

1579. Nelson, James Gordon; Butler, R.; and Wall, Geoffrey. eds. *Tourism and sustainable development : monitoring, planning, managing.* Waterloo, Ont.: Univ. of Waterloo, 1993. 284 p. 0921083440
Contains 20 chapters by various authors on all aspects of ecotourism, focusing on Canada.

1580. Newman, Steven M. *Worldwalk.* NY: Avon, 1990, c1989. 560 p. : ill. 0380711508
This personal account of a four year, 15,000 mile walk around the world describes his adventures and trials, emphasizing the encouragement and assistance he received from strangers.

1581. Outside Magazine. *Out of the noosphere : adventure, sports, travel and the environment : the best of Outside magazine.* NY: Simon & Schuster, 1992. 480 p. : ill. 0671754461
Includes articles and short stories by Edward Abbey, Rick Bass, Tim Cahill, William Kittredge, Barry Lopez, Thomas McGuane, Peter Matthiessen, Bob Shacochis, Randy Wayne White, and others. HR AOT. #

1582. Perkins, Robert F. *Into the great solitude : an Arctic journey.* NY: Dell, 1992, c1991. 222 p. 0440212484
AR: *Against straight lines.*
A modern adventurer retraces a canoe journey made in 1834. Facing physical and emotional challenges, Perkins found courage, greater unity with nature, and a deeper understanding of the world and his place in it.

1583. Rennicke, Jeff. ed. *River days : travels on western rivers : a collection of essays.* Golden, CO: Fulcrum, 1988. 174 p. : photos. 1555910297
Contains 27 personal accounts of river running through western wilderness areas. Includes observations on riparian ecology, water use issues, and nature.

1584. Rice, Larry, and Rice, Judy. *Baja to Patagonia : Latin American adventures.* Golden, CO: Fulcrum, 1993. 217 p. : photos, maps. 1555911137
The Rices relate their experiences backpacking, canoeing, and kayaking throughout Latin America, and offer advice on planning, natural history, ecotourism, and establishing friendly relations with local residents. Recommended for adventurous people who wish to avoid the beaten path.

1585. Richards, Mary Bradshaw. *Camping out in the Yellowstone,* 1882. Salt

Lake City: Univ. of Utah, 1994, c1882. 139 p. : photos. 0874804493
A rare and intimate look at Yellowstone shortly after it was established as a
Park, before commercial development and landscape degradation. Includes
28 historic photos and descriptions of natural wonders, such as the Excel-
sior Geyser, which no longer exist. Recommended for history and wilderness buffs.

1586. Roorbach, Bill. *Summers with Juliet*. Boston: Houghton Mifflin, 1992.
292 p. 0395573238
Describes an eight year courtship which took place largely in the wild as
they traveled and camped cross-country, learning the ways of wildlife and
each other. AOT. #

1587. Ross, Cindy, and Gladfelter, Todd. *The hiker's companion : 12,000 miles
of trail-tested wisdom*. Seattle: Mountaineers, 1993. 207 p. : ill. 0898863538
Tips and advice on hiking including health and safety isssues, physical conditioning, environmental ethics, low-impact camping, and much more. #

1588. Starkell, Don. *Paddle to the Amazon*. NY: Ivy, 1992, c1989. 317 p. : photos, maps. 0804109338
The incredible saga of a two year, 12,000 mile, two man canoe trip from Win-
nipeg to the heart of the Amazon. Combines a thrilling adventure story, a
nature chronicle, and a story of love and conflict between a father and son. #

1589. Stensaas, Mark. *Canoe country wildlife : a field guide to the Boundary
Waters and Quetico*. Duluth: Pfeifer-Hamilton, 1993. 221 p. : ill. 0938586653
Midwest Independent Publishers Association Award, 1993.
Includes wildlife checklists, a calendar to record natural events, and activi-
ties designed to help both adults and children learn by discovery. HR AOT,
and for parents to use with children. #

1590. Strickland, Ron. ed. *Shank's mare : a compendium of remarkable walks*.
NY: Paragon House, 1988. 420 p. 1557780749 1557780978 (pbk)
Hit the trail with selections from 77 authors including Edward Abbey,
Robert Marshall, Dian Fossey, Clin Fletcher, Gary Snyder, Loren Eiseley,
Winston Churchill, Aleksander Solzhenitsyn, Norman Mailer, Evelyn
Waugh, Anais Nin, and Ernest Hemingway. HR.

1591. Sutherland, Audrey. *Paddling my own canoe*. Honolulu: Univ. of
Hawaii, 1978. 136 p. 0824806182
During this adventurous woman's solo expeditions around the wild and
lonely northeast coast of Molokai she rediscovered a sense of wonder and
unity of nature she had known as a child.

1592. Walton, Izaak, and Cotton, Charles. *The compleat angler*. NY: Oxford, 1983, c1653. 435 p. 0198123132
The classic work on catching and cooking trout and grayling, selecting bait, and tying flies. Offers lyrical descriptions of pastoral England and interesting philosophical and literary comments. Only the *Bible* and the *Book of common prayer* have been reprinted more often than this delightful book. Essential for anglers. HR.

1593. Weidensaul, Scott. *Seasonal guide to the natural year : a month by month guide to natural events* : Mid-Atlantic. Golden, CO: Fulcrum, 1992. 342 p. : ill., maps. 1555911056
Describes not only where to go and what to look for, but the best times of the year for viewing animals, birds, and natural phenomena in this highly populated, but environmnentally diverse region. Includes significant natural history and ecotourism information.

1594. ___. *Seasonal guide to the natural year : a month by month guide to natural events, New England & New York*. Golden, CO: Fulcrum, 1993. 328 p. : ill., maps. 1555911358
Describes not only where to go and what to look for, but the best times of the year for viewing animals, birds, and natural phenomena in upstate New York and New England. Includes significant natural history and ecotourism information.

1595. Whelan, Tensie. *Nature tourism : managing for the environment*. Washington: Island, 1991. 223 p. : maps, photos. 155963037X 1559630361 (pbk)
Includes ten articles exploring the pros, cons, and various aspects of ecotourism including economics, marketing, local participation, and its relation to sustainable development and to family farming.

1596. White, Randy Wayne. *Batfishing in the rainforest : strange tales of travel & fishing*. NY: Lyons & Burford, 1991. 250 p. 155821187
Lively and highly personalized observations on travel, fishing, nature, animals, and politics from a columnist for *Outside* magazine.

1597. Wilber, Jerry. *Wit & wisdom of the great outdoors*. Duluth, MN: Pfeifer-Hamilton, 1993. 1 v. (unpaged) : ill. 0938586726
Midwest Independent Publishers Association Award, 1993.
Provides a wealth of nature lore, including tips for fishing, hints for campers, advice for parents, hunting secrets, ecological reminders, and recipe suggestions in 365 entries covering the entire year. Each page includes space for the reader's observations. AOT. #

1598. Wood, Douglas. *Paddle whispers*. Duluth: Pfeifer-Hamilton, 1993. 176

p. : ill. 0938586734
Through poetic text and drawings, interwoven with quotes by John Muir, Walking Buffalo, Sigurd F. Olson, Henry David Thoreau, and others, Wood traces a canoe journey by paddle that renews the spirit. HR, especially for Olson fans and upper Midwesterners.

Books for young adults:

1599. Paulsen, Gary. *Woodsong*. NY: Viking Penguin, 1991, c1990. 132 p. : ill. 0140349057
For a rugged outdoorsman and his family, life in northern Minnesota is a wild experience involving wolves, deer, and the sled dogs that make their way of life possible. Includes an account of the author's first Iditarod, a dogsled race across Alaska. Good family reading. #

Greener Homes and Gardens

Some of the countercultural and alternative Christian trends of the 1960s and 70s have entered the mainstream in the 80s and 90s, as reflected in a proliferation of books on living a more ecologically responsible lifestyle. Books on voluntary simplicity are provided by Devall, Johnson, Levering, Longacre, Mitchell, the Nearings, Nichols, and Shi. Also check "Simplicity" in the subject index for related books.

Some authors focus on fairly specific subjects: Altman, Dadd, Garland, and Shoemaker on reducing toxic substances in the home; the Earth Works Group, and Kimball on recycling; Landau and Merilees on backyard wildlife; Donahoe and Wilkinson on green Christian homes; Makower on commuting; and Robbins and Steinman on diet. Also see Lappe's classic *Diet for a Small Planet* (653). Organic or low impact gardening and pest control are covered by Bowers, Christopher, Cox, Hamilton, Hansen, Kitto, Minter, Ocone, Pest Publications, Stein, Swain, and Tilgner.

The other books are of a general nature, offering everything from a few general practices to hundreds of specific tips on energy use, water conservation, shopping, housing, furniture, health care, cosmetics, cleaning products, child care, and the whole gamut of home activities. The authors are as varied as novelist Nichols, activist Rifkin, and the ever helpful and hint-ful Heloise. Lowery and Lobieki furnish a good guide for middle-grade students.

Books for adults:

1600. Altman, Roberta. *The complete book of home environmental hazards*. NY: Facts on File, 1990. 290 p. : ill. 0816020957
Includes practical and technical information on discovering and eliminating toxic and other environmental hazards in the home. Contains an extensive list of consumer-oriented government agencies.

1601. Anderson, Bruce. *Ecologue : the environmental catalogue and consumer's guide for a safe Earth*. NY: Prentice Hall, 1990. 255 p. : photos. 0130845183
Describes 27 product categories (cleaners, pet care, water purification, appliances, groceries, office supplies, etc.), testing and recommending over 600 specific products. Includes advice on evaluating products yourself and other useful consumer education information.

1602. Bowers, Janice Emily. *A full life in a small place : and other essays from a desert garden*. Tucson: Univ. of Arizona, 1993. 166 p. 0816513457 0816513570 (pbk)
Describes her desert-wise garden in Arizona and provides ecologically based gardening tips. She then extrapolates to consider desert ecology, humanity's place in the balance of nature, and the effect of nature on the human soul.

1603. Capone, Lisa. *The Conservationworks book : practical conservation tips for the home and outdoors*. Boston: Appalachian Mountain Club, 1992. 96 p. : ill. 1878239112
A Books Change Lives selection.
A brief and practical guide for the family that shows both adults and children how to reduce waste, save energy, and protect the environment around the home and in the outdoors. HR. #

1604. Christensen, Karen. *Home ecology : simple and practical ways to green your home*. Golden, CO: Fulcrum, 1990. 334 p. : ill. 1555910629
A common sense guide that unites ecology and home economics, demonstrating how to evaluate and make changes in the home.

1605. Christopher, Thomas. *Water-wise gardening : America's backyard revolution*. NY: Simon & Schuster, 1994. 295 p. : col. photos. 0671738569
Shows how drought and water pollution is provoking a revolution in garden design. Provides a comprehensive guide to the priciples of water-conserving gardening.

1606. Co-op America Foundation. *Co-op America's national green pages*. Washington: the Foundation, annual. ca. 120 p. : ill.
Lists over 90 categories of products and services, including body care prod-

ucts, books, bed & breakfasts, clothing, financial services, food, furniture, health, housing, jobs, restaurants, travel, and socially and environmentally responsible businesses. Get this free with a Co-op America membership by calling 202-872-5307. HR.

1607. Cox, Jeff. *Your organic garden with Jeff Cox*. Emmaus, PA: Rodale, 1994. 344 p. : ill. 0875966233 0875966241 (pbk)
Rodale publishes many similar books including *Growing fruits and vegetables organically*, and the six volume *Organic gardening series*. This companion volume to a PBS series includes facutal and philosophical background material on organic gardening, but is primarily a compendium of practical advice on all aspects of gardening. Recommended for beginning to advanced gardeners.

1608. Dadd, Debra Lynn. *The nontoxic home and office : protecting yourself and your family from everyday toxics and health hazards*. Rev. ed. LA: Tarcher, 1992, c1986. 240 p. : ill. 0874776767
Explains how to avoid toxins in water, food, cleaning products, and other household and office products and how to evaluate product safety. Recommended for families and small businesses.

1609. Devall, Bill. *Living richly in an age of limits : exploring voluntary changes toward a green lifestyle*. Salt Lake City: Gibbs Smith, 1993. 256 p. 0879055596 0879055642 (pbk)
Explores how to change the way we live, to live more authentic lives and to reduce our impact on biological diversity. Changes in jobs, housing, transportation, gardening, consumption, landscapes, and built environments are all considered. HR.

1610. Donahoe, Sydney L. *Earth keeping : making it a family habit*. Grand Rapids: Zondervan House, 1990. 144 p. : ill. 0310538017
Offers a Christian perspective encouraging families not just to pray together, but to take practical action to conserve water and energy, reduce waste in the home, be wise consumers, and act more responsibly toward nature. Upbeat and encouraging.

1611. Earth Works Group. *50 simple things you can do to save the Earth*. Berkeley: Earth Works, 1991, c1989. 96 p. : ill. 0816151288
Demonstrates a variety of practical things you can do in your home and elsewhere to reduce waste, conserve resources and energy, and improve personal and planetary health without major sacrifices and inconveniences. This influential volume led to a spate of similar books. HR. #

1612. ___. *The Next step : 50 more things you can do to save the Earth*. Berkeley: Earth Works, 1991. 120 p. : ill. 0836223020

More practical advice on alternative light bulbs, non-aerosol sprayers, cloth diapers, solar power, bicycles, butterfly gardens, "affinity" credit cards which support environmental causes, and alternative transportation. HR. #

1613. ___. *The student environmental action guide.* Berkeley: Earth Works, 1991. 96 p. 0665004329
"25 simple things we can do" to effect positive environmental changes in homes and schools. Written by and for college students, but also recommended for high schools. #

1614. Garland, Anne Witte. *For our kid's sake : how to protect your children against pesticides in food.* SF: Sierra Club, 1990. 87 p. : ill. 0871566133
Contains descriptions of hazardous pesticides in fruits and vegetables, specific instructions on eliminating or limiting them, and suggestions for political action to limit their use.

1615. Gershon, David, and Gilman, Robert. *Household ecoteam workbook.* Woodstock, NY: Global Action Plan, 1992. 145 p. 0963032704
Home economics from a green perspective. Contains a six month program "to bring your household into environmental balance"

1616. Gorder, Cheryl. *Green Earth resource guide : a comprehensive guide about environmentally-friendly products.* Tempe, AZ: Blue Bird, 1990. 256 p. : ill. 0933025238
Provides information on products and services relating to environmental restoration, clean air and water, ecotourism, energy conservation, household products, organicaly grown products, recycling baby and child care, and investments.

1617. Hamilton, Geoff. *The organic garden book : the complete guide to growing flowers, fruit, and vegetables naturally.* NY: D. Kindersley, 1994, c1987. 288 p. : col. photos. 156458528X
Describes how to garden with, rather than against, nature through chemical-free techniques to grow healthy, better tasting fruits and vegetables in healthy soil. AOT. #

1618. Hansen, Michael. *Pest control for home and garden : the safest and most effective methods for you and the environment.* Yonkers, NY: Consumer Reports, 1993. 372 p. : ill. 0890434239
Demonstrates how to use integrated pest management to identify pests, cut off their food supply, and use light, temperature, humidity, or their natural enemies to eradicate them. Covers insects, slugs and snails, rodents, fleas and ticks. Provides lists of relatively safe insecticides, herbicides, and fungicides.

1619. Harland, Edward. *Eco-renovation : the ecological home improvement guide.* Post Mills, VT: Chelsea Green, 1994. 288 p. : ill. 0930031660
A British architect's advice on paths toward ecological and energy conserving living and housing. Includes the benefits of houseplants, detecting and removing radon and other harmful substances, water conservation, solar energy, insulation, organic gardening, and lower consumption levels. Somewhat lacking in practical detail.

1620. Harms, Valerie. ed. *The National Audubon Society almanac of the environment : the ecology of everyday life.* NY: Putnam, 1994. 290 p. : ill. 0399139427
Demonstrates the extent of various activities on the environment. Contains five sections: body, home, community, land and ocean, and cultural ecology. Not really a true almanac since it contains more text than statistical information. HR.

1621. Harris, D. Mark, and Pusztai, Lyn. *Embracing the Earth : choices for environmentally sound living.* Chicago: Noble, 1990. 162 p. : ill. 0962268321
A user-friendly guide to evaluating and modifying personal lifestyles, including consumption, reuse and recycling, travel, water use, heating and lighting. Includes a directory of environmental organizations.

1622. Heloise. *Heloise's hints for a healthy planet.* NY: Perigee, 1990. 160 p. : ill. 0399516255
The syndicated wizard of home ec provides basic energy and water saving tips, advice on ecologically prudent use of cleaning products, and methods of reducing trash by sensible shopping and reuse of packing materials.

1623. Hynes, H. Patricia. *EarthRight : what you can do in your home, workplace, and community to save the environment.* Rocklin, CA: Prima, 1990. 236 p. : ill. 1559580275 (pbk) 1559580283
Focusing on five specific areas—pesticides, solid waste, drinking water, ozone depletion, and global warming—Hynes defines the problems and their scope, suggesting things to do in the home to help mitigate these problems. Provides advice on community-wide action.

1624. Johnson, Warren A. *Muddling toward frugality.* SF: Sierra Club, 1978. 256 p. 0871562146
Johnson proposes a slow but significant shift from affluent, material oriented lifestyles to more responsible modes focusing on quality of life. Frugality is defined here not as poverty, but as making the maximum wise use of existing resources while preserving them for the future.

1625. Kimball, Debi. *Recycling in America : a reference handbook.* Santa Barbara: ABC-CLIO, 1992. 254 p. 0874366631

Library Journal Best Reference Source, 1992.
Covers the basics of recycling and its relation to the reduction of global warming, deforestation, and resource depletion. Reviews materials being recycled, those being overlooked, and the creation of products from recycled materials. HR AOT. #

1626. Kitto, Dick. *Planning the organic vegetable garden for healthy crops throughout the year.* Golden, CO: Fulcrum, 1993, c1986. 160 p. : ill. 1555911099
An excellent guide for beginning or intermediate organic gardeners that covers planning, crop selection, scheduling, manure, composting, crop rotation, and ways to get the greatest output form minimal space. Particularly for gardeners in the northeast and other cool to moderate climates.

1627. Lamb, Marjorie. *Two minutes a day for a greener planet.* SF: Harper & Row, 1990. 243 p. : ill. 0062505076
The CBC's "Two Minute Ecologist" provides a collection of easy, environmentally sound practices which can be done in a short time in the home or office (e.g. cleaning and reusing antifreeze, eating citrus fruits as natural insect repellents, and using bicycle couriers.)

1628. Landau, Diana, and Stump, Shelley. *Living with wildlife : how to enjoy, cope with, and protect wild creatures around your home and theirs.* SF: Sierra Club, 1994. 352 p. : ill. 0871565471
Provides information on range, habitat, enemies, defenses, and behavior of over 100 wildlife species. Describes how to make our homes and society more hospitable to wildlife, humane human-wildlife interactions, and dealing with unwelcome household visitors. HR.

1629. Levering, Frank, and Urbanska, Wanda. *Simple living : one couple's search for a better life.* NY: Penguin, 1993, c1992. 272 p. 014012393
A journalist and a screenwriter leave Los Angeles to live on a rural orchard. Intended as a guide for others who "also choose to make their lives contain less and matter more."

1630. Longacre, Doris Janzen. *Living more with less.* Scottdale, PA: Herald, 1980. 294 p. 0836119304
"More with less" doesn't mean poverty, but rather living more creatively, finding new ways to put basic resources to use, conserving, and wasting less. Combines Mennonite philosophy with many practical tips.

1631. Makower, Joel. *The green commuter : a guide to driving . . . and not driving for a healthier, cleaner world.* Washington: Tilden, 1992. 171 p. 0915765950
Covers selecting the "greenest" cars and cleanest gasolines; maintenance, tires and accessories; and recycling batteries, tires, and the entire car.

Includes a chapter on alternatives such as mass transit, cycling, car or van-pooling, and telecommuting. HR.

1632. Makower, Joel; Elkington, John; and Hailes, Julia. *The green consumer.* Rev. ed. NY: Penguin, 1993. 339 p. : ill. 0140177116
Provides criteria for rating the environmental "friendliness" of products and evaluates a variety of types of common products. Also covers boycotts, investing, and challenging polluters. Includes a directory. HR.

1633. Marlin, Alice Tepper. ed. *Shopping for a better world : a quick and easy guide to socially responsible supermarket shopping.* Rev. ed. NY: Ballantine, 1992, c1991. 433 p. 034537083X
Evaluates about 200 companies and over 1,000 products in terms of corporate philanthropy, equal opportunity, military contracts, animal testing, environmental records, and related issues. HR.

1634. Merilees, Bill. *Attracting backyard wildlife : a guide for nature-lovers.* Stillwater, MN: Voyageur, 1989. 159 p. : ill. 0896581306
Concerned with rapid destruction of natural habitats and limited government efforts to provide green space, Merilees shows how individuals can form small wildlife habitats in their own back yards. Contains practical advice on how yards and gardens can accomodate birds, reptiles, amphibians, and small animals.

1635. Minter, Sue. *The healing garden : a natural haven for body, senses and spirit.* Boston: Tuttle, 1993. 160 p. : col. ill. 0804819750
Use your garden not just to grow food and flowers, but health and happiness, too. Provides historical background and practical advice on the marriage of botany and medicine, nature's pharmacy, awakening the senses, and gardens as spiritual havens. Includes planting charts, recipes, meditations, stories, and much more.

1636. Mitchell, John Hanson. *Living at the end of time.* Boston: Houghton Mifflin, 1990. 223 p. 0395445949
AR: *A field guide to your own backyard.*
A naturalist explains how he developed a more natural lifestyle both in the wilderness and in suburbia. Also ponders the origins of Thoreau's mysticism and how human experience can be enhanced by internalizing the wild.

1637. Nearing, Helen. *Loving and leaving the good life.* Post Mills, VT: Chelsea Green, 1992. 197 p. : ill. 0930031636
A memoir of Helen and Scott Nearing and an evocation of the ideals they stood for and practiced: self-sufficiency, voluntary simplicity, social justice, conservation, and world peace.

1638. Nearing, Helen, and Nearing, Scott. *Living the good life : Helen and Scott Nearing's sixty years of self-sufficient living.* NY: Schocken, 1989. 411 p. 0805209700
A reprint of *Living the good life,* and *Continuing the good life.* The Nearings have been "a voice in the wilderness" to three generations of Americans by promoting a simpler, less wasteful, more spiritually fulfilling lifestyle. They provide both a philosophical overview of voluntary simplicity and a practical guide to it. HR.

1639. Needleman, Herbert L., and Landrigan, Philip J. *Raising children toxic free : how to keep your child safe from lead, asbestos, pesticides, and other environmental hazards.* NY: Farrar, Straus & Giroux, 1994. 400 p. 0374523924
Written by doctors for the general public, this book covers radiation, pesticides, pollutants in the air, lead, mercury, tobacco, and other common hazards. Includes questions about home safety and directory listings. HR for home use and environmental health educators.

1640. Nichols, John Treadwell. *Keep it simple : a defense of the Earth.* NY: Norton, 1992. 14 p., 68 plates : col. photos. 0393309010 (pbk) 0393033864
Nichols describes how he has benefited from a radically simplified lifestyle. Includes photos chiefly of the natural world around his New Mexico home.

1641. Ocone, Lynn, and Pranis, Eve. *The National Gardening Association guide to kids' gardening.* NY: Wiley, 1990. 148 p. : ill. 0471520926
Rev. ed. of *The youth gardening book.*
Provides practical and philosophical advice along with over 100 garden activities, projects, and games. "Gardening grows good kids" is the unique perspective of this book which is HR for parents, teachers, and teens. #

1642. Pest Publications (Firm). *Shepherd's purse : organic pest control handbook.* Rev. ed. Summertown, TN: Book Co., 1993. 80 p. : col. photos. 0913990981
Provides detailed descriptions of 50 common garden pests and how to control them using non-chemical, organic methods such as integrated pest control.

1643. Pohanish, Richard P., and Greene, Stanley A. eds. *Hazardous substances resource guide.* Detroit: Gale, 1993. 624 p. 0810384946
American Libraries Outstanding Reference Source, 1994.
Provides background for the layperson; explains over 1,000 hazardous substances found in the home, workplace, and community; lists over 1,500 organizations, government agencies, services, and publications. HR.

1644. Rifkin, Jeremy. ed. *The Green lifestyle handbook : 1001 ways to heal the Earth.* NY: Holt, 1990. 220 p. 0805013695

Practical information on home energy use, toxics in the home, shopping, traveling, recycling, environmentally friendly offices, tree planting, genetic diversity, and community organizing. A.k.a.: *The Green lifestyle guide.* HR.

1645. Robbins, John. *Diet for a new America.* Walpole, NH: Stillpoint, 1987. 440 p : ill. 0913299545
Considers negative changes in the 20th century American diet, effects of meat and dairy consumption on the individual and the environment, elements of a healthy diet, evaluating the nutritional content and environmental consequences of all types of food, and the rewards of a reduced-meat or vegetarian diet. HR.

1646. Sanders, Scott R. *Staying put : making a home in a restless world.* Boston: Beacon, 1993. 220 p. 0807063401
The author expounds upon the concept of staying in one place be it home, marriage, or landscape.

1647. Schoemaker, Joyce M., and Vitale, Charity Y. *Healthy homes, healthy kids : protecting your children from everyday environmental hazards.* Washington: Island, 1991. 221 p. 1559630574
A comprehensive, non-technical guide to identifying and remedying toxic and other environmental hazards in the home. Includes information on heightened sensitivities of young and growing children to toxic substances.

1648. Schultz, Warren. ed. *Natural pest control : the ecological gardener's guide to foiling pests.* Brooklyn: Brooklyn Botanic Garden, 1994. 112 p. : col. ill. 0945352832
Depicts over 50 common garden pests, describing life cycle, symptoms, and controls for each. Provides information on cultural and physical controls, natural predators, natural pesticides, and how to read a pesticide label.

1649. Seymour, John, and Girardet, Herbert. *Blueprint for a green planet : your practical guide to restoring the world's environment.* NY: Prentice Hall, 1987. 192 p. 0317565621
Analyzes the causes and effects of common environmental problems, individual and household practices that are injurious to the environment, and alternative, Earth-friendly lifestle practices. This was one of the first citizen action guide and is still one of the best. HR.

1650. Shi, David E. *The simple life : plain living and high thinking in American culture.* NY: Oxford, 1986, c1985. 332 p. 019563475
AR: *In search of the simple life.*
Analyzes both historic and contemporary theory and practice on enriching

personal experience through voluntary simplicity. "Enlightened material restraint" is traced to colonial times.

1651. Stein, Sara Bonnett. *Noah's garden : restoring the ecology of our own back yards*. Boston: Houghton Mifflin, 1993. 294 p. 0395653738
Critiques modern gardens and lawns as antiecological anachronisms, recommending organic gardening, cultivating a wide variety of local flora, encouraging the return of local wildlife, and letting natural processes like leaf mulching occur. This lively and lyrical book is HR.

1652. Steinman, David. *Diet for a poisoned planet : how to choose safe foods for you and your family*. NY: Ballantine, 1992, c1990. 392 p. 0517575124
Describes food contamination by pesticides, additives, irradiation, and processing; reviews and rates over 500 foods and beverages; makes recommendatios on nontoxic homes; and provides a safe food shopping guide, a directory of safe food sources and relevant activist groups, and a glossary.

1653. Swain, Roger B. *Groundwork : a gardener's ecology*. Boston: Houghton Mifflin, 1994. 162 p. : ill. 0395684005
Swain's basis for ecologically sound practices in gardening is grounded in a broad interpretation of what it means to garden in harmony with nature. This practical book bridges the gap between strict organic gardening and chemically-based gardening. HR.

1654. Teitel, Martin. *Rain forest in your kitchen : the hidden connection between extinction and your supermarket*. Washington: Island, 1992. 112 p. 1559631538
Demonstrates how supermarkets rely on monoculture factory farming which is pesticide and herbicide intensive, thus hastening extinctions and loss of biodiversity. Encourages consumers to demand a greater diversity of products such as brown eggs and local produce strains.

1655. Tilgner, Linda. *Let's grow! : 72 gardening adventures with children*. Pownal, VT: Storey Communications, 1988. 208 p. : photos. 0882664719 0882664700 (pbk)
Covers organic gardening, botanical lore, tree planting, wild foods, indoor gardening, and related topics. HR for parents and teachers.

1656. Wilkinson, Loren, and Wilkinson, Mary Ruth. *Caring for creation in your own backyard : over 100 things Christian families can do to help the Earth*. Ann Arbor: Vine, 1992. 268 p. 0892837519
Details how families can practice Christian stewardship of nature through using less, driving less, recycling, and buying Earth-friendly products.

Books for young adults:

1657. Lowery, Linda, and Lorbiecki, Marybeth. *Earthwise at home : a guide to the care & feeding of your planet.* Minneapolis: Carolrhoda, 1993. 48 p. : ill. 0876147295
Suggests activities that can help save our planet, including recycling, power conservation, and smart shopping. GR 5-8.

Green Business and Industry

Like the books on sustainable development, the ones in this chapter by Davis, Hoffman, Pearson, the Millers, and Romm demonstrate that the false dichotomy between environmentalism and success in business has harmed both the environment and the economy.

Bennett, Callenbach, Hawken, Makower, Marguglio, Sanders, Silverstein, and Westerman offer practical guides to general business reforms. Others are more specific: Kazis on labor; Cairncross and Manne on how pollution taxes and market forces may be more effective than regulations; Moavenzadeh on the international construction business. These books all seem honest, perceptive, upbeat, and well-intentioned.

Coddington, Miller, and Ottman promote green marketing. When major polluters like Coors portray themselves as environmentally friendly, oil companies claim solar power is far in the future, tobacco executives swear under oath that smoking is a harmless habit, etc. ad nauseum, they create well-founded suspicions about public relations and the veracity of industrial leaders and advertisers. The Plants' *Green Marketing, Hope or Hoax?* is required reading to separate true green marketing from corporate hype.

Books for adults:

1658. Bennett, Steven J.; Freierman, Richard; and George, Stephen. *Corporate realities and environmental truths : strategies for leading your business in the environmental era.* NY: Wiley, 1993. 232 p. : ill. 0471530735
Suggests that what began primarily as a public relations tool has given way to a genuine recognition of environmental responsibilities. Covers organizational, economic, and legal concerns.

1659. Bennett, Steven J., and Freierman, Richard. eds. *Save the Earth at work! : how you can create a waste-free, non-polluting, non-toxic workplace.* Holbrook, MS: Bob Adams, 1991. 144 p. 1558500294
Primary coverage is identifying and correcting environmental hazards and

problems at work, with secondary coverage of producing Earth-friendly products.

1660. Cairncross, Frances. *Costing the Earth : the challenge for governments, the opportunities for business.* Boston: Harvard Business School, 1992. 341 p. 0875843158
Demonstrates that energy efficiency, pollution reduction, and lower waste is both profitable and in the public interest. Contends that taxes and other market mechanisms are more effective than regulations in convincing industries to be environmentally responsible.

1661. Callenbach, Ernest; Capra Fritjof; Goldman, Lenore; Lutz, Rudiger; and Marburg, Sandra. *EcoManagement : the Elmwood guide to ecological auditing and sustainable business.* SF: Berrett-Koehler, 1993. 206 p. 1881052273
AR: *The ecotopian encyclopedia for the 80s.*
Shows businesses how to move from the negative mode of merely complying with regulations to a positive eco-audit which examines every aspect of operations for long-term ecological sustainability. Provides a practical program of checklists. HR.

1662. Coddington, Walter, and Florian, Peter. *Environmental marketing : positive strategies for reaching the green consumer.* NY: McGraw-Hill, 1993. 252 p. : ill. 0070115990
Explains how to develop effective environmental marketing strategies that involve every unit of the organization while promoting products, supporting environmental stewardship, and educating consumers.

1663. Davis, John. *Greening business : managing for sustainable development.* Cambridge: Blackwell, 1994. 220 p. 0631172025 0631193154 (pbk)
Demonstrates how both the developed and developing world can profitably engage in sustainable development through directed innovation for businesspeople and industrialists. HR.

1664. Harrison, E. Bruce. *Going green : how to communicate your company's environmental commitment.* Homewood, IL: Irwin, 1993. 344 p. 15562339459
Explains how "going green" effects customer loyalty and higher profits and how to create a corporate communication program to raise public awareness. Useful to businesses wishing to improve their "green market" and to activists who suspect that "green business" is sometimes little more than manipulative P.R.

1665. Hawken, Paul. *The ecology of commerce : doing good business.* NY: Harper-Collins, 1993. 250 p. 0887306551
Argues that a "restorative economy" is our only opportunity to avoid plane-

tary ecological and economic collapse. Combines background information with practical suggestions on how to achieve both environmental and economic sustainability. HR.

1666. Hines, Lawrence Gregory. *The market, energy, and the environment.* Boston: Allyn & Bacon, 1988. 305 p. : ill 0205105688
Considers the market concept as an analytical framework to examine approaches to energy and environmental problems, how the approaches have worked, and in what respects they may be modified. The role of the market is also considered in areas such as federal benefit-cost analysis, natural gas pricing, air emissions trading, etc.

1667. Hoffman, W. Michael; Frederick, Robert; and Petry, Edward S. eds. *Business, ethics, and the environment* ; with *The corporation, ethics, and the environment.* NY: Quorum, 1990. 2 v. 0899305504
The first volume of these conference papers focuses on the role of corporations and business in protecting the environment. The second goes beyond the typical cost/benefit arguments to consider the ethical responsilbilites of businesses.

1668. Kazis, Richard, and Grossman, Richard Lee. *Fear at work : job blackmail, labor, and the environment.* Philadelphia: New Society, 1991. 336 p. 0865712069
Rejects economic myths that have put labor and environmentalists at odds through anecdotes, history, and social analysis. Argues for greater cooperation between the labor and environmental movements. Analyzes co-option and compromise by major environmental organizations. HR.

1669. Makower, Joel. *The E-factor : the bottom-line approach to environmentally responsible business.* NY: Dutton, 1994, c1993. 304 p. 0452271908
Shows how companies can minimize waste while maximizing profits, how to influence suppliers to create environmentally sound products and services, and how to work with customers to create win/win situations. Describes market-oriented approaches to regulation such as green taxes, smog futures, etc. HR.

1670. Manne, Alan S., and Richels, Richard G. *Buying greenhouse insurance : the economic costs of carbon dioxide emission limits.* Cambridge: MIT, 1992. 182 p. 026213280X
Presents a model to determine the economic costs of limiting CO^2 emissions. Provides background on the greenhouse effect, various scenarios regarding the use of fossil fuels with and without carbon constraints, and region by region costs and benefits of an international agreement. Fairly technical.

1671. Marguglio, B. W. *Environmental management systems.* NY: Dekker, 1991. 194 p. 0824785231
Provides managers, scientists, engineers, and regulators with specific features of environmental policies and procedures that should exist at the plant level. Very useful for activists as a baseline to assess the environmental controls used by local companies.

1672. Moavenzadeh, Fred. *Global construction and the environment : strategies and opportunities.* NY: Wiley, 1994. 324 p. : ill. 0471012890
Describes how environmental regulations are creating international markets for environmentally sensitive construction and engineering firms. Identifies issues and regulations, describes new technologies, targets specific markets, and provides other practical advice.

1673. Moore, Curtis, and Miller, Alan S. *Green gold : Japan, Germany, the United States, and the race for environmental technology.* Boston: Beacon, 1994. 288 p. 0807085308
Investigates the environmental and economic successes in Germany and Japan attained by using technologies developed by U.S. taxes. Analyzes how the Reagan/Bush administrations and many current industrial leaders have damaged America environmentally and economically. Recommends policy changes. HR.

1674. Ottman, Jacquelyn A. *Green marketing.* Lincolnwood, IL: NTC Business, 1993. 188 p. : ill. 0844232505
Contends that we are on the verge of a fundamental realignment of our mindset and expectations, toward an enduring public ethic that expects every product to be designed to minimize its environmental impact.

1675. Pearson, Charles S. *Down to business : multinational corporations, the environment and development.* Washington.: World Resources Inst., 1985. 107 p. 091582504X
Demonstrates how multinationals can be involved in environmentally responsible development, citing specific cases. Considers the difficulties of environmental-economic assessment and suggests the use of indicators which would measure sustainability. Reviews Third World conditions.

1676. Plant, Christopher, and Plant, Judith. eds. *Green business : hope or hoax?* Philadelphia: New Society, 1991. 136 p. : ill. 086571195X 0865711968 (pbk)
Articles from *The new catalyst* distinguishing between deceptive shallow green consumerism such as "dolphin-safe" tuna and "recyclable" plastics vs. deep green practices such as land trusts, green tax, and sustainable agriculture. HR for consumers, businesspeople, and activists.

1677. Romm, Joseph J. *The once and future superpower : how to restore America's economic, energy, and environmental security.* NY: Morrow, 1992. 320 p. 0688118682
Debunks these myths: wise energy policy hurts economic growth, the free market can revitalize the economy on its own, combating global warming will be costly, cutting the military budget will endanger national security, and higher taxes are needed. Provides a strategy to redirect American industry to sustainable development. HR.

1678. Saunders, Tedd, and McGovern, Loretta. *The bottom line of green is black : strategies for creating profitable and environmentally sound businesses.* SF: Harper, 1993. 282 p. 0062507524 0062507532 (pbk)
Proposes that balancing economic viability with ecological responsibility is imperitive to sustain the planet, meet government regulations, and satisfy consumer demand. Presents detailed, pragmatic, business action plans.

1679. Silverstein, Michael. *The environmental economic revolution : how business will thrive and the Earth survive.* NY: St. Martin's, 1993. 216 p. 0312097972
Describes a set of already well-advanced largely market-based rather than regulation-based changes, which collectively constitute a new economic reality that aligns environmentally and economically sound behavior.

1680. Westerman, Marty, and Joachims, Melody. *The business environmental handbook.* Grants Pass, OR: Oasis/PSI Research, 1993. 283 p. 1555711634 (pbk) 1555713041
Shows how businesses can profit by being environment-friendly. This practical volume includes worsheets, lists of waste exchanges and trade groups, etc. Particularly good for creating comprehensive environmental plans for businesses. HR.

Vocational Guidance

Although a few of the books in the Conservation or the Activism and Direct Action sections offer some vocational guidance, these provide it in greater detail and depth. Some are very specific: Paradis covers nonprofit organizations, one of Basta's books is for scientists and engineers, and Lee, Miller, and Shorto focus on working with wildlife or other animals. The rest are general in scope. VGM's *Resumes for Environmental Careers* is unique in its coverage and is highly recommended as a companion to any of the other guides.

As with directories and almanacs, currency is a crucial factor in evaluating vocational guidance books. Readers are encouraged to supplement these books with more recent information from periodicals, newsletters, and online databases.

Books for varying levels:

1681. Basta, Nicholas. *The environmental career guide : job opportunities with the Earth in mind.* NY: Wiley, 1991. 195 p. 0471534161
Describes the new "green collar" work force: 3.5 million jobs in 1990, expected to double by the year 2000. Surveys types of jobs available, offering an overview, an historical and current assessment, educational requirements, and environmental impact for each type. Lists nonprofit organizations, environmental publications, and federal and state agencies.

1682. ___. *Environmental jobs for scientists and engineers.* NY: Wiley, 1992. 228 p. 047154034X
Describes many jobs for scientists and engineers in the new "green collar" work force. Surveys types of work available, offering an overview, an historical and current assessment, educational requirements, and environmental impact for each type. Lists nonprofit organizations, environmental publications, and federal and state agencies.

1683. Cohn, Susan. *Green at work : finding a business career that works for the environment.* Washington: Island, 1992. 223 p. 1559631740
Discusses general job skills and knowledge, advising that management, marketing, finance, and accounting skills are particularly valued. Correlates traditional job skills with types of "green jobs" and provides a lengthy corporate directory listing each firm's environmental program and contact person. AOT. #

1684. Contessa, Mike. ed. *Environmental employment and volunteer directory.* Phoenix: Environmental Networking, 1994. 48 p.
Provides some background on job hunting and types of environmental jobs. Lists agencies, organizations, and companies in Arizona with environmental job and volunteer opportunities.

1685. Fanning, Odom. *Opportunities in environmental careers.* Lincolnwood, IL: VGM, 1991. 146 p. 0844281611
Examines educational requirements and types of jobs in the life sciences and ecology, environmental protection and public health, natural resources, and land use and human settlements. Provides a good, brief overview. Recommended mainly to supplement more detailed books. #

1686. Kaplan, Andrew. *Careers for outdoor types.* Brookfield, CT: Milbrook, 1991. 65 p. : photos. 1562940228

Includes interviews with a variety of people who work outdoors in land and park management, agriculture, recreation, and related fields. Each one tells what they do, how they got started, and what preparation is required for each career. AOT. #

1687. Kinney, Jane, and Fasulo, Michael. *Careers for environmental types and others who respect the Earth.* Lincolnwood, IL: VGM, 1993. 152 p. 0844241024 0844241032 (pbk)
Describes careers in law, public relations, forestry, journalism, engineering, business and environmental quality control. Includes information on environmental education and job hunting and a directory of state agencies. Provides a good, brief overview. Recommended mainly to supplement more detailed books. #

1688. Lee, Mary Price, and Lee, Richard S. *Opportunities in animal and pet care careers.* Lincolnwood, IL: VGM, 1994. 147 p. 0844240818
Describes careers in veterinary medicine, zoos, animal welfare, working with wildlife, dogs or horse specialists, and related fields. Describes how these careers have changed over time, aptitudes and attitudes, volunteer and apprentice opportunities, education, and future outlook. AOT. #

1689. Miller, Louise. *Careers for animal lovers & other zoological types.* Lincolnwood, IL: VGM, 1991. 158 p. 0844281255
Provides information on job duties, education required, career paths, and making a living in a variety of animal related jobs in veterinary medicine, pet stores, animal shelters, wildlife refuges, wildlife protection, pet grooming, etc. AOT. #

1690. ___. *Careers for nature lovers & other outdoor types.* Lincolnwood, IL: VGM, 1992. 123 p. 0844281328
Examines jobs that allow you to spend a great deal of time outdoors such as agronomy, ornithology, soil conservation, geology, oceanography, toxicology, forestry, and pollution control. Provides a good, brief overview. Recommended mainly to supplement more detailed books. #

1691. Morgan, Bradley J., and Palmisano, Joseph M. eds. *Environmental career directory : a practical, one-step guide to getting a job preserving the environment.* Detroit: Visible Ink, 1993. 348 p. 0810394472
Contains "Advice from the pros" on 17 different fields, the job search process, an extensive job opportunities directory including both internships and professional positions, and guides to career resources such as associations, search agencies, periodicals, and books. HR AOT. #

1692. Paradis, Adrian A. *Opportunities in nonprofit organization careers.* Lincol-

nwood, IL: VGM, 1994. 151 p. 0844240893
Describes opportunities in a variety of nonprofits including associations, cultural organizations, health care, social service, education, environmental organizations, special interest groups, hobbies, and fund-raising. #

1693. Sharp, Bill. *The new complete guide to environmental careers.* 2d ed. Washington: Island, 1993. 364 p. : photos. 1559631783
An introduction to career choice and job hunting covering planning, environmental education, solid waste, hazardous waste, air quality, water quality, land and water conservation, fishery and wildlife management, parks and recreation, and forestry. Includes history of each field, issues and trends, opportunities, education, salary ranges, profiles of professionals, and case studies. HR AOT. #

1694. Shorto, Russell. *Careers for animal lovers.* Brookfield, CT: Millbroook, 1992. 64 p. : photos. 1562941607
Includes interviews with a variety of people who work with animals including a pet store owner, illustrator, dairy farmer, and zoo biologist. Tells what they do, how they got started, and what preparation is required for each career. #

1695. VGM Career Horizons. *Resumes for environmental careers.* Lincolnwood, IL: VGM, 1994. 151 p. 0844241598
Covers the elements of successful resume writing, assembly and layout, and cover letters. Provides dozens of sample resumes and cover letters. HR AOT. #

1696. Warner, David J. *Environmental careers : a practical guide to opportunities in the 90s.* Boca Raton: Lewis, 1992. 267 p. 0873715241
Covers natural resources management, environmental protection, health and safety, environmental education, allied environmental careers, and non-degree technical careers. Includes salaries, types of employers, and employment trends. For students, guidance counselors, and mid-career changers. HR AOT. #

Bibliographies and other Sources

Even the most pacifistic activist should be armed with the facts. This section contains several notable bibliographies which were useful in compiling *Earth Works* and are appropriate for virtually all but the smallest academic and public libraries. Most are general in scope and some are quite comprehensive, particularly the five volume *Beacham's Guide to Environmental Issues & Sources.*

Bibliographies

Three new bibliographies which are also of interest to individual readers are Meredith's *The Environmentalist's Bookshelf*, Stein's *The Environmental Sourcebook*, and Sinclair's *E for Environment*. The EPA's *Guide to Selected National Environmental Statistics in the* U.S. is a clear guide to the sometimes confusing statistical treasure trove of government documents.

Books from the Big Outside and the *Island Press Environmental Sourcebook* provide both descriptions of hundreds of books from various publishers and mechanisms to order the books directly. These two excellent sources can fill most environmental book order needs. Good sources in other chapters include the Acorn *Environmentalists Catalog* (1481) Makower's *Nature Catalog* (1527) and Baldwin's *Whole Earth Ecolog* (1407) Several books in the Literary Studies chapter and *Sisters of the Earth* (1703) include extensive bibliographies.

Books for adults:

1697. Beacham, Walton. *Beacham's guide to environmental issues & sources*. Washington: Beacham, 1993. 5 v. (3350 p.) 0933833318
Contains 39 long chapters on a broad range of topics, providing overviews; subtopics; and lists of current books, journal articles, and videos for each subtopic. HR.

1698. *Bibliographic guide to the environment*. Boston: G.K. Hall, annual. ca. 400 p. 0783821768 (1994)
Describes books and periodicals on conservation, energy, pollution, atmospheric trends, endangered species, waste management, public policy, and law. Interdisciplinary and international in scope.

1699. Davis, Donald Edward. *Ecophilosophy : a field guide to the literature*. San Pedro, CA: Miles, 1989. 137 p. 0936810181
Second ed. due in 1995. Provides over 300 detailed abstracts. Invaluable in the study of the philosophical basis of ecology. HR.

1700. Fletcher, Marilyn. ed. *Reader's guide to twentieth century science fiction*. Chicago: American Library Assn., 1989. 272 p. 0838905048
Provides biographical, critical, and plot summary information on over 125 prominent science fiction and fantasy authors. Includes a bibliography of sci-fi magazines and journals and lists of major award winners. HR.

1701. Foreman, Dave. *Books of the big outside*. Tucson: Ned Ludd, quarterly. 80 p. : ill.
Describes over 400 books, maps, and recordings on a wide variety of subjects emphasizing wilderness, conservation, preservation, and environmental action. Foreman's annotations are highly personal and often

entertaining. Books can all be ordered directly from Ned Ludd. For a free catalog call 1-505-867-0878. HR.

1702. Ingham, Zita, and Steffens, Ron. eds. *Association for the Study of Literature and the Environment bilbliography*, 1990-1993. Jonesboro, AR: ASLE, annual. 120 p.
Contains nearly 700 brief, annotated entries for books, articles, and dissertations written 1990-1993. Covers literary criticism, fiction, poetry, gender, teaching, visual arts, politics, history, philosophy, rhetoric, and theology relevant to environmental studies. HR

1703. *Island Press environmental sourcebook : books for better conservation and management*. Washington: Island, annual. 48 p. : ill.
Contains fairly detailed entries for over 200 recent books on biodiversity, climate change, business and the environment, water, wetlands, forestry, policy, public lands, regulation, solid waste, conservation, and sustainable development. Includes books from Island and other publishers which can be ordered through the catalog. Call 1-800-828-1302 for a free copy. HR.

1704. Jansma, P. E. *Reading about the environment : an introductory guide*. Englewood, CO: Libraries Unlimited, 1993. 252 p. 0872879852
Contains 800 annotated entries for books about ecology, environmental issues, overpopulation, pollution, acid rain, global warming, greenhouse effect, toxic wastes, energy and resource depletion, pesticides, radiation, land management, health problems, and potential solutions. Scientific and technical issues are emphasized.

1705. Meredith, Robert W. *The environmentalist's bookshelf : a guide to the best books*. NY: G. K. Hall, 1993. 272 p. 0816173591
Survey results from 263 noted environmentalists ranking the top books in the field. Provides detailed annotations for the 100 core books and 250 strongly recommended books and briefer listings for an additional 150 books. HR.

1706. Miller, Joseph Arthur. compiler *The Island Press bibliography of environmental literature*. Washington: Island, 1993. 414 p. 1559631899
Includes over 3,000 well indexed, annotated entries describing books, government publications, and journals. Recommended for all public and academic libraries and environmental organizations.

1707. Pringle, David. *Science fiction, the best 100 novels : an English language selection*, 1949-1984. NY: Caroll & Graf, 1985. 224 p. 08881842591
Provides a two page analysis for each novel selected. The high number of great science fiction novels with environmental themes selected by Pringle is noteworthy.

1708. Sheldon, Joseph Kenneth. *Rediscovery of creation : a bibliographical study of the church's response to the environmental crisis.* Metuchen, NJ: Scarecrow, 1993. 282 p. 0810825392
Lists hundreds of books and articles on the history of Christianity and the environment, environmental ethics, creation care, other religions, Christian organizations focusing on creation, and related subjects. Please note that "creation" does not refer to the fundamentalist theory of "creationism."

1709. Sinclair, Patti K. *E for environment : an annotated bibliography of children's books with environmental themes.* New Providence, NJ: Bowker, 1992. 292 p. 0835230287
Contains well written, evaluative entries describing over 500 books written for primary through junior-high level readers. Essential. HR.

1710. Stein, Edith Carol. *The environmental sourcebook.* NY: Lyons & Burford, 1992. 264 p. 1558211640 (pbk) 1558211659
Contains background information followed by annotated lists of books, periodicals, and organizations on these subjects: overview, population, energy, climate and atmosphere, biodiversity, water, oceans, solid waste, hazardous substances, endangered lands, and development. HR.

1711. U.S. EPA. *A Guide to selected national environmental statistics in the U.S. government.* 2nd ed. NY: Diane, 1994. 92 p. 0788102982
Over 30 offices and bureaus of the federal government publish statistics about the environment. This guide describes each program, listing data coverage, collection methods, publications, databases, and contact people. The USGPO ed. can be found in depository libraries.

Fiction

8. Literary Anthologies

Anthologies may be the best introduction for readers unfamiliar with ecofiction and those wishing to expand their knowledge of various writers, subjects, or regions. They often include a wide variety of authors, and sometimes combine short stories, poetry, novel extracts, journal entries, and non-fiction essays in a single volume. Due to this mix one is also advised to check the Anthologies section of the Nature Writing chapter.

The following list correlates anthologies with relevant fiction chapters in *Earth Works* . General: Baron, Begiebing, Haslam, McNair, McNamee, Nunley, Salkeld, and Tanner. Animals: Beard, Downs, Rosen, West, and Wilbur. Ecofeminist: Allen, Anderson, Brant, Piekarski, Thomas, Zahara. Environmental Crisis: Frost, Stovall. Frontier: Piekarski. Native American: Allen, Brant, Edmonds, Evers, King, Lerner, Lesley, Peyer, Sarris, Trafzer, and Velie. New West: Blackburn, Hedin, Kittredge, Martin, Miller, and Vinz. Science Fiction: Dozois and Robinson.

Books for adults:

1712. Allen, Paula Gunn*. ed. *Spider Woman's granddaughters : traditional tales and contemporary writing by Native American women.* Boston: Beacon, 1989. 242 p. 080708100
AR: *American Indian literature,* 1900-1970.
Contains 24 selections by Allen, Zitkala-Sa, Louise Erdrich, Ella Deloria, Mary Tallmountain, Linda Hogan, Leslie Silko, and others. Emphasizes the unity and interdependence of humanity, animals, plants, the Earth itself and the spirit world. AOT. #

1713. Anderson, Lorraine. ed. *Sisters of the Earth : women's prose and poetry about nature.* NY: Vintage, 1991. 426 p. 0679733825
Essays, poems, and stories written by 90 American women naturalists and literary authors, including Emily Dickinson, Laura Ingalls Wilder, Terry Tempest Williams, Joy Harjo, May Swenson, Willa Cather, Alice Walker, and Marge Piercy. Reveals important parallels between nature and women as nurturers, healers, and victims and those between feminist and ecological thought. HR AOT. #

1714. Baron, Robert C., and Junkin, Elizabeth Darby. eds. *Of discovery & destiny : an anthology of American writers and the American land*. Golden, CO: Fulcrum, 1986. 414 p. 1555910041
Poetry, short stores, and excerpts about the American experience of the land and nature by a wide variety of authors including Parkman, Teale, Longfellow, Whitman, Mowat, Carson, Thoreau, Dillard, Lopez, Cooper, Cather, London, Muir, McPhee, Austin, Twain, Burroughs, and Abbey. HR AOT. #

1715. Beard, Patricia. ed. *The Voice of the wild : an anthology of animal stories*. NY: Viking, 1992. 245 p. 0670842931
Contains 19 entertaining factual and fictional encounters with the animal world stressing the development of environmental consciousness based on our common natural heritage. Authors include Tim Cahill, Pat Conroy, Isaac Dinesen, Kipling, London, McGuane, Seton, and Jean Stafford. #

1716. Begiebing, Robert J., and Grumbling, Owen. eds. *The Literature of nature : the British and American traditions*. Medford, NJ: Plexus, 1990. 730 p. 0937548170 (pbk) 0937548162
Prose and poetry by Blake, Wordsworth, Coleridge, Keats, Tennyson, Darwin, Hopkins, Jefferson, Bartram, Whitman Thoreau, Emerson, Dickinson, Muir, Marsh, Jewett, Lawrence, Faulkner, Fowles, Snyder, Dillard, Burroughs, Yeats, Frost, London, and many others. HR AOT. #

1717. Blackburn, Alexander. ed. *The Interior country : stories of the modern West*. Athens, OH: Swallow, 1987. 333 p. 0804008876 0804008884 (pbk)
Emphasizes the personal connection between these writers and a Western landscape threatened by overdevelopment. Authors include Frank Waters, Ed Abbey, Walter Van Tilburg Clark, Craig Leslie, Rudolfo Anaya, John Nichols, Max Schott, William Kittredge, Jean Stafford, Gladys Swan, Raymond Carver, Clark Brown, Barry Lopez, Leslie Marmon Silko, Wallace Stegner, James B. Hall, Joanne Greenberg, William Eastlake, and David Kranes. HR.

1718. Brant, Beth*. ed. *A Gathering of spirit : a collection by North American Indian women*. Ithaca, NY: Firebrand, 1988. 238 p. 0932379567 0932379559 (pbk)
Poetry, journal entries, stories, novel excerpts, letters, and drawings from a wide variety of native women. Dispells the "squaw" myth of dependence and secondary status, presenting native women as strong and creative, in touch with the Earth, and active in the process of revitalizing traditional Indian cuture.

1719. Downs, Robert Bingham. ed. *The bear went over the mountain : tall tales of American animals*. Detroit: Omnigraphics, 1994, c1964. 358 p. 155889310
Contains 60 short, humorous selections by E.B. White, Davy Crockett, Jesse

Stuart, H.L. Mencken, Joel Chandler Harris, Mark Twain, Vardis Fisher, Josh Billings, James Thurber, Ring Lardner, Robert Benchly, William Saroyan, and others. #

1720. Dozois, Gardner R., and Williams, Sheila. eds. *Isaac Asimov's Earth.* NY: Ace, 1992. 210 p. 0441373771
Originally published in *Isaac Asimov's science fiction magazine.* Stories about various environmental problems by Frederik Pohl, D. A. Smith, Marta Randall, Dave Smeds, Nancy Kress, Mary Rosenbkum, Michael Swanwick, Kim Stanley Robinson, and Charles Sheffield.

1721. Elder, John, and Wong, Hertha Dawn. eds. *Family of Earth and sky : indigenous tales of nature from around the world.* Boston: Beacon, 1994. 323 p. 0807085286
Contains 70 legends from a wide variety of indigenous cultures divided into four themes: origins, animal tales and transformations, tricksters, and tales to live by. Intended to reinforce the environmental awareness in Western nature writing by placing it within a more global context. HR.

1722. Evers, Larry. ed. *The South corner of time : Hopi, Navajo, Papago, Yacqui tribal literature.* Ann Arbor: Books on Demand, 1993, c1980. 240 p. : photos. 0783750463
Short prose selections and poems revealing a diverse group of modern authors, artists, and photographers whose work is grounded in Earth-centered tribal traditions.

1723. Frost, Helen. ed. *Season of dead water.* Portland: Breitenbush, 1990. 113 p. 0932576834 (pbk) 0932576826
Stories, poems, and essays about the *Exxon Valdez* oil spill by over 50 authors.

1724. Haslam, Gerald W. ed. *Many Californias : literature from the Golden State.* Reno: Univ. of Nevada, 1992. 388 p. 0874171822 0874171830 (pbk)
Benjamin Franklin Award.
AR: *California heartland.*
This multi-region, multi-period, and multi-cultural collection rebuts California stereotypes by presenting a varied physical, cultural, and literary landscape. Authors include Harte, Twain, Bierce, London, Norris, Saroyan, Jeffers, Muir, Kerouac, Snyder, Didion, Stegner, Soto, Ina Coolbrith, Joaquin Miller, Helen Hunt Jackson, and Lawrence Clark Powell. HR.

1725. Hedin, Robert, and Stark, David. eds. *In the dreamlight : twenty-one Alaskan writers.* Port Townsend, WA: Copper Canyon, 1984. 161 p. 0914742752 (pbk) 0914742760

The shamanic tradition and the unity of the material and spiritual realms is featured in 21 short stories and poems by Hedin, Stark, John Haines, Mary Tallmountain, Mary Baron, and others.

1726. **King, Thomas*. ed.** *All my relations : an anthology of contemporary Canadian native fiction.* Norman: Univ. of Oklahoma, 1992. 220 p. 0806124296
A wide variety of prose and poetry from 19 current authors including King, Harry Robinson, Peter Blue Cloud, and Beth Brant. Explores the essential relationships that exist in traditional cultures between people, animals, the land, reality, and imagination.

1727. **Kittredge, William, and Smith, Annick. eds.** *The last best place : a Montana anthology.* Seattle: Univ. of Washington, 1988. 1158 p. 0295969741
Contains stories, novel excerpts, poetry, journal entries, and essays beginning in native times, but with more emphasis on Montana as a modern literary retreat. Authors include George Catlin, Lewis & Clark, Audubon, Norman MacLean, William Kittredge, Richard Brautigan, Thomas McGuane, Ivan Doig, and Richard Hugo. Despite a dearth of female authors it is HR.

1728. **Lerner, Andrea. ed.** *Dancing on the rim of the world : an anthology of contemporary Northwest native American writing.* Tucson: Univ. of Arizona, 1990. 266 p. 0816512159
These poems and stories by Northwest Indian authors explore how "land and literature shape each other and how the landscape itself become a character and a force in our lives." Authors include James Welch, Mary Tallmountain, Duane Niatum, Chrystos, R. A. Swanson, and Jo Whitehorse Cochran. HR.

1729. **Lesley, Craig, and Stavrakis, Katheryn. eds.** *Talking leaves : contemporary Native American short stories.* NY: Laurel, 1991. 385 p. 0440503442
Stories implicitly or explicitly conveying respect for nature by Paul Gunn Allen, Joseph Bruchac, Michael Dorris, Louise Erdrich, Joy Harjo, Linda Hogan, Thomas King, N. Scott Momaday, Duane Niatum, Mary Tallmountain, Gerald Vizenor, James Welch and others. HR.

1730. **Martin, Russell. ed.** *New writers of the Purple Sage : an anthology of contemporary Western writers.* NY: Penguin, 1992. 381 p. 0140169407
Includes 24 selections about the contemporary West by William Kittredge, Gretel Ehrlich, Rudolpho Anaya, N. Scott Momaday, James Welch, Pam Houston, Charles Bowden, Lisa Sandlin, Barbara Kingsolver, Linda Hogan, Gary Nabhan, Tim Sandlin, Terry Tempest Williams, and others. HR.

1731. **Martin, Russell, and Barasch, Marc. eds.** *Writers of the purple sage : an anthology of recent western writing.* NY: Penguin, 1984. 340 p. 0140073701

Includes 20 selections about the contemporary West by N. Scott Momady, Leslie Marmon Silko, Ivan Doig, Norman Maclean, Rudolpho Anaya, David Quammen, Gretel Ehrlich, Ed Abbey, Richard Ford, William Kittredge, Thomas McGuane, and others. HR.

1732. McNair, Wesley. ed. *The Quotable moose : a contemporary Maine reader.* Hanover, NH: Univ. of New England, 1994. 253 p. 0874516730
Fiction, poetry, and essays by 40 Maine writers including Carolyn Chute, Richard Gillman, and Elizabeth Cooke.

1733. McNamee, Gregory. ed. *Named in stone and sky : an Arizona anthology.* Tuscon: Univ. of Arizona, 1993. 196 p. 0816513481
A diverse collection of stories, poems, and essays from native traditions or by modern writers including Van Dyke, Elman, Ferlinghetti, Kingsolver, Austin, Cather, Bowden, Powell, Momaday, Lummis, Leopold, Lawrence, Kerouac, and Abbey. AOT. #

1734. Miller, John, and Anderson, Genevieve. eds. *Southwest stories : tales from the desert.* SF: Chronicle, 1993. 210 p. 0811802167
Stories, poems, letters, and legends by Larry McMurtry, John Muir, Sandra Cisneros, Henry Miller, Georgia O'Keefe, Edward Abbey, Barbara Kingsolver, D.H. Lawrence, Sam Shepard, Leslie Marmon Silko, and others.

1735. Nunley, Richard. ed. *The Berkshire reader : writings from New England's secluded paradise.* Stockbridge, MS: Berkshire House, 1992. 530 p. : ill. 0936399333
Stories, essays, poems, journals, and misc. writings about the land and people from the 17th century to the present day by Thoreau, Emerson, Hawthorne, Melville, Wharton, Millay, Wilbur, and many others. HR.

1736. Peyer, Bernd. ed. *The Singing spirit : early short stories by North American Indians.* Tucson: Univ. of Arizona, 1989. 175 p. 0816511144
Contains short stories written 1881-1936, a period when Native American literature was moving from oral to written form. Authors include Susette Le Flesche, Pauline Johnson, Angel DeCora, William Jones, Francis Le Flesche, Gertrude Bonnin, Charles A Eastman, Alexander Posey, John M. Oskison, John Joseph Mathews, and D'Arcy McNickle.

1737. Piekarski, Vicki. ed. *Westward the women : an anthology of Western stories by women.* Albuquerque: Univ. of New Mexico, 1988, c1984. 179 p. 082631063X
Stories by women writers from the late 19th to mid 20th century presenting a different point of view of the West from male writers of that era. Authors include Cather, Austin, Sandoz, Eustis, Silko, and seven others.

1738. Robinson, Kim Stanley. ed. *Future primitive*. NY: TOR, 1994. 384 p.
0312854749
Prose and poetry depicting simple, environmentally responsible, low-tech
societal options instead of the standard, mechanistic sci-fi models.
Authors include Terry Bisson, Gary Kilworth, Robert Silverberg, Ursula Le
Guin, and others. AMT. #

1739. Rosen, Michael J. ed. *The Company of animals : 20 stories of alliance and
encounter*. NY: Doubleday, 1993. 272 p. : ill. 0385468172
AR: *The company of dogs*, *The company of cats*.
Features short stories about wild and domestic animals by a diverse array
of authors including Maxine Kumin, J.F. Powers, Leslie Marmon Silko, Italo
Calvino, Robley Wilson, Jr., and Haruki Murakami. HR. #

1740. Salkeld, Audrey, and Smith, Rosie. eds. *One step in the clouds : an
omnibus of mountaineering novels and short stories*. SF: Sierra Club, 1991. 1056 p.
: ill. 0871566389
Includes 31 short stories by Guy de Maupassant, Anne Sauvy, John Daniel,
John Long, and others; *The ice chimney*, a play by Barry Collins; the novellas
Mother Goddess of the world by Kim Stanley Robinson, and *Like water and like
wind* by David Roberts; and the novels *One green bottle* by Elizabeth Coxhead,
North Wall by Roger Hubank, *Solo faces* by James Salter, and *Vortex* by David
Harris. Essential for mountaineers.

1741. Sarris, Greg*. ed. *The Sound of rattles and clappers : a collection of new California Indian writing*. Tucson: Univ. of Arizona, 1994. 161 p. 0816512809
0816514348 (pbk)
Poems, stories, and essays by Janice Gould, James Luna, Stephen Meadows, William Oandasan, Wendy Rose, Georgiana Valoyce-Sanchez, Greg
Sarris, Kathleen Smith, and Darryl Babe Wilson. They evince a strong love
for their native land and people and express anger over their degradation.

1742. Stovall, Linny. ed. *Extinction*. Hillsboro, OR: Blue Heron, 1992. 149 p.
: ill. 0936085509
Issue #2 of *Left bank* magazine. Stories, essays, poems, interview, and
reviews about vanishing species and habitat loss by Barry Lopez, Tess Gallagher, Sallie Tisdale and others. This varied anthology is of the highest literary quality. HR.

1743. Tanner, Tony. ed. *The Oxford book of sea stories*. NY: Oxford, 1994. 410 p.
0192142100
Includes 27 sea stories by Conrad, Crane, Faulkner, Fitzgerald, Forster,
Hemingway, Kipling, London, Masefield, Melville, Poe, H. G. Wells, and
others. HR AOT. #

1744. Thomas, Sue. ed. *Wild women : contemporary short stories by women cele-brating women*. Woodstock, NY: Overlook, 1994. 368 p. 0879515147
This ecofeminist anthology contains 47 short stories celebrating the kinship of women to nature, the wild nature within women, and the wild woman archetype. Authors include Isabel Allende, Erica Jong, Andrea Dworkin, Fay Weldon, Alice Walker, Margaret Atwood, Anne Rice, and Pam Houston. HR.

1745. Trafzer, Clifford E. ed. *Earth song, sky spirit : short stories of the contempo-rary native American experience*. NY: Doubleday, 1993. 495 p. 0385469594
Stories and excerpts from novels by Louise Erdrich, Michael Dorris, Leslie Marmon Silko, James Welch, Gerald Vizenor, Joseph Bruchac, Paula Gunn Allen, Duane Niatum, Diane Glancy, and others. HR.

1746. Velie, Alan R. ed. *The lightning within : an anthology of contemporary American Indian fiction*. Lincoln: Univ. of Nebraska, 1991. 161 p. 0803246595
Selections by N. Scott Momaday, Louise Erdrich, Michael Dorris, Simon Ortiz, James Welch, Leslie Marmon Silko, and Gerald Vizenor. HR.

1747. Vinz, Mark, and Tammaro, Thom. eds. *Inheriting the land : contemporary voices from the Midwest*. Minneapolis: Univ. of Minnesota, 1993. 335 p. 0816623031
Poetry, stories, and essays from 84 contemporary authors including Robert Bly, Meridel LeSueur, Jon Hassler, Bill Holm, Louise Erdrich, Garrison Keilor, Dan O'Brien, Linda Hasselstrom, and Larry Woiwode. "Explores the quiet, idyllic beauty of the region, the cohesive spirit of its people, and the diversity and power of the land itself."-Intro.

1748. West, Mark I. ed. A *wondrous menagerie : animal fantasy stories from Amer-ican children's literature*. Hamden, CT: Archon, 1993. 139 p. 0208023836
Includes 17 stories, mostly from the late 19th and early 20th century by Harriet Beecher Stowe, Louisa May Alcott, Mark Twain, Joel Chandler Har-ris, Zitkala-Sa, L. Frank Baum, Henry Beston, and others. Good family read-ing. HR. #

1749. Wilbur, Richard. comp. A *bestiary*. Drawings by Alexander Calder. NY: Pantheon, 1993, c1955. 74 p. : ill. 0679428755
Stories, tales, novel excerpts, and poems about all manner of creatures. Authors include Plato, Homer, Shakespeare, Thoreau, Vachel Lindsay, Mari-anne Moore, Elizabeth Bishop, T. E. Lawrence, Sherwood Anderson, Jose Ortega y Gassett, Gerard Manley Hopkins, Richard Eberhart, William Faulkner, W. B. Yeats, Franz Kafka, and many others. HR AOT. #

1750. *Writing & fishing the Northwest*. Hillsboro, OR: Blue Heron, 1991. 144 p. 0936085193

Issue #1 of *Left bank* magazine. Stories, poems, essays, interviews, and reviews about writing, fishing, and writing about fishing by Wallace Stegner, Craig Lesley, William Gibson, and others. HR.

1751. Zahava, Irene. ed. *Hear the silence : stories by women of myth, magic & renewal.* Trumansburg, NY: Crossing, 1986. 194 p. 0895942127 0895942119 (pbk)
Stories by various authors relating the need to communicate with the Earth in order to live a balanced life.

9. General Fiction

This chapter contains novels by classic and significant contemporary authors which do not fit into a particular genre such as science fiction or mystery or have a specific theme such as environmental action. It includes such distinguished novelists as Bellow, Defoe, Faulkner, Gardner, Huxley, D. H. Lawrence, Melville, Steinbeck, and Twain.

The dearth of female authors in this chapter is not intended to imply that they should continue to be excluded from the literary canon. I would argue the contrary, noting the overall quality of Ecofeminist fiction and the major contribution made by women to a variety of genres. The canon might be expanded to include Atwood, Austin, Cather, Erdrich, Le Guin, Piercy, Leslie Silko, and Wilhelm at the least. They are not listed in this chapter because some write in specific genres and because after consultation with ecofeminist scholars I decided to provide separate chapters for Ecofeminist and Environmental Action fiction to draw greater attention to these increasingly significant forms.

Apppropriate Literary Anthologies include *Of Discovery and Destiny*, *The Literature of Nature*, *Many Californias*, *The Quotable Moose*, *Named in Stone and Sky*, *The Berkshire Reader*, *One Step in the Clouds*, *The Oxford Book of Sea Stories*, and *Writing and Fishing in the Northwest*.

The teen section includes adaptations of adult classics, two young adult books by Steinbeck, and general or adventure books by Dickinson, White and others. The classic works of Stratton-Porter (2209-12) and most novels by Paulsen (2198-2207) can be found in Country Life.

Books for adults:

1752. Barth, John. *The sot-weed factor.* Garden City, NY: Doubleday, 1987, c1960. 756 p. 0385240880
An early American settler and tobacconist learns the disparities between the heroic images and hardscrabble realities of settlement in this comic picaresque. HR.

1753. Bellow, Saul. *Henderson, the rain king.* NY: Penguin, 1984. 342 p. 0140072691
An American millionaire travels to Africa to search for himself in confronta-

tion with the wilderness, and finds a man he never knew inside. After making rain, he is chosen for an experiment with a lion. Will his new powers suffice? HR.

1754. Berger, Thomas. *Robert Crews*. NY: Morrow, 1994. 240 p. 0688119204
AR: *Little Big Man*.
In this modern variation of *Robinson Crusoe* a middle-aged alcoholic is stranded after a plane crash and forced to face the specters of his past in isolation. He develops new physical and psychological resources to survive and meets Friday, a woman escaping a violent husband, in this tale of the restorative powers of nature and love.

1755. Boyle, T. Coraghessan. *The road to Wellville*. NY: Viking, 1993. 476 p. 0670837660
AR:*World's end*.
In 1907, Dr. John Harvey Kellogg's attempts to improve America's eating habits and operate a world-class "temple of health" resort are undermined when his adopted son is used by unscrupulous rivals to market a spurious "Kellogg" health food called Perfo.

1756. Brautigan, Richard. *Richard Brautigan's Trout fishing in America* ; *The pill versus the Springhill mine disaster* ; *and, In watermelon sugar*. Boston: Houghton Mifflin, 1989, c1968. 360 p. : ill. 0395500761
Three of Brautigan's classic novellas published together.
AR: *A confederate general from Big Sur*. All three books contrast the ruins of modern industrial society with the faint hope of hippie pastoralism. In *Trout fishing* he presents once pristine trout streams as stairs, water faucets, telephone booths, and wrecking yard scrap. HR AMT. #

1757. Defoe, Daniel. *Robinson Crusoe*. NY: Knopf, 1991. 256 p. 0679405852
The 1991 Children's Classics ed. recommended for younger readers. The diary of an Englishman shipwrecked for thirty years on a deserted island, where he uses his wits and inner resources not only to survive, but to build a new life with his new companion, Friday. This classic is HR AOT, despite the obsolescence of its vision of enlightened human despotism over nature. #

1758. Dickey, James. *Deliverance*. NY: Dell, 1992, c1970. 236 p. 0440318688
When four city men test their manhood on a wilderness canoe trip, they become the prey of vicious rednecks and the rampaging river. HR AMT. #

1759. ___. *To the white sea*. Boston: Houghton Mifflin, 1993. 275 p. 0395475651
An American flyer shot down over Tokyo during WW II calls on the knowl-

edge and skills he used in the Alaskan wilderness to work his way across Japan in an attempt to escape back into nature on the northern island of Hokkaido.

1760. Faulkner, William. B*ig woods : the hunting stories.* NY: Vintage, 1994, c1955. 198 p. 0679752528
Faulkner is hunting for far more than bear as he considers the conflict between wilderness and society and the effect of nature on the human mind. Includes his classic novella, *The Bear,* which focuses on the conflict between human exploitation and wilderness. HR.

1761. ___. *Go down, Moses.* NY: Vintage, 1990, c1942. 365 p. 0679732179
Considers the destruction of the southern wilderness and its consequences for Indian, black, and white people and for the human soul. HR.

1762. Gardner, John. *Grendel.* NY: Vintage, 1989, c1971. 174 p. : ill. 0679723110
HR: *The sunlight dialogues.*
A retelling of *Beowulf* from the "monster's" perspective. Grendel's early separation from his mother prefigures his separation from both nature and humanity, and his intense sense of alienation. In death, he is reunited with nature and transformed: "Is it joy I feel?" HR.

1763. Hawkes, John. *Adventures in the Alaskan skin trade.* NY: Penguin, 1986. 396 p. 0140092838
Hawkes uses Alaska as a metaphor for the wildness within people and uses the story's action to explore eroticism, writing, relationships, human nature, and nature.

1764. ___. *The beetle leg.* London: Chatto & Windus, 1967, c1951. 159 p. 0811200620
The memory of a construction worker buried alive in a dam haunts its builders and affects its completion. An interesting theme, but the obtuse avant-garde style makes this a very uneven book.

1765. Hawthorne, Nathaniel. *The Blithedale romance.* NY: Oxford, 1991, 1852. 256 p. 0192825984
A group of dreamers flees the city to form a utopian socialist farm and live on the land, but their idealism is tempered by the hard work and their own flaws and ambitions. Based on the actual histories of some communes of the time, it foreshadows the fate of some similar experiments in the 1930s and 1960s.

1766. Hersey, John. B*lues.* NY: Vintage, 1988, c1987. 205 p. 0394757025

AR: *Hiroshima diary, The marmot drive, My petition for more space.*
Hersey reflects on fishing, the natural history of the bluefish, the wonders of the ocean, and the web of interdependence of all living things.

1767. ___. *Key West tales.* NY: Knopf, 1994. 227 p. 0679429921
The Pulitzer Prize winning author's final book contains 15 stories set in and around Key West, with some stories dealing with nature or environmental themes, particularly "Did you ever have such sport?" wherein Audubon visits Key West and slaughters birds for sport.

1768. Huxley, Aldous. *Ape and essence.* Chicago: Dee, 1992, c1948. 216 p. 0929587782
In 1948 a movie director finds a script set in 2108 about a scientific expedition from New Zealand to North America. America was ravaged by nuclear war and is now controlled by devil worshipping mutants. Another warning from Huxley that technologically enhanced aggression uncurbed by greater wisdom or spirituality is self-destructive.

1769. ___. *Brave new world and* Brave new world revisited. NY: Harper & Row, 1990, c1932. 199, 97 p. 0060901012
In an age of socially engineered babies, a rare naturally born young man known as the Savage initially delights in the superficial wonders of civilization. Then he becomes increasingly alienated by the shallowness, artificiality, and meaningless of such a controlled existence. HR AMT. #

1770. ___. *Island.* NY: Collins, 1989, c1962. 304 p. 006680985X
When an oilfield developer is shipwrecked on the Indian Ocean island of Pala, he tries to overthrow the antitechnological native government for one that will allow him to drill. He slowly realizes that the Pala is an idyllic Utopia and that he has lived an empty, miserable life.

1771. Lawrence, D. H. *Lady Chatterley's lover.* NY: Grove, 1993. 384 p. 0802133347
When her aristocratic husband is unable to impregnate her, Connie Chatterly falls in love with a sensual, nature-oriented gamekeeper. The furor over the book's unbridled sexuality has obscured Lawrence's usual contrasting of the power of the natural world and the destructiveness of the class system and modern industrialism. HR.

1772. ___. *The plumed serpent (Quetzalcoatl).* NY: Vintage, 1992, c1926. 445 p. 0679734937
A Irishwoman who falls under the spell of the Mexican landscape and the myths of Quetzalcoatl marries a mixed-blood Indian and tries to cope with the contradictions between an ancient culture and the modern industrial world. HR.

1773. ___. *The white peacock.* NY: Penguin, 1990. 432 p. 0140182195
Lettie and George search of their authentic selves, but are thwarted by
mind-body dualism. Lawrence rejects "idealism," or living according to
other people's ideas, in favor of a more unified and naturalistic life.

1774. Matthiessen, Peter. *Far Tortuga.* NY: Vintage, 1988, c1975. 408 p.
0394756673
Describes the mythic voyage of the schooner *Lillias Eden* from Grand Cay-
man to the turtle fishing grounds of the Carribean. The crew must overcome not just a cruel and domineering captain, but hurricanes, shipwrecks,
and witchcraft.

1775. McClanahan, Ed. *The natural man.* Frankfort, Ky: Gnomon, 1993,
c1983. 229 p. 0917788567
A coming-of-age book for adults about what it means to be "natural" in
modern America and its implications. AMT. #

1776. McGuane, Thomas. *The bushwhacked piano.* NY: Random House, 1994,
c1971. 190 p. 0394258864
Nicholas Payne goes against the grain of American life, pursuing out-
landish dreams such as "bat towers" in an attempt to forge a way of life
which is free, imaginative, and respectful of nature.

1777. ___. *Ninety-two in the shade.* NY: Penguin, 1987, c1973. 197 p.
0140099077
A marine biology student, fleeing the "Hotcakesland" of consumerism, pollution, and development, winds up in Key West and dreams of being an
ecologically conscious fishing guide. After burning another guide's boat in
retaliation for a prank, he pursues the dream despite a death threat. This
violent but sensitive novel which some critics compared to Hemingway's
best work is HR.

1778. ___. *The sporting club.* NY: Buccaneer, 1994, c1968. 220 p. 1568494017
When the Centennial Club holds its centenary celebration, one of the
younger men decides it is time to "purify" the membership which is limited
to those bred to rule: the American Royalty. His plan takes shape as the
opening of a time capsule sets off a wild, abandoned party, and men revert
to a murderous level.

1779. Melville, Herman. *Moby-Dick, or, The whale.* NY: Vintage, 1991. 660 p.
0679725253
A sailor recounts the ill-fated voyage led by the fanatical Captain Ahab in
search of the white whale that had crippled him. This exploration of the
dark side of human nature, pitting human compulsiveness against the

power of nature, is perhaps the greatest American novel and is a key precursor of ecofiction. HR AOT. #

1780. ___. *Typee : a peep at Polynesian life ; Omoo : a narrative of adventures in the South Seas ; Mardi, and a voyage thither.* NY: Viking, 1982. 1333 p. 0940450003
Known collectively as *The Polynesian trilogy.* In *Typee* the narrator and another sailor desert on a South Pacific island, are overcome by the dense cloud forest, and are taken in by a supposedly savage tribe which is actually gentle and Earth-oriented. In *Omoo* the now more feral narrator meets a "renegado" wilder than himself. When the sailors attempt to escape in Tahiti they are placed in a stockade where they, too, reach a wilder state. *Mardi* is a more complex inquiry into the themes raised in the first two. All are HR for adults, the first two HR AOT. #

1781. Miller, Henry. *Big Sur and the oranges of Hieronymus Bosch.* NY: New Directions, 1964, c1957. 484 p. 0811201074
A literary bohemian finds paradise in an isolated, physically primitive, intellectually sophisticated community. Although not as powerful or erotic as his *Tropic* books, this evocation of nature and beat culture is still recommended for mature adults.

1782. Rothschild, Michael. *Wondermonger.* NY: Viking, 1990. 230 p. 0670833266
AR: *Rhapsody of a hermit.*
Rothschild combines nature, myth, psychological and spiritual insights, and lyrical prose in nine short stories which some reviewers likened to the work of Faulkner or Garcia Marquez. Includes three short stories from *Rhapsody of a hermit.* HR.

1783. Shacochis, Bob. *The next new world.* NY: Penguin, 1990, c1989. 209 p. 0140121056
Winner, Prix de Rome.
Short stories from *Outside* magazine columnist, approximately half of which have a salty southern maritime flavor. Some violence and sexual conduct AMT. #

1784. ___. *Swimming in the volcano.* NY: Scribner, 1993. 519 p. 0684192608
An American expatriate becomes embroiled in political and romantic intrigue. Subplots include the nature vs. development controversy, the lives and livelihood of fishers and divers, and a conflict between a young man and his father at sea. AMT. #

1785. Simpson, Thomas William. *Full moon over America.* NY: Warner, 1994. 416 p. 0446518085
A descendent of robber barons on his father's side and Mohawk Indians on

his mother's is elected president after he captures the mood of the disgusted electorate by mooning the incumbent. He plans to be the first ecologically-oriented president, but can a true reformer and outsider hope to even survive his inauguration?

1786. Sinclair, Upton. *Oil!* Cambridge: Bentley, 1981, c1927. 527 p. 0837604443
This classic muckraking novel exposes the environmental destruction and political corruption created by the oil industry. HR.

1787. Steinbeck, John. *Cannery Row.* NY: Penguin, 1994, c1945. 185 p. 0140187375
AR sequel: *Sweet Tuesday.*
Tales of a group of working class and unemployed men living in Monterey in the 1930s. The depictions of the Central California coast and tide pool ecology are very vivid. The central character, a marine biologist, is based on Ed Ricketts who co-authored *Sea of Cortez* with Steinbeck. HR AOT. #

1788. ___. *The grapes of wrath.* NY: Penguin, 1992, c1939. 640 p. 0140186409
Pulitzer Prize.
A family of Oklahoma farmers escapes the devastation of the dust bowl only to find that in the "golden land" of California the workers work the land while the owners get the gold. HR AOT. #

1789. ___. *To a God unknown.* NY: Penguin, 1986, c1933. 264 p. 0140042334
A homesteader is driven to extreme action by his frustration over a withering drought. AOT. #

1790. Theroux, Paul. *Millroy the Magician.* NY: Random House, 1994. 437 p. 067940247
An unusual magician enlists a tiny fourteen year old girl in his attempts to improve America's health through vegetarianism. He finds his perfect platform as a TV evangelist promoting the pure foods mentioned in the Bible.

1791. ___. *The Mosquito Coast.* NY: Buccaneer, 1994, c1982. 374 p. 1568493487
An outcast inventor moves his family to Central America for a simpler and happier life in the jungle. His attempts to introduce technology by building an ice plant for Indians who have never even seen ice bring unforseen consequences for the jungle and for his family. HR.

1792. Tobias, Michael. *Voice of the planet.* NY: Bantam, 1990. 388 p. 0553283677
AR: *Deva.*

An ecologist travels to a Buddhist monestary in the Himalayas where he meets Gaia, herself, speaking through a computer. Their relationship moves from the intellectual and spiritual to the emotional realm, where their powerful attraction but tumultuous relationship mirrors humanity's life on Earth. HR AMT. #

1793. Twain, Mark. *Adventures of Huckleberry Finn.* NY: Dover Thrift, 1994, c1885. 224 p. 0486280616
Modern Library ed. (1993) recommended for heavy use.
AR: *Life on the Mississippi.*
A young man and an escaped slave who desire freedom and adventure "light out for the territories" on a log raft down the Mississippi River. This classic American novel is HR. #

1794. ___. *Roughing it.* Berkeley: Univ. of Cal., 1992, c1872. 1100 p. : ill. 0520084985
Twain presents his traveling and mining experiences in Nevada and California as a series of fictionalized tall tales. Although he sometimes praises nature and mildly protests the worst mining excesses, his bemused reaction to a massive forest fire is a powerful example of the myth of the bottomless bounty of the west. AMT. #

Books for young adults:

1795. Dickinson, Peter. *A bone from a dry sea.* NY: Delacorte, 1993, c1992. 199 p. 0385308213
In two parallel stories, an intelligent female member of a prehistoric tribe becomes instrumental in advancing the lot of her people, and the daughter of a paleontologist is visiting him on a dig in Africa when important fossil remains are discovered. Recommended mainly for teens, but also for adults. #

1796. ___. *Tulku.* NY: Dell, 1993. 286 p. 0440214890
A 13 year-old boy escapes from slaughter by the Boxers in China and joins forces with an English botanist and her escort, traveling with them to Tibet where the power of Buddhist monks transforms the lives of all of them. #

1797. Melville, Herman. *Moby Dick.* Adapted and ill. by Bill Sienkiewicz, Dan Chichester, and Willie Schubert. NY: Berkley, 1990. 41 p. : ill. 0425120236
Ishmael, a sailor, recounts the ill-fated voyage of a whaling ship led by the fanatical Captain Ahab in search of the white whale that had crippled him. Presented in comic book format. Recommended for upper elementary and reluctant teen readers.

1798. Steinbeck, John. *The pearl*. NY: Penguin, 1992, c1945. 90 p.
014017737X
An adaptation of a Mexican folktale wherein a fisherman finds a huge pearl, which causes unexpected misfortunes for his family when it is removed from the sea. This parable about the relation between nature and society contains lovely depictions of nature and rural life in Mexico and is HR. #

1799. ___. *The red pony*. NY: Penguin, 1992, c1945. 100 p. 0140177361
The story of a conflict between a nine year old boy, his father, and a hired hand. Young Tom's relationship with his pony teaches him to love and respect animals and himself. HR. #

1800. Twain, Mark. *The adventures of Huckleberry Finn*. NY: Children's Classics, 1992. 244 p. : ill. 0517081288
Recounts the adventures of a young boy and an escaped slave as they travel down the Mississippi River on a raft. GR 4-10. Joanne Gise's adaptation (Troll, 1990) recommended for younger and reluctant readers. HR.

1801. White, Robb. *Deathwatch*. NY: Dell, 1973. 220 p. 0440917409
AR: *No man's land*.
Needing money for school, a college boy accepts a job as guide on a desert hunting trip and nearly loses his life. #

10. Environmental Crisis

Jeffries' *After London*, published in 1885, may be the first modern English language disaster novel. Orwell's *Coming Up for Air* appeared in 1939, while the first major American eco-thriller, Stewart's *Earth Abides*, was published ten years later. Stone had already written the ultra-realistic novels *Storm* and *Fire*, which featured the natural forces themselves as protagonists, but *Earth Abides* was about human caused ecocatastrophe.

Although novels about the consequences of nuclear war or testing were common in the 1950s and 60s, the proliferation of all manner of environmental crisis books has occured in the last fifteen years. The prevailing mood is one of doom, gloom, fear, and dystopia. Nearly a quarter deal with the ecological and social consequences of nuclear war, another quarter depict a general environmental crisis in the 21st century or beyond, and an equal number insist that it's already happening now or will in the very near future. The lethal effects of air pollution are featured in approximately ten books as are the consequences of earthquakes, particularly upon dams. Threats as diverse as volcanos, drought, killer bees, rising sea levels, oil spills, nerve gas, and fires are scattered through the rest.

In disaster books people are fortunate to merely survive. Usually thses novels serve as Jeremiads, but sometimes they seem to signal submission to doom. The Environmental Action chapter covers similar challenges, but those books feature people actively engaged in trying to do something other than flee in terror or close the door to the shelter. Many other disaster novels can be found in the Science Fiction chapter. Similar books by Atwood, Charnas, Elgin, Piercy, Wilhelm and Winterson are located in Ecofeminist Fiction. *Season of Dead Water* (1723) contains 50 selections about the *Exxon Valdez* oil spill.

Books for adults:

1802. Abbey, Lloyd Robert. *The last whales.* NY: Ivy, 1991, c1989. 340 p. : map. 0804107475
A bull blue whale escapes mercury poisoning in the North Atlantic, migrates, and mates with a South Atlantic blue. Their descendents then try to survive the ravages of nuclear winter and the greenhouse effect. A wealth of information about habitats, natural history, and challenges to

cetacean survival are couched in a moving adventure story. AOT. #

1803. Amis, Martin. *London fields*. NY: Harmony, 1989. 470 p. 0517577186
Life in early 21st century Britain becomes horrific due to ozone layer depletion, as human emotions become as ravaged as the environment.

1804. Auster, Paul. *In the country of last things*. NY: Penguin, 1988, c1987. 188
p. 0140097058
A grim vision of a desperate future where nature is unbalanced, unpredictable,
and brutal. Resources have been used up, food is scarce, and people live in
sprawling cities of ruins where morals and faith are as depleted as the planet.

1805. Barth, John. *The Tidewater tales*. NY: Putnam, 1987. 655 p. 0399132473
Want a relaxing day on the water? Stay away from Chesapeake Bay where
one encounters spies floating in the water, toxic wastes dumped by the
mafia, and espionage stations in a place where nature is fighting a losing
battle with an increasingly brutal "civilization."

1806. Baxter, Charles. *A relative stranger*. NY: Penguin, 1991, c1990. 223 p.
0140156283
This collection of stories includes an excellent one called "The Disap-
peared" about a Swedish engineer who becomes intrigued with an unusual
young Detroit woman with a "radioactive soul." The toxic, smoky, dying city
of Detroit is used as a harbinger of the probable fate of the industrialized
world.

1807. ___. *Shadow play*. NY: Norton, 1993. 399 p. 0393034372
Publishers Weekly Best Book, 1993.
After allowing a chemical plant to open in his economically depressed
town, bureaucrat Wyatt Palmer receives his payoff in the Faustian bargain
when his cousin gets terminal cancer at the plant. Palmer then faces the
moral dillemas of whether to mercy-kill his cousin and whether and how to
exact revenge on the plant operator. HR.

1808. Becker, Mary Kay, and Coburn, Patricia. *Superspill : an account of the
1978 grounding at Bird Rocks*. Seattle: Fremont, 1989. 161 p. 0962430226
An account of the grounding of an oil tanker on its voyage from Alaska to
Seattle; the effects of the spill on wildlife, fisheries and the economy; and
the subsequent coverup and whitewash by Octagon Oil and the government. The many parallels to the *Exxon Valdez* grounding seem almost
prophetic.

1809. Bishop, Michael. *Count Geiger's blues : a comedy*. NY: T. Doherty, 1992.
374 p. 0312851995

After swimming in a pond containing radioactive waste, Xavier Thaxton goes through many changes, culminating in his transformation to Count Geiger, the personification of a cartoon superhero he previously scorned. He is untouched by bullets and beatings, but realizes that the radiation is killing him. Far–fetched but amusing for AOT, especially comic book fans. #

1810. Castillo, Ana. *So far from God*. NY: Norton, 1993. 251 p. 0393034909
Despite their relative isolation in Tome, NM, Hispanic families are affected by AIDS, industrial pollution, poverty, and government neglect. Some turn to nature, alternate spirituality, and natural medicine to help overcome their plight.

1811. Cussler, Clive. *Sahara*. NY: Simon & Schuster, 1992. 541 p. 0671681559
An adventurer crosses the Sahara and uncovers a conspiracy involving a deadly toxic compound which threatens the lives of thousands and is killing marine life.

1812. DeLillo, Don. *White noise*. NY: Penguin, 1986. 326 p. 0140077022
The "white noise" of radio transmissions, microwaves, ultrasonic appliances, television, and the babble of consumerism has significantly degraded the quality of life, but most people seem oblivious to the techno-cultural buzz. Then a chemical cloud released from an industrial accident provides a threat that is more difficult to ignore. This black comedy is HR.

1813. Ehrenreich, Barbara. *Kipper's game*. NY: Farrar Straus Giroux, 1993. 310 p. 0374181551
This eco-medical novel features ecological and social breakdown, animal rights, genetic engineering, corruption, fetal tissue research and various related themes. The somewhat jumbled plot can be confusing, but mirrors the chaos described very well.

1814. Ephron, Amy. *Biodegradable soap*. Boston: Houghton Mifflin, 1991. 159 p. : ill. 0395572274
This satire of glitzy modern life in Los Angeles features a heroine who worries about pollution, toxic waste, and war, but uses her concerns as an escape from her divorce and other personal failures. A potentially powerful theme is undermined by a lack of substance and a style as superficial as the lives it ridicules. Not recommended.

1815. Franzen, Jonathan. *Strong motion*. NY: Farrar Straus Giroux, 1992. 508 p. 0374271054
An anti-abortion minister claims that a series of unusual earthquakes in

New England are an expression of God's wrath, but the heir of a chemical company fortune and a seismologist suspect that the company's dumping of toxic wastes into a deep well is the real cause. HR.

1816. Gaiman, Neil, and Pratchett, Terry. *Good omens : the nice and accurate prophecies of Agnes Nutter, witch.* NY: Workman, 1990. 354 p. 0894808532
After "the totally reliable witch" predicts that the environmental Armageddon will occur next Saturday, a devil and an angel try to find "The Four Motorcyclists of the Apocalypse" and kill the Antichrist. The Antichrist is a twelve year old boy who loves his dog, which unfortunately happens to be a Hellhound. HR AMT. #

1817. Harrison, Payne. *Thunder of Erebus.* NY: Ivy, 1993, c1991. 495 p. 0804108773
When a Soviet-American geological team discovers massive oil fields in Antarctica, the two oil-starved countries go to war. Primarily a war story with an underlying environmental message.

1818. Herzog, Arthur. *Heat.* NY: Tudor, 1989, c1977. 277 p. 0944276512
AR: *Earthsound, The Swarm, Orca, Make us happy.*
An engineer's warnings that too much carbon dioxide in the atmosphere will lead to imminent disaster are ignored by the government, industry, and consumers. Massive tornados, hurricanes, droughts, hailstorms, and waterspouts threaten to eradicate humanity, but there is still one small hope for survival. AMT. #

1819. Hill, Russell. *The edge of the earth.* NY: Ballantine, 1992, c1986. 151 p. 0345331249
Previous title: *Cold Creek cash store.*
When inflation, materialism, social decay, and environmental degradation bring American society to a low ebb of anarchy, Evan Walker escapes to the Sierra where he finds a "family" of other refugees nonviolently living off the land. If Evan can find them, will the violent, greedy hordes find them too?#

1820. Hoban, Russell. *Riddley Walker.* NY: Summit, 1990, c1980. 220 p. 0671701274
One of *Science fiction's 100 best novels* .
A poetic nomad describes a barbaric future in England one millenium after nuclear war has ravaged the planet, when gunpowder is being rediscovered and used despite a widespread fear of science, particularly "the Little Shynin Man the Addom." The unusual language will discourage lazy readers but delight most wordsmiths. HR

1821. Houston, James D. *Continental drift.* NY: McGraw-Hill, 1987, c1976. 337

p. 0070304882
A California family copes with internal conflict, crime, and the stress of living along a fault line. Expecting a land of abundance and promise, they find one of fear and degradation.

1822. Jeffries, Richard. *After London : or, Wild England.* NY: Oxford, 1980, c1885. 248 p. 0192812661
A British naturalist uses the device of a cataclysm which has essentially destroyed civilization to describe what England would be like if returned to a natural, unspoiled state. Unusually lyrical for a disaster novel, this is a fascinating precursor of contemporary ecofiction.

1823. Kilian, Crawford. *Icequake.* Vancouver: Douglas & McIntyre, 1979. 229 p. : ill. 088942311
In 1985 a team of scientists becomes stranded on an antarctic iceflow after a massive Earthquake. With their radio destroyed, they manage to drift to New Zealand where they discover that the Earth's magnetic field has been weakened and that solar flares are destroying the ozone layer and atmosphere.

1824. ___. *Tsunami.* Toronto: McClelland & Stewart, 1984, c1983. 218 p. 0770418570
Solar flares, loss of Earth's electromagnetic field, ozone burnoff, cancer from ultraviolet rays, antarctic Earthquakes, rising sea levels, massive tidal waves, social collapse, and that's just the first fifty pages! Trying to cope with continuing disasters, anarchy, local warlords, and skirmishes over remaining resources takes up the rest. A veritable orgy for disaster and dystopia readers.

1825. Knighton, Gary. *Isles of Omega.* Hungington, WV: Univ. Editions, 1993. 117 p. 1560022205
After a periodic Earth "wobble" from an imbalance at the poles, much of the world is flooded, and the remaining human inhabitants live on islands where they regroup in primitive bands to protect themselves from pirates, storms, plague, and each other.

1826. Maron, Monika. *Flight of ashes.* London: Readers Int'l., 1986. 188 p. 0930523229 0930523237 (pbk)
Translation of *Flugasche.*
When an East German reporter stands by her story that the continued operation of an antiquated power plant has made the local environment unlivable, she faces both subtle and overt efforts to change or suppress it.

1827. Masters, Dexter. *The accident.* NY: Penguin, 1985, c1955. 336 p. 0140077766

Depicts the last eight days of an atomic scientist's life before he dies from the radiation he was exposed to at Los Alamos. This groundbreaking cautionary tale about the effects of nuclear testing and development is still powerful and relevant 40 years later. HR.

1828. Miller, Walter M. A *canticle for Leibowitz*. NY: Lightyear, 1993, c1961. 320 p. 0899683533
Hugo Award, 1961.
The Leibowitz Abbey has preserved the scraps of the world's knowledge after a nuclear war, but will the knowledge that could revive technology also lead to further wars? HR AMT. #

1829. Moran, Richard. *Cold sea rising*. NY: Arbor House, 1986. 352 p. 087795755X
When volcanic eruptions break the massive Ross Ice Shelf away from Antarctica, the rising sea level threatens to engulf the world's seaports and coastlines. Soviet hardliners see an opportunity to turn it to their strategic advantage, with only an American admiral, a Russian scientist, and a respected journalist standing in the way of panic and disaster.

1830. ___. *Dallas down*. NY: Arbor House, 1988. 289 p. 0877959099
In 1999 the Southwest is crippled by a killer drought. A greater threat is caused by overpumping groundwater and oil, creating a sinkhole that could swallow Dallas. The intrigue includes a Texas oilman who knows the secret to save Dallas but wishes to use the knowledge for his own gain, a geologist who must overcome cave phobia, and a pact between the Soviets and the new Marxist rulers of Mexico.

1831. ___. *Empire of ice*. NY: Forge, 1994. 351 p. 0312855273
The British Isles and Europe are threatened by a new ice age when the volcanoes of the Mid-Atlantic Ridge erupt in the year 2000. A geophysicist hopes to sink geothermal wells to supply Britain with heat and energy in the face of an IRA-inspired civil war.

1832. Murphy, Pat. *The city, not long after*. NY: Doubleday, 1989. 244 p. 038524925X
After a plague decimates the human population, a nearly empty San Francisco is starting to develop into a near ecotopia when it is invaded. How can the peaceful, unarmed, and seemingly defenseless people defend themselves?

1833. Nichols, Robert. *From the steam room : a comic fiction*. Gardiner, ME: Tilbury House, 1993. 232 p. 0884481298
AR: *Daily lives in Nghsi-Altai*.

After New York's infrastructure collapses, race and class relations reach an all time low and the noxious environment becomes a constant threat. Financiers and politicians attempt to rescue the city. Recommended only for sophisticated adults due to its complex style.

1834. ___. *In the air.* Baltimore: Johns Hopkins Univ., 1991. 161 p. 080184195X 0801841968 (pbk)
Bodies from the Bhopal chemical cloud appear on a Maine beach, a customer receives an electric bill surcharge for dead Nicaraguans, and various other surreal events occur in 13 stories. Makes a powerful connection between First World consumption and Third World poverty, dehumanization, environmental ruin, and war. HR.

1835. Norwood, Warren. *Shudderchild.* NY: Bantam, 1987. 350 p. 0553264559
Massive Earthquakes have killed millions and devastated the planet. A Texan must cope with paramilitary insurgents in his efforts to protect the Earth from the nuclear radiation escaping from thousands of sites. #

1836. Orwell, George. *Coming up for air.* NY: Penguin, 1990, c1939. 246 p. 0140182284
AR: 1984, *Animal farm.*
A middle aged Englishman returns to his childhood home to find that it has been ravaged by industrial waste and pollution.

1837. Palmer, David R. *Emergence.* NY: Bantam, 1990. 302 p. 0553255193
Portions of *Emergence* nominated for the Nebula, Hugo, and John W. Campbell Awards.
After bionuclear war destroys most of the human population, a new species (Homo post hominem) emerges. Candy Smith-Foster makes a quest across the ravaged land seeking others of her own new kind. AOT. #

1838. Perry, Elaine. *Another present era.* NY: Farrar Straus Giroux, 1990. 244 p. 0374105286
In a near–future New York City plagued by violent storms, floods, and brown-outs, a young mulatta architect tries to improve the quality of life and public morale through beautiful, natural interior designs.

1839. Preuss, Paul. *Core.* NY: Morrow, 1993. 350 p. 068809662X
When a shift in the Earth's magnetic field produces fatal levels of ozone depletion, a geologist plans to drill a tube to the center of the Earth to learn how to prevent further shifting. Corporate criminals steal his new technology for their own gain, threatening the lives of millions. This complex, slow starter will reward patient readers.

1840. Roessner, Michaela. *Vanishing point*. NY: T. Doherty, 1993. 348 p. 0312852134
A scientist joins a small group of survivors who are trying to discover why 90% of the Earth's human population has suddenly vanished. They must also find safe haven from "bounders" and other survivors who have degenerated into hostile, marauding cults.

1841. Roshwald, Mordecai. *Level 7*. Chicago: L. Hill, 1989. 192 p. 1556520654
The last inhabitants of a fallout shelter rebel against crowding, fear, and the spectre of a degraded environment. HR.

1842. Sanders, Scott R. *Terrarium*. NY: T. Doherty, 1985. 272 p. 0812553802
By the mid 21st century almost all of humanity has escaped ecological disasters by inhabiting a global "Enclosure," a system of totally enclosed cities and travel tubes. A small group of dissidents returns to the wilderness despite fears that they will be found and killed. Then they discover the surprising origins and purpose of the Enclosure. HR AOT. #

1843. Shakespeare, L. M. *Poisoning the angels*. NY: St. Martin's, 1993. 176 p. 0312098952
The EPA tries to fine Santhill Chemicals $25 million for dumping toxic waste; Santhill uses a forged form to make Lloyd's of London liable; Lloyd's investigates; and a string of double-crosses, blackmailings, beatings, and a murder occur. Everyone "pays" in the end, but who will fork over the $25 mill?

1844. Smith, Martin Cruz. *Nightwing*. Boston: Hill, 1987, c1977. 224 p. 0940595052
A plague of infected vampire bats strikes a Hopi-Navajo reservation in the Southwest. An Indian sheriff and a mad scientist are the last line of defense.

1845. ___. *Stallion Gate*. NY: Random House, 1986. 321 p. 0394530063
The natural wilds of New Mexico are the setting for a most unnatural act: the development and testing of the atom bomb. Smith's version includes espionage, intrigue, and conflict between nature and traditional society on one side and atomic physicists and generals on the other.

1846. Smith, Wilbur. *Hungry as the sea*. NY: Dutton, 1993, c1978. 395 p. 0385136056
After a supertanker built to minimum specifications creates a giant oil spill, a rescue crew must overcome deadly conditions, corporate irresponsibility, and cynicism to rescue the crew and limit the spill.

1847. Stewart, George Rippey. *Earth abides*. NY: Lightyear, 1993, c1949. 317 p. 0899683073
One of *Science fiction's 100 best novels.*
A vacationer returns from his remote cabin to discover that almost all other human life has been exterminated. Can the few city-softened survivors cope with the hardships of returning to the land and the loss of their culture? A cautionary tale about the overspecialization of Homo sapiens and our separation from the natural world. HR.

1848. ___. *Fire*. Lincoln: Univ. of Nebraska, 1984, c1948. 336 p. : map. 0803291388
AR: *Sheep rock, Not so rich as you think..*
The protagonist of this book is the wildfire itself which eventually grows from a small spark to a massive conflagration. Only eleven days of back-breaking, dangerous work can finally extinguish it. #

1849. ___. *Storm*. Lincoln: Univ. of Nebraska, 1983, c1941. 349 p. 0803291353
Follows the development of tropical storm Maria as it builds over the Pacific and wreaks havoc across California, and the struggles of scientists and rescue personnel to understand and cope with it. Maria itself is the central character, a twist which was ahead of its time, and which led to wildly mixed reviews in 1941.

1850. Strieber, Whitley, and Kunetka, James W. *Nature's end : the consequences of the twentieth century*. NY: Warner, 1986. 418 p. 044651344X
In 2025 America is an overpopulated environmental wasteland whose natural resources have been squandered. A "depopulationist" movement would give everyone a pill, one third of which would be poisioned, to reduce population, but there is a more sinister purpose behind the movement. HR.

1851. ___. *Warday and the journey onward*. NY: Warner, 1988, c1984. 524 p. 0446357278
After a limited nuclear war and subsequent ecological catastrophe, two friends wander through the country and encounter anarchy, violence, extreme poverty, and famine. A powerful, disquieting novel portraying the environmental and social consequences of nuclear winter.

1852. Strong, Jonathan. *An untold tale*. Cambridge: Zoland, 1993. 228 p. 0944072321
Primarily an exploration of human sexuality, this book also depicts the degradation of the natural and human environment in a New Hampshire town.

1853. Sucharitkul, Somtow. *Starship & Haiku*. NY: Ballantime, 1988, c1981. 210 p. 0345338669
In 2023, Japan is the last outpost of civilization after global devastation, but a mysterious Death Lord now stalks the land. Only one woman can save her people, the whales, and possibly life on Earth. Will he find and kill her first?#

1854. Theroux, Paul. *O-Zone*. NY: Ivy, 1987, c1986. 544 p. 0804101515
After several wealthy New Yorkers travel to the forbidden O-Zone, a site contaminated by nuclear waste, they find themselves haunted by images of devastation and the rebels they meet there.

1855. Tobias, Michael. *Fatal exposure*. NY: Pocket, 1991. 281 p. 0671725726
A hole in the ozone layer opens over the Arctic and shifts southward, bringing blinding sun, deadly heat, and murderous swarms of insects. The government employs a cover-up, leaving millions of potential victims in its approaching path. (The cover-up is similar to the ones denying the impact of nuclear and other toxic wastes, leaks in the Alaska pipeline, oil spills, acid rain, etc.)#

1856. Turner, George. *The destiny makers*. NY: Morrow, 1993. 321 p. 068812187X
AR: *Beloved son*.
By 2069 Earth is so overpopulated and environmentally ravaged that procreation and the treatment of terminal illness are capital crimes. When an old man is cured of Alzheimer's disease and rejuvenated, an intrigue begins that could lead to the end of the world. #

1857. ___. *Drowning towers*. NY: Arbor House, 1987. 318 p. 1557100381
Original title: *The sea and summer*.
Arthur C. Clark Award, 1987.
A boy struggles to deal with a future in a world which is overpopulated, hyperautomated, and being flooded by the rising oceans. Turner's warning of the potentially disasterous effects of pollution is skillfully laid between the lines. AOT. #

1858. Weston, Susan B. *Children of the light*. NY: St. Martin's, 1987, c1985. 262 p. 0312903057
A college student's canoe trip unexpectedly brings him into a future, post-cataclysmic world where the survivors struggle against hostile climatic conditions to raise crops in contaminated soil.

1859. Wohl, Burton. *The China Syndrome*. Mattituck, NY: Amereon, 1990, c1979. 200 p. 0848812247

A TV news reporter and a cameraman uncover a startling "incident" while investigating a nuclear power plant. The real meaning of the incident is revealed by a veteran engineer whose integrity makes him a murder target. #

1860. Wolf, Christa. *Accident : a day's news.* NY: Farrar Straus Giroux, 1989. 113 p. 0374100462
Translation of *Storfall.*
An East German writer is abruptly forced to confront the effects of modern technology on her own life as her brother undergoes a delicate operation for a brain tumor, while in Chernobyl radiation fills the sky.

1861. Wren, M. K. A *gift upon the shore.* NY: Ballantine, 1990. 375 p. 0345363418
After nuclear war and nuclear winter have ravaged the Northwest, violence and looting are the law of the land. Two women try to preserve civilization by saving the last library, but the fanatical Flock believes all books should be destroyed. AOT. #

11. Animals

This chapter lists books about animals, and relations between people and other animals. Human records as ancient as cave drawings depict humanity's fascination with animals. They have frequently appeared in art, folklore, and literature ever since, and are well represented in many collections of legends in the Native North Americans chapter. Animals, humans, and plants are united in the Australian dreamtime as described in the works of Wongar (1970-73) and Narogin (1965-66)

Some books are incredibly accurate in their depictions of wildlife, notably those by naturalists R. D. Lawrence and Eckert. Other authors, such as Adams, create highly anthromorphic beasts in order to examine the sometime bizarre species known as Homo sapiens. Clark, Boyd, London, Le Guin, Popham, Schaefer, Quinn, Stafford, Sterchi, and others have written very perceptively about human-animal relations. Engel, Hoffman, Lerman, McClure, Mueller, and Wharton make this connection as intimate as possible.

Many of the best animal books written for young adults are also appropriate for adults or for family reading: the realistic works of naturalists Clarkson and Eckert; the more fanciful novels by Dickinson, L'Engle, and Hughes; and the family classics The Jungle Books, The Wind in the Willows, Bambi, and The Yearling. Literary anthologies for all ages include The Voice of the Wild (1715), The Bear Went over the Mountain (1719), The Company of Animals (1739), A Wondrous Menagerie (1748), and A Bestiary (1749).

Relevant science fiction titles include Wells' Island of Dr. Moreau (2425), Brin's Startide Rising and The Uplift War, (2369-70) McIntyre's Dreamsnake (2262) and Star Trek IV (2400), Norton's Fur Magic (2408) and Vonnegut's Galapagos (2424). Important mainstream authors such as Faulkner (1760-61), Melville (1779), and Gardner (1762) consider many aspects of human-animal nature. Dodge's Fup (2161) is a cult classic. Jewett (2247-52) and Walker (2279) offer Ecofeminist perspectives. Environmental Action authors such as Hoban (2514), Cameron (2499), Matthee (2518), and Smith (2534) present animal protection and conservation messages. Platt takes preservation one step further when several animals unite to save the Earth (2526).

Although it isn't technically a work of fiction, Schaefer's fascinating American Bestiary (897) is a fine complement to his Conversations with a Pocket Gopher. Both are highly recommended for adults and mature teens as are related novels by Lopez (2135-7)

Books for adults:

1862. Adams, Richard. T*he plague dogs.* NY: Fawcett, 1986, c1977. 480 p. : ill. 0449211827
AR: T*he unbroken web.*
Two dogs escape the horrors of animal experimentation to roam the heaths, where they kill chickens and sheep. An unethical reporter rumors that they are spreading bubonic plague. AMT. #

1863. ___. *Watership Down.* NY: Avon, 1993, c1972. 478 p. 0380004283
After their warren is bulldozed for a new development, a group of rabbits encounters both natural and human threats while searching for a safe place to establish a new home. HR. #

1864. Boyd, William. *Brazzaville Beach.* NY: Morrow, 1990. 316 p. 0688103634
A primate ecologist suspects that the chimps she is studying have become warlike and cannibalistic. She is captured by guerillas (not gorillas!) while seeking more evidence. Meanwhile, her mathematician husband's attempt to create a formula to "reproduce the magical, infinite variety of the natural world" drives him insane.

1865. Bradley, Will. *Ark liberty.* NY: ROC, 1992. 336 p. 0451451325
In this ecological future saga, some people are building arks to preserve as many animal species as possible while others are seeking to destroy them.

1866. Burnford, Sheila Every. T*he incredible journey.* Boston: G.K. Hall, 1989, c1961. 211 p. : ill. 0816147248
A Siamese cat, an old bull terrier, and a young Labrador retriever travel together 250 miles through the Canadian wilderness to return to their home. Good family reading. #

1867. Clark, Walter Van Tilburg. T*he track of the cat.* Reno: Univ. of Nevada, 1993, c1949. 409 p. 0874172306
His oft-anthologized owl story, H*ook,* is also HR. Rancher Curt insists on killing the panther he suspects of killing his cattle, over the objections of Art and Joe Sam who believe it is a symbol which arises when people dishonor the land. By hunting the cat, Curt begins to destroy the real killer: himself. HR AOT. #

1868. Clarkson, Ewan. *Ice trek.* NY: St. Martin's, 1987. 192 p. 031201046X
AR: T*he badgers of Summercombe, Halic: the story of a grey seal, The shadow of the falcon, The many-forked branch, The running of the deer, Syla the mink, The wake of the storm; Wolf country, a wilderness pilgrimage.*
Two men stranded in Alaska's rugged, remote Brooks Range after a plane cash

are pursued by a grizzly as they try to make their way back to civilization.

1869. Engel, Marian. *Bear.* Boston: Godine, 1987, c1976. 141 p. 0879236671
A woman develops a very close relationship with a bear, changing her life
in unforseen ways. This controversial novel is recommended only for
mature, sophisticated adults.

1870. Galloway, Les. *The forty fathom bank.* SF: Chronicle, 1994, c1984. 118
p. : ill. 081180342
When North Sea fishing is interrupted by WWII, a northern California commercial fisherman discovers that the indigenous gray nurse shark is loaded
with the now scarce vitamin A. Prices soar, and a human feeding frenzy on
sharks begins.

1871. Harrigan, Stephen. *Aransas.* Houston: Gulf, 1986, c1980. 259 p.
0877190577
A young man returns to his home in Texas to work in a friend's porpoise
show. His work with dolphins and respect for them as intelligent, feeling
creatures leads him to the verge of the most important decision in his life.
AOT. #

1872. Hawkes, John. *Sweet William : a memoir of Old Horse.* NY: Simon &
Schuster, 1993. 269 p. 0671740571
Observations on life, nature, and human-animal relationships from an
aging horse's point of view.

1873. Hill, Lloyd E. *The village of Bom Jesus.* Chapel Hill, NC: Algonquin,
1993. 227 p. 0945575882
Chronicles of life in an Amazon village featuring villagers, a witch doctor,
wild animals, and a very unusual calico cat. Depicts rain forest ecology,
human-animal interactions, and the importance of rain forests on the
human psyche. HR. #

1874. Hoffman, Alice. *Second nature.* NY: Putnam, 1994. 254 p. 0399139087
Robin Moore frees a young, feral man called the "Wolf Man" from a wolf
trap he was caught in while escaping transfer to a mental hospital. She
falls in love with him, but when animals and a teenage girl are murdered,
the Wolf Man becomes a key suspect. HR.

1875. Kilworth, Garry. *The foxes of first dark.* NY: Doubleday, 1990. 371 p.
0385264275
Follows a clan of foxes in their struggle for survival against natural predators, hard winters, fox hunters, and habitat loss. Explores the baffling love-
hate attitudes of humans to foxes. #

1876. Kiteley, Brian. *Still life with insects*. St. Paul: Graywolf, 1989. 114 p. 155971898
An amateur entomologist suffers a nervous breakdown, but uses writing in his journal to regain his equilibrium. He intersperses insect sightings with observations on family, life, and nature. This book considers the restorative power of nature to the psyche and generational cycles in human families. HR AMT. #

1877. LaFarge, Tom. *The crimson bears*. LA: Sun & Moon, 1993. 272 p. 1557130744
A brother and sister explore Bargetown, the central meeting place in the Commonwealth of Bears for every known species. The denizens of Barge-town attempt to resurrect their heroic, naturalistic past while also trying to forge a future with the "speakable" humans. This fable was written for adults, but will appeal to some older teens. #

1878. Lawrence, R. D. *Cry wild*. Toronto: HarperCollins, 1991, c1970. 146 p. 0006470483
The story of a young wolf raised in the Canadian wilderness: his flight from a forest fire, capture and mistreatment, escape, and pursuit by enraged farmers and trappers. This book is so realistic that it was initially cataloged as nonfiction natural history by the Library of Congress. HR. #

1879. ___. *The white puma*. NY: Holt, 1990. 329 p. 0805006850
The respected naturalist offers an environmentally accurate and exciting novel told from the perspective of a rare white puma in British Columbia. Includes courtship, mating, family relations, hunting, and flight from wealthy hunters. The encounter between the puma and an unarmed female conservationist is both chilling and inspirational. HR AMT. #

1880. Le Guin, Ursula K. *Buffalo gals and other animal presences*. Santa Barbara: Capra, 1987. 196 p. 0884962709
A highly imaginative and varied collection of essays, stories, and poems about animals and about the proper relations between people, other animals, plants, and the land itself. All of nature speaks to us, and Le Guin tantalizes us to listen. HR. #

1881. Lerman, Rhoda. *Animal acts*. NY: Holt, 1994. 263 p. 0805014187
A middle-aged woman escapes her husband and lover for a more primal lover who turns out to be a gorilla. Beneath the raucous comedy is a more serious consideration of male-female and human-animal relations.

1882. London, Jack. *The call of the wild*. NY: Children's Classics, 1991, c1903. 254 p. : ill. 0517060035

The adventures of a St. Bernard-Scotch shepherd dog who is forcibly taken to the Klondike gold fields where he faces human cruelty and eventually becomes the leader of a wolf pack. London's implicit message that all "children" whether human or animal require respect and love is a foreshadowing of the animal rights movement. HR. #

1883. ___. *White fang.* Boston: G.K. Hall, 1993, c1907. 311 p. 0816158894
The themes of nature, domestication, respect, and human-animal relationships that emerged in *Call of the wild* are reconsidered through a plot reversal: this time we follow the slow domestication of a wolf. Although not as strong as its great predecessor, it is still HR. #

1884. McClure, Michael. *Scratching the beat surface.* SF: North Point, 1989, c1982. 175 p. : photos. 085647074X
Beat poet and playwright McClure presents a "new perception of art as a living bio-alchemical organism." He also explores the animal nature of human beings, the environmental costs of rejecting that animality and the place of humans in the ecological web.

1885. Michener, James A. *Creatures of the kingdom : stories of animals and nature.* NY: Random House, 1993. 281 p. : ill. 0679413677
Includes excerpts about a variety of different animals drawn from his many novels. The lives, habitats, survival or extinction, and (where appropriate) relations with humans, of dinosaurs, mastodons, beavers, birds, snakes, hyenas, salmon, crabs, and other animals are imaginatively depicted here. #

1886. Mueller, Ruth. *The eye of the child.* Philadelphia: New Society, 1985. 225 p. 0865710465
A six-year-old gypsy girl who can 'speak bird' narrates a fantasy novel about the consequences of humanity's separation from nature and its ecological and pscyological costs.

1887. Peak, Michael. *Catamount.* NY: ROC, 1992. 282 p. 0451451414
Two lonely predators, a puma and an eagle, receive dream warnings from the animal gods of the coming of the ultimate predator: human hunters. #

1888. Popham, Melinda Worth. *Skywater.* St. Paul: Graywolf, 1990. 205 p. 155597127X
Edward Abbey Award for Ecofiction, 1990.
Two old settlers in the Sonoran Desert and a small band of coyotes struggle to maintain their traditional ways of life. An allegory about the relationship between animals, humans and the Earth, told mainly from a coyote's perspective. HR AOT. #

1889. Quinn, Daniel. *Ishmael*. NY: Bantam, 1992. 266 p. 0553078755
Turner Tomorrow Fellowship Award, 1992.
A telepathic gorilla engages the human narrator in a Socratic dialogue
about the Takers (humans), their greed, their refusal to share with the other
animals, and the inevitable disaster to come unless the Takers change their
ways. HR.

1890. Schaefer, Jack. *Conversations with a pocket gopher, and other outspoken
neighbors*. Santa Barbara: Capra, 1992, c1978. 128 p. 0884963489
AR: *The canyon*, *Mavericks*.
A gopher, shrew, bat, kangaroo rat, jaguar, and puma chide humans on misuse of the land, two dimensional thinking, the arms race, and ecological
ruin. This darkly humorous work rejects technological quick-fixes in favor of
a long-range ecological orientation. HR AMT. #

1891. Searls, Hank. *Sounding*. NY: Ballantine, 1985, 1982. 280 p.
0345325265
Two interweaving stories, about the sperm whales' struggle for survival and
a Russian submarine crew, consider the interrelationship of all living
things.

1892. Siegel, Robert. *Whalesong*. SF: Harper, 1991, c1981. 143 p.
0062507982
A whale escapes from the pollution and ecological degradation of the
northern Atlantic only to discover that the southern seas have been poisoned by radiation. Includes detailed descriptions of whale habits and
habitats. #

1893. ___. *White whale*. SF: Harper, 1991. 228 p. 0062507974
A young beluga whale feeds, frolics with playmates, and listens to the
songs of the adult whales. #

1894. Stafford, Jean. *The mountain lion*. Austin: Univ. of Texas, 1992, c1947.
231 p. 0292751362
During the last of several summers spent on their uncle's ranch in the Colorado Rockies, a teenage brother and sister become estranged. They and
their uncle all want to see the elusive mountain lion, but for different reasons. AMT. #

1895. Stebel, S. L. *Spring thaw*. NY: Walker, 1989. 236 p. 0802710689
During a blizzard the captain of a seal hunting expedition encounters a
phantom who protects the seals. The struggle against the blizzard is more
than matched by the one between his conscience and the greed of his first
mate.

1896. Sterchi, Beat. *Cow*. NY: Pantheon, 1988. 353 p. 0394584511
Translation of *Blosch* .
A worker expects his new job on a dairy farm will be a bucolic joy, but is shocked by the horrors of the slaughterhouse. His humane treatment of the cows and friendship with one in particular makes him an outcast. HR.

1897. Stover, Laren. *Pluto, animal lover*. NY: HarperCollins, 1994. 160 p. 0060171111
Stover's "animal loving" includes misanthropic treatment of other humans, sadistic sexual relationships, and "euthanizing" pets. This immature work is listed to warn potential readers that the title is utterly misleading. Not recommended for anyone except Stover's much needed psychiatrist.

1898. Wharton, William. *Birdy*. NY: Vintage, 1992, 1978. 309 p. 0679734120
When a young soldier becomes insane during WW II he retreats to the solace of nature by assuming the personality of a canary. A war-weary childhood friend tries to bring him back to reality, but is true sanity possible in a crazed world?

1899. ___. *Franky Furbo*. NY: Holt, 1989. 228 p. 0805011579
An aging American author living in Italy becomes confused about his past: the seeming miracle of a fox saving him and a German soldier during WWII, his physical and psychological recovery, and his unusual wife and children. Is he actually returning from the future to save the Earth from ecological disaster? HR for adventurous adults.

1900. Williams, Heathcote. *Sacred elephant*. NY: Harmony, 1989. 175 p. : photos. 0517575477
Verse, prose, and photographs depicting a gentle, sensitive animal which is being forced into labor or slaughtered for ivory.

1901. Wilson, A. N. *Stray*. NY: Orchard, 1989, c1987. 247 p. 053108440X
AR: *Tabitha*.
A proud old alley cat tells his life story to his grandson, including his adventures in a convent, in a feline commune, in laboratory experiments, and with his hearer's grandmother. Wilson presents a plea for the careful and ethical treatment of animals between the lines. #

1902. Wood, Douglas. *Old Turtle*. Duluth, MN: Pfeifer-Hamilton, 1992. 1 v. (unpaged) : watercolors. 0938586483
The animals, rocks, trees, birds and fish use Old Turtle's teachings to help humans learn to see God in one another and in the beauty of all the Earth. #

Books for young adults:

1903. Alexander, Rosanne. *Selkie*. London: Deutsch, 1991. 223 p.
0233986545
A few days after a seal is born an oil slick engulfs remote Skomer Island,
killing seals, birds, and other wildlife. He is saved by wildlife workers, but
grows too trusting and is injured in a later encounter with people. #

1904. Blathwayt, Benedict. *Stories from Firefly Island*. NY: Greenwillow, 1993.
120 p. : ill. 0688124879
The animals of Firefly Island listen as Tortoise explains such mysteries as
why frogs croak at night. GR 5-9.

1905. Clough, B. W. *An impossumble summer*. NY: Walker, 1992. 144 p. : ill.
0802781500
After years of living abroad, ten-year-old Rianne and her family move to
Reston, Virginia, where the acquaintance of a talking possum seems to influence the luck in their lives, sometimes for better and sometimes for worse.

1906. Curwood, James Oliver. *Baree, the story of a wolf-dog*. NY: Newmarket,
1990, c1917. 241 p. 1557040753 1557040745 (pbk)
Original title: *Baree, son of Kazan*.
A half tame wolf-dog pup is separated from its parents and must learn to
survive in the Canadian wilderness. He learns lessons of compassion and
loyalty from otters, rabbits, owls, bears, beavers and humans. For dog
lovers of all ages. #

1907. ___. *The bear*. South Yarmouth, ME: Curley, 1992, c1916. 205 p.
0792710533 (pbk) 0792710525
Original title: *The grizzly king*.
An orphaned cub is befriended by a giant wounded grizzly. Threatened by
hunters, they embark on a desperate quest for survival. #

1908. Dickinson, Peter. *Eva*. NY: Dell, 1990, c1988. 219 p. 0440207665
ALA Notable Children's Book, 1989.
After a terrible accident, a young girl wakes up to discover that her brain
has been transplanted into the body of a chimpanzee. She then discovers
what it is like to be treated like an animal. HR AOT#

1909. Eckert, Allan W. *Incident at Hawk's Hill*. NY: Bantam, 1987, c1971. 191
p. 0553266969
Newbery Medal, 1971.
AR: *The crossbreed*, *The dark green tunnel*, *The dreaming tree*, *The great auk*, *The
HAB theory*, *Savage journey*, *The silent sky*, *Song of the wild*, *The wand*.
A shy, lonely six-year-old wanders into the Canadian prairie and spends a

summer under the protection of a badger. HR. #

1910. Fisher, R. L. *The prince of whales : a fantasy adventure*. NY: TOR, 1988, c1985. 160 p. : ill. 0812566378
Toby, a young whale attempting to carry on the ancient tradition of underwater singing, must struggle for survival against pollution and exploding harpoons.

1911. Fox, Michael W. Dr. *Fox's fables : lessons from nature : twenty-three fables in which animals talk about how they really feel and live*. Washington: Acropolis, 1980. 157 p. 0874912911 0874915163 (pbk)
AR: *Ramu and Chennai, Sundance coyote, Vixie*.
These fables, written by a veterinarian and animal rights activist, explore relationships among animals and between them and people. #

1912. Grahame, Kenneth. *The wind in the willows*. Ill. by Arthur Rackham. NY: Knopf, 1993, c1908. 249 p. : col. ill. 0679418924
On one level, this is an animal adventure for children and young adults featuring the highly anthromorphic Rat, Toad, Mole, and Badger. On a deeper level it is about the despoiling of the British countryside and rural way of life by the forces of turn-of-the-century industrialism. This classic is HR for all ages. #

1913. Holling, Holling Clancy. *Minn of the Mississippi*. Boston: Houghton Mifflin, 1992, c1952. 85 p. : ill. 039517578X
Newbery Honor Book, 1952.
The adventures of a three-legged snapping turtle as she travels from the headwaters to the mouth of the Mississippi River illustrate the life cycle of the turtle and the geography, history, geology, and climate of the river. HR GR 4-8.

1914. ___. *Pagoo*. Boston: Houghton Mifflin, 1990. 96 p. : ill. 0395539641
An intricate study of tide pool life is presented through the story of Pagoo, a hermit crab. HR GR 4-8.

1915. ___. *Seabird*. Boston: Houghton Mifflin, 1978, c1975. 58 p. : ill. 0395266815
Newbery Honor Book, 1975.
The history of America at sea is present through the travels of Seabird, a carved ivory gull. HR GR 4-8.

1916. Hughes, Monica. *Hunter in the dark*. NY: Avon, 1984, c1982. 144 p. 0380677024
A teenage boy goes on a secret hunting trip alone, in an effort to come to

terms with his leukemia and to test his strength and resourcefulness in battling the elements and stalking the white-tailed deer that is his quarry.

1917. **Hughes, Ted.** *Tales of the early world.* Ill. by Andrew Davidson. NY: Farrar Straus Giroux, 1991, c1988. 121 p. : ill. 0374373779
One of England's premier nature poets provides fancifully humorous prose vignettes on the creation of various species. HR for all ages. #

1918. **Johnson, Annabel.** *I am Leaper.* NY: Scholastic, 1990. 105 p. : ill. 0590434004
A kangaroo rat who can communicate with humans enlists the aid of a boy, to help defeat a "monster" that has been terrorizing the desert where she lives. GR 4-8.

1919. **Katz, Welwyn Wilton.** *Whalesinger.* NY: Dell, 1993, c1990. 212 p. 044021419X
A scientific field trip near the possible site of a sunken treasure ship from Francis Drake's expedition, brings together two teenagers, an emotionally isolated boy, and a girl who is discovering she shares an empathic bond with two gray whales in the area.

1920. **Kipling, Rudyard.** *The jungle books.* NY: Oxford, 1992. 373 p. 0192829017
Presents the adventures of Mowgli, a boy reared by a pack of wolves and the wild animals of the jungle. Also includes other short stories set in India, including "Rikki-Tikki-Tavi." HR all ages. #

1921. **Kjelgaard, Jim.** *Big Red : the story of a champion Irish setter and a trapper's son who grew up together, roaming the wilderness.* NY: Buccaneer, 1993, c1945. 218 p. : ill. 1568491115
AR: *Outlaw Red, Irish Red.*
A wealthy sportsman hires a backwoods French Canadian orphan to exercise Big Red, his champion Irish setter. The bonds of loyalty and understanding of nature and each other that arise between the three lead them thorough a succession of adventures. HR GR 4-9.

1922. ___. *Desert dog.* NY: Bantam, 1984. 129 p. : ill. 0553154915
After his master dies, a champion greyhound chooses the freedom of the desert, where his ready intellect is continually challenged in the battle against thirst, hunger, and natural enemies. GR 4-9.

1923. ___. *Snow dog.* NY: P. Smith, 1992, c1948. 163 p. : ill. 0844665940
The adventure of a dog born in the wilderness, left to fend for himself when a black wolf kills his mother and brothers, and finally won over by the trap-

per his mother had deserted. GR 4-9.

1924. ___. *Wild trek*. NY: P. Smith, 1992, c1950. 242 p. : ill. 0844665940
AR: *A nose for trouble, Haunt Fox, Lion hound, Snow dog, Stormy.*
A trapper and his dog travel into the Caribou Mountains to rescue a naturalist and a pilot whose plane has been downed in the rugged wilderness
of northern Canada. GR 4-9

1925. Langton, Jane. *The fledgling*. NY: Scholastic, 1990, c1980. 182 p.
0590434519
Newbery Honor Book, 1980.
Georgie's fondest hope, to be able to fly, is fleetingly fulfilled when she is
befriended by a Canada goose. HR.

1926. Lawson, Robert. *Robbut : a tale of tails*. Hamden, CT: Linnet, 1989,
c1948. 94 p. 0208022368
Dissatisfied with his own tail, a young rabbit tries out that of a cat, a garter
snake, and a fox. GR 4-8.

1927. L'Engle, Madeleine. *A ring of endless light*. NY: Farrar Straus Giroux,
1990, c1980. 324 p. 0440972329
Newbery Honor Book, 1981.
During the summer her grandfather is dying of lukemia and death seems
all around, 15-year-old Vicky finds comfort with the pod of dolphins with
which she has been doing research. HR. #

1928. Mazer, Anne. *The oxboy*. NY: Knopf, 1993. 109 p. 0679841911
A young boy who is half-human and half-ox struggles to survive in a society
where animals are hated and contact with them is prohibited. GR 3-7.

1929. McClung, Robert M. *Samson, last of the California grizzlies*. Hamden, CT:
Linnet, 1992. 94 p. : ill. 0208023275
During the first nine years of his life a California grizzly struggles for survival
against the ever increasing threat of humans in the Sierra Nevada. GR 3-7.

1930. ___. *Shag*. Hamden, CT: Linnet, 1991. 96 p. : ill. 0208023135
Relates the daily struggle of a buffalo against famine, drought, and death
by the hunter's bullet, in the days when the bison moved in mighty herds
on America's plains. GR 3-7.

1931. ___. *The true adventures of Grizzly Adams : a biography*. NY: Morrow, 1985.
200 p. : ill. 0688057950
Recounts the adventures of the 19th-century frontier hunter, with an
emphasis on his experiences with bears. GR 5-9.

1932. Monson, A. M. *The deer stand.* NY: Lothrop, Lee & Shepard, 1992. 171 p. 0688110576
When her family moves from Chicago to the wilds of Wisconsin and Bits has trouble making new friends at school, she spends her time trying to tame a deer in the woods near her house. GR 7-10.

1933. Morey, Walt. *Death walk.* Hillsboro, OR: Blue Heron, 1991. 166 p. 0936085785
ALA Best Books for Young Adults nominee, 1991.
Seventeen year old Joel is stranded in the Alaskan wilds, where he survives shootouts, ordeals in the ice and snow, and a mine cave-in (when he is saved by two wolves.) This survival and coming of age novel is HR GR 5-12.

1934. ___. *Gentle Ben.* NY: Puffin, 1992. 191 p. : ill. 0140360352
AR: *Operation blue bear.*
Traces the friendship between a boy and a bear in the rugged Alaskan Territory. GR 5-12.

1935. ___. *Gloomy Gus.* Hillsboro, OR: Blue Heron, 1989, c1970. 182 p. 0936085177
Because of an agreement his alcoholic father makes with a circus, 15-year-old Eric begins a long journey south from Alaska with the huge Kodiak bear he raised from a cub. GR 4-8.

1936. ___. *Home is the North.* Hillsboro, OR: Blue Heron, 1990, c1967. 162 p. 0936085118
A portrait of the land and people of wilderness Alaska presented through the experiences of an orphan whose year of decisions, responsibilities, and growth help him to accept the future. GR 4-9.

1937. ___. *Kavik the wolf dog.* NY: Dutton, 1977, c1968. 192 p. : ill. 0525450181
A wolf-dog instinctively travels 2000 miles from Washington to Alaska to return to the boy who once saved his life. GR 5-9.

1938. Mowat, Farley. *Owls in the family.* NY: Bantam, 1985, c1961. 107 p. : ill. 0553155857
AR: Sequel: *The dog who wouldn't be* (GR 5-adult).
Two owls take over the household, in this autobiographical tale based on naturalist Mowat's boyhood on the Canadian prairies. GR 4-8.

1939. Muir, John. *Stickeen.* Berkeley: Heyday, 1990, c1909. 96 p. : ill. 0877971951
A fictionalized version of Muir's adventures in Alaska with a mongrel dog.

One particularly vivid section involves being lost on a massive glacier. HR, especially for dog lovers. #

1940. North, Sterling. *Rascal*. NY: Puffin, 1990, c1963. 189 p. : ill. 0140344454
The author recalls his carefree life in a small midwestern town at the close of World War I, and his adventures with his pet raccoon, Rascal.

1941. ___. *The wolfling : a documentary novel of the* 1870s. NY: Puffin, 1992. 223 p. : ill. 0140361669
In the 19th-century midwest, a young boy adopts a wolf whelp and gains the attention and friendship of the Swedish-American naturalist Thure Kumlien.

1942. Parnall, Peter. *Marsh cat*. NY: Macmillan, 1991. 128 p. : ill. 0027701204
AR: *The rock*.
A huge, wild cat faces new dangers when he moves from his marshland home to a barn during the winter months. He is almost killed by a trap, saved by a young girl and a vet, and returns to the marsh in the spring. HR GR 3-8.

1943. Rawlings, Marjorie Kinnan. *The yearling*. 50th anniversary ed. ill. by Edward Shenton. NY: Collier, 1988, c1938. 428 p. : ill. 0020449313
Pulitzer Prize, 1939.
A young boy living in the Florida backwoods is forced to decide the fate of a fawn he has lovingly raised as a pet. This classic is HR for all ages. Good family reading. #

1944. Rounds, David. *Cannonball River tales*. SF: Sierra Club, 1992. 97 p. : ill. 0871565773
Tall tales about Tom Terry, whose home by the banks of the magical Cannonball River is populated by a talking rabbit, a flying silver dragon, and other unusual characters who all offer amusing stories about caring for wildlife and living in harmony with the earth. GR 6-9.

1945. Ryden, Hope. *Bobcat year*. NY: Lyons & Burford, 1990, c1981. 205 p. : ill. 1558210555
Companion to her nonfictional *Bobcat*. Follows the life of a family of bobcats as they progress through the year. Criticized by some reviewers for the anthropomorphized bobcats and one-dimensional hunters, but praised by others. #

1946. Salten, Felix. *Bambi : a life in the woods*. Adapted by Joanne Ryder. Ill. by Michael J. Woods. NY: Simon & Schuster, 1992, c1928. 158 p. : ill. 0671739379

Another ed. for elementary schools is titled Walt Disney's Bambi. The adventures of a young deer in the forest as he grows into a beautiful stag. The forest fire and the death of Bambi's mother are classic scenes. HR GR 5-8.

1947. Seton, Ernest Thompson. *Animal heroes : being the histories of a cat, a dog, a pigeon, a lynx, two wolves & a reindeer and in elucidation of the same, over 200 drawings.* Berkeley: Creative Arts, 1987, c1905. 362 p. : ill. 0887390552
Eight stories detailing the struggle for existence of such animals as a slum cat, a homing pigeon, a wolf, a lynx, and a reindeer. #

1948. ___. ed. *Animal stories.* NY: Derrydale, 1991, c1902. 383 p. 0517037610
Original title: *The animal story book.*
A collection of 59 animal tales and essays by writers including Aesop, Rudyard Kipling, and Washington Irving. HR. #

1949. ___. *The biography of a grizzly and 75 drawings.* Lincoln: Univ. of Nebraska, 1987, c1900. 169 p. : ill. 0803291841
AR: *The biography of a silver fox.*
The fortunes and misfortunes of a lone grizzly bear who learns early that people are his enemies and that he must fight for survival. #

1950. ___. *Collected novels.* Secaucus, NJ: Castle, 1985. 454 p. 0890099367
Contents: *Wild animals I have known* — *Lives of the hunted* — *Animal heroes* — *Monarch the big bear.* Readers are warned again that Seton is entertaining, but that his animals are less than accurately drawn from nature. #

1951. Seton, Ernest Thompson, and Steilen, Mark. *Lobo, the wolf : King of Currumpaw.* Seattle: Storytellers Ink, 1991. 64 p. : ill. 0962307246
An adaptation of Seton's *Lobo* by Mark Steilen for GR 3-8.

1952. Shachtman, Tom. *Beachmaster : a story of Daniel au Fond.* NY: Holt, 1988. 166 p. : ill. 080500498X
Sequels: *Driftwhistle, Wavebender.*
An account of the adventures of a sea lion, including his experiences with humans in a sea laboratory and his assumption of the role as head of the tribe. GR 7-10.

1953. ___. *Driftwhistler : a story of Daniel au Fond.* NY: Holt, 1991. 176 p. : ill. 0805012850
At the head of a band of 13 different species of sea mammals, Daniel the sea lion seeks to fulfill a legend and find Pacifica, the long-drowned, ancient home cove of their race. GR 7-10.

1954. ___. *Wavebender : a story of Daniel au Fond.* NY: Holt, 1989. 166 p. : ill.

0805008403
Sequel: *Driftwhistle*.
Further adventures of a sea lion as he wisely leads his tribe out of various dangers. GR 7-10.

1955. Thomas, Jane Resh. *Fox in a trap*. NY: Clarion, 1987. 78 p. : ill. 0899194737
Daniel looks forward to helping his uncle set traps for the foxes that have been plaguing the family farm until the discovery of a severed fox paw makes him seriously question what they are doing. GR 4-8.

1956. Williamson, Henry. *Tarka the otter : his joyful water-life and death in the country of the two rivers*. Boston: Beacon, 1990, c1927. 279 p. : ill. 0807085073
AR: *Salar the salmon*.
Tarka the otter pursues an active life, sometimes playful and sometimes dangerous, in the Devonshire countryside. #

12. Australia and Oceania

A wide variety of contemporary ecofiction is being written in and about Australia and Oceania. Nearly two-thirds is by or about Australian Aboriginal and other people indigenous to the region, whose Earth-centered cosmology is similar to that of Native Americans.

Mudrooroo Narogin (aka Colin Johnson) wrote the first major published novel by an Aborigine, *Wild Cat Falling*, in 1965. More recently he has written *The Kwinkan* about a shady land developer/politician and the brilliant *Master of the Ghost Dreaming* wherein two displaced Aborigines and an African slave use their animal familiars to try to save their people from an oppressive missionary. B. Wongar was born in Yugoslavia, but moved to Australia in his youth and has lived in the outback with his native wife Djumula for 40 years. His *Nuclear Trilogy* incorporates tribal myth and metafictional techniques. Other writers who employ Aboriginal themes or issues include Bird, Fox, Wrightson and the controversial Morgan.

Concern about native cultures and/or the environment is expressed in works situated across the Pacific: Davenport and Christensen in Hawaii; Peel in Melanesia; Clarke and young adult authors Bosse and the Myers in Borneo.

D. H. Lawrence's *Kangaroo* provides a fascinating early 20th century view of Australian life. Other fiction from or about the region includes the satire of Schutte, Astley's and Winton's environmental action novels, and the young adult adventures of Thiele.

Books for adults:

1957. Astley, Thea. *Vanishing points*. NY: Putnam, 1992. 234 p. 039913770X
Contains two "slyly linked" novellas: *The Genteel Poverty Bus Company*, and *Inventing the Weather*. MacIntosh Hope thinks he's found paradise on a small island but then must confront Clifford Truscott who is building a giant tourist lodge nearby. Julie Truscott becomes disgusted with Clifford's greed and moves to a small mission on Bukki Bay. Clifford tracks her down and decides to develop that area too, but again faces opposition. AOT. #

1958. Bird, Carmel. *The Bluebird Cafe*. NY: New Directions, 1991. 180 p. 0811211568

This novel set in modern Tasmania includes the interaction of white residents, natives, and animals. AOT. #

1959. Christensen, Mark. *Aloha.* NY: Simon & Schuster, 1994. 288 p. 0671870238
In this apocalyptic satire a land developer plans to tap an underwater volcano to create a new Hawaiian island to serve as a haven for illlegal activity and money laundering.

1960. Clarke, Terence. *The King of Rumah Nadai.* SF: Mercury House, 1994. 192 p. 1562790609
When a U.S. AID official is ordered stateside from Borneo for allowing one of his workers to "go native," he rebels and goes upriver to live with the Iban. In the deep rainforest he experiences both the wonders and terrors of nature and native society. HR.

1961. Davenport, Kiana. *Shark dialogues.* NY: Atheneum, 1994. 469 p. 0689121911
AR: *Pacific woman.*
This tale of native Hawaiians striving for sovereignty revolves around four modern women and their spiritualist grandmother. In flashbacks we learn the native perspective: anger over commercial debasement and environmental degradation by mainlanders. Polynesian shark legends are used effectively. " . . . much better written than Michener's *Hawaii.*"-Nancy Pearl, *Library Journal,* April 1, 1994, p. 131. HR.

1962. Fox, Stuart. *The back of beyond.* NY: Forge, 1994. 352 p. 0312853661
AR: *Black fire.*
When an Australian developer plans to build a city on the sacred hunting grounds of the Ularis, disemboweled bodies are found on the site. The Ularis get the blame, but they credit the work to ancient spirits. An unusual team of investigators and journalists travels through the "the back of beyond" to seek out the truth and then to protect Ulari culture.

1963. Lawrence, D. H. *Kangaroo.* Corrected ed. by Richard Southall. NY: Cambridge Univ., 1994, c1923. 418 p. 0521384559
Author Richard Somers (Lawrence) travels to Australia. He finds fault with the structured programs of both socialists and facists compared to what he considers a free and open form of democracy arising from the Aboriginal peoples, the settlers, and even the land itself. A strong evocation of people, place, and politics in early 20th century Australia.

1964. Morgan, Marlo. *Mutant message down under.* NY: HarperCollins, 1994. 187 p. 0060171928

An American tourist sets out on a walkabout of the outback with a tribe of Aborigines. She learns how they live in harmony with the land and returns with their "message." Originally self-published as nonfiction, it was marketed by HarperCollins as fiction. Ethically questionable at best. Not recommended.

1965. Narogin, Mudrooroo. *The kwinkan*. Pymble, Aus.: Angus & Robertson, 1993. 130 p. 0207179441
A real estate speculator loses a parliamentary election and is sent on a "special mission" to an island about to be granted independence. He becomes involved in intrigue with the ruling family, an ambitious Japanese executive, and a mysterious aboriginal special agent. Incorporates Australian aboriginal myths to demonstrate the psychic toll of government-corporate skullduggery. HR.

1966. ___. *Master of the ghost dreaming*. Pymble, Aus.: Angus & Robertson, 1991. 148 p. 0207169527
AR: *Wild cat falling* , *Wild cat screaming*.
A group of Aborigines is exiled to an island without enough food or water, decimated by disease, and tyranized by an arrogant missionary/anthropologist. A male and female aborigine and an escaped African slave use their shamanistic powers and animal familiars to try to return to their homeland and restore the balance of nature. This poetic, moving novel is HR AMT. #

1967. Peel, Colin D. *Atoll*. NY: St. Martin's, 1992. 211 p. 0312076460
AR:*Snowtrap*.
After 30 years of nuclear testing, the radiatiation level of the Mururoa atoll has reached alarmingly destructive levels. Two Islanders and a mining engineer discover that there is a Russian nuclear sub in the area, and that the French government plans to blame the rising radiation on the Russians.

1968. Schutte, James E. *The bunyip archives*. Dallas: Baskerville, 1992. 282 p. 0962750980
The bunyips (a gentle, beer drinking, marsupial Austalian version of Bigfoot) are mainly a scientific curiosity until a rumor that bunyip blood is a veritable fountain of youth makes them the target of poachers. A research biologist, his ex-girlfriend, a beer swilling outbacker, and Wild Wendy of Woolloona team up to save the Bunyips in this humorous satire.

1969. Winton, Tim. *Shallows*. Saint Paul: Graywolf, 1993. 260 p. 1555971938
A young travel guide from an Australian whaling family risks estrangement from her family and community when she joins a group of foreign anti-whaling activists. AMT. #

1970. Wongar, B. *Gabo Djara*. NY: Dodd Mead, 1987. 232 p. : ill. 0396088619
In order to save the Australian Aborigenes and their land from total devastation by uranium mining, a large green ant emerges from the Dreamtime. He encounters the Australian Parliament, the Queen of England, the President of the U.S., and even the white man's God (imprisoned in a drum of uranium ore) as state, church, and business conspire in nuclear development schemes. HR.

1971. ___. *Karan : a novel of Australia*. NY: Braziller, 1991, c1985. 248 p. 0807612421
A modernized Aborigine serves as an unwitting flunky in a "Welfare Center" until he is taken on an "boong hunt" and realizes the massive environmental and genocidal effects of nuclear testing. Acting on a tip from a tribal visionary, he learns that the hospital patients are human guinea pigs who never emerge alive, so he tries to re-establish his aboriginal roots and flee. Very powerful. HR.

1972. ___. *The last pack of dingoes*. Pymble, Aus.: Angus & Robertson, 1993. 136 p. : ill. 0207171475
AR: *Babaru*.
A collection of stories about the extinction of Australian aboriginal peoples and culture and the degradation of nature which combines traditional myths (especially human-animal metamorphoses) with modern settings and problems. HR.

1973. ___. *Walg*. NY: Dodd Mead, 1983. 213 p. 0396081894
Concerned that uranium mining and nuclear testing are sterilizing people, Australian scientists attempt to breed radiation-immune people from the local Aborigines. A young, pregnant woman flees the compound to return to her home in the North, hoping that the birth of her baby will be an act of renewal for all her people and the land itself. HR.

Books For Young Adults

1974. Bosse, Malcolm J. *Deep dream of the rain forest*. NY: Farrar Straus Giroux, 1993. 179 p. 0374317577
While on an expedition in the jungles of Borneo in 1920, 15-year-old Harry Windsor is captured by Bayang, a young Iban tribesman. Bayang believes that Harry has some power to help him and an outcast Iban girl on their dream quest. #

1975. Myers, Christopher A., and Myers, Lynne Born. *Forest of the clouded leopard*. Boston: Houghton Mifflin, 1994. 128 p. 0395674085

The death of his grandfather forces a 15-year-old Iban boy to face the conflict between the traditions and beliefs of his tribe and the modern ways he has learned at a Western-style boarding school. GR 5-9.

1976. Thiele, Colin. *Jodie's journey*. NY: Harper & Row, 1990. 169 p. 0060261331
AR: *Blue fin, Fight against Albatross Two, The hammerhead light, Shatterbelt*.
A 12-year-old disabled by rheumatoid arthritis and no longer able to ride her beloved horse faces a crisis when the two of them are alone at her remote Australia home and a devastating fire approaches.

1977. ___. *Rotten Egg Paterson to the rescue*. NY: HarperCollins, 1991, c1989. 133 p. : ill. 0060261056
Original title: *Danny's egg*.
Having rescued an emu egg from the clutches of a goanna in the Australian scrublands, a 12-year-old is determined to hatch it despite a series of near-disasters and the interference of the school bully.

1978. ___. *Shadow shark*. NY: Harper & Row, 1988, c1985. 214 p. 006026179X
Original title: *Seashores and shadows*.
Two 14-year-old cousins join a group of fishermen in pursuit of a massive shark off the coast of Southern Australia. #

1979. Wrightson, Patricia. *Moon-dark*. NY: McElderry, 1988, c1987. 169 p. : ill. 0689504519
As people build and settle on Australian land formerly occupied by wildlife, an old fisherman's dog named Blue and all the other animals of the territory carry on nocturnal activities to make their homes safer. Told from Blue's perspective. GR 4-9.

1980. ___. *The Nargun and the stars*. NY: Puffin, 1988, c1970. 184 p. : ill. 014030780X
A rock from the beginning of time moves from South America to Australia in search of the stars.

13. Native American Fiction

This chapter includes novels about Native North Americans written by both native and non-native authors. It is limited to books which either depict indigenous values or contrast them with those of settlers and developers. Many western writers present stereotypical images and a one-sided view of the struggle over the land. Others romanticize the "noble savage." The novels listed here are more complex or ambiguous, particularly the work of Waters, whose entire body of fiction and non-fiction can be viewed as an attempt to reconcile and combine the best of his mixed white and Indian heritage.

The debate about whether white authors practice culture appropriation when they write about the indigenous experience is too complex to summarize here. I have placed an asterisk after the names of enrolled members of tribes, but that hardly settles the issue. One might argue that some white writers such as Eastlake, Richter, or Manfred provide a truer picture of both settlers and indigines than some Native American works, e.g. Storm's *Seven Arrows*. Sensitivity to stereotypes, tokenism, distortions of history, phony dialogue, simplistic depictions of women and elders, and effects on a reader's self-image are more useful than a scholarly blood test.

North America has been the home of a rich oral tradition for millenia. Several anthologies of legends are included in the Native Americans non-fiction section. Also see *Family of Earth and Sky* (1721) for legends from other indigenous cultures.

The Native American novel is a fairly recent form. Some scholars trace its inception to Mourning Dove, Mathews, and Standing Bear in the late 1920s and 30s. In the 60s and early 70s a re-emergence of Native identity was reflected in the work of Chief Eagle, Nasnaga, Momaday, and Pierre. Alexie, Cook-Lynn, Dorris, Erdrich, Hale, Hogan, King, Owens, Silko, Welch, and Vizenor provide a wide variety of first-rate contemporary fiction.

Related novels can be found in other chapters: Guthrie (2088-91), Jackson (2095), Manfred (2101-2), Richter (2104-8), Davenport (1961), Narogin (1965-6), Wongar (1970-3), Cather's *The Professor's House* (2228), Laurence's *The Diviners* (2258), Sojourner's *Sisters of the Dream* (2272), Wilhelm's *Juniper Time* (2282), Bishop's *Unicorn Mountain* (2365), Norton's *Fur Magic* (2408), Doss' *The Shaman Sings* (2290), Straley's *The Woman Who Married a Bear* (2327), Eastlake's *Portrait of an Artist with Twenty-Six Horses* (2123), Henderson's *Native* (2130), Eastlake's *Dancers in the Scalp House* (2508), Matthiessen's *At Play in the Fields of the Lord*, (2519), and Metz's *King of the Mountain* (2520). Good Environmental Action novels in this chapter include Doane's *Bullet Heart*, Lesley's *River Song*, Owens' *Wolfsong*, and King's *Green Grass, Water Running*.

Books for adults:

1981. Alexie, Sherman*. *The Lone Ranger and Tonto fistfight in heaven*. NY: Atlantic Monthly, 1993. 223 p. 0871135485
Short stories set primarily in the present, on and around the Spokane Indian reservation. Alexie often alludes to tradtional times when the Spokanes had a strong, Earth-centered religion, mutually supportive social order, and sense of identity and purpose. In one story a contemporary Indian stands trial for an event which took place in 1858. HR AMT. #

1982. Brant, Beth*. *Food & spirits*. Ithaca, NY: Firebrand, 1991. 125 p. 0932379931 0932379923 (pbk)
AR:*Mohawk Trail*.
Poignant stories about modern Mohawk Indians who are trying to return to traditional values. HR.

1983. Cady, Jack. *Inagehi*. Seattle: Broken Moon, 1994. 258 p. 0913089508
In 1957 30 year-old Cherokee Harriette Johnson inherits 700 acres of timberland and learns that her father, who had been clearcutting the mountain, was murdered. Johnson is faced with two mysteries, who killed her father and the more puzzling one of why her father was behaving contrary to traditional Cherokee values.

1984. Cameron, Anne*. *Daughters of Copper Woman*. Erie, KS: Inland, 1988, c1981. 150 p. 0889740224
AR: *The journey*, Dzelarhons.
"Weaving together the lives of mythic and imaginary characters, Daughter of Copper Woman offers a shining vision of womanhood, of how the spiritual and social power of women...can endure and survive." This book is also about women using their powers to help the Earth survive. #

1985. Cook-Lynn, Elizabeth*. *From the river's edge*. NY: Arcade, 1991. 147 p. 1559700513
The action revolves around a cattle rustling trial, the building of a dam, and the subsequent flooding of Indian land. Contrasts the Indians' and settlers' attitudes about use of the land demonstrating how degrading nature also destroys indigenous cultures. #

1986. Craven, Margaret. *I heard the owl call my name*. NY: Buccaneer, 1991, c1973. 159 p. 089668542
An episcopal priest sent to a remote British Columbia tribe learns from the Indians to live closely to the land. He becomes a sad witness to the gradual disintegration of a culture which has provided him with the courage and insight to fearlessly face his own death. HR for adults, particularly for senior citizens.

1987. Doane, Michael. *Bullet heart*. NY: Knopf, 1994. 316 p. 0679425071
When golf course construction unearths a frontier cemetary the whites are reburied, but the bones of the one Sioux woman are sent to a museum, beginning a 20 year "bone war" over Indian lands and rights. HR.

1988. Dorris, Michael*. *A yellow raft in blue water*. NY: Warner, 1988, c1987. 372 p. 0446387878
Dorris uses three generations of Native American women as narrators, thus presenting both a complex tale of the changing Indian experience, and revealing how the changing values between generations has contributed to the decline of traditional culture. AMT. #

1989. Dorris, Michael*, and Erdrich, Louise*. *The crown of Columbus*. NY: HarperCollins, 1991. 382 p. 0060160799
Two college professors pursue the lost diary of Columbus. This complex novel attempts to combine the personal and the scholarly while it contrasts European and Indian views of American history. An interesting but flawed book recommended primarily for Erdrich and Dorris fans.

1990. Eastlake, William. *The bronc people*. LA: Seven Wolves, 1991, c1958. 254 p. 0962738751
A would-be bronc rider and his Black foster brother roam the high country of northern New Mexico where they learn many lessons from an old Indian fighter, a former missionary who has been coverted to Indian values, and a wide variety of Navajo people. HR. #

1991. ___. *Go in beauty*. LA: Seven Wolves, 1991, c1956. 279 p. 0962738735
A young novelist whose books about "Indian Country" were just becoming successful elopes to Europe with his brother's wife. Removed from hozro (beauty and harmony) and alienated from his brother, he eventually loses his inspiration and sense of meaning. He makes a desperate journey to Mexico to try to regain them.

1992. Erdrich, Louise*. *The bingo palace*. NY: HarperCollins, 1994. 274 p. 0060170808
AR: *The beet queen*.
Lipsha Morrissery and his boss Lyman Lamartine become rivals first over

the love of Shawnee Ray and then over tribal land surrounding a lake where Lyman wants to build a huge bingo hall. On a vision quest, Lipsha is sprayed by a skunk whose message is that the sacred land "ain't real estate." HR AMT. #

1993. ___. *Love medicine : new and expanded version.* NY: HarperPerennial, 1993, c1984. 367 p. 0060975547
National Book Critics Circle Award, 1984.
In Erdrich's first novel, she introduces the Kashpaw and Lamartine families and tells an intergenerational tale of Indian assimilation and resistance from the mid 1930s to the 80s. The character Nanapush, in particular, represents a link to traditional ways and love for the land. HR.

1994. ___. *Tracks.* NY: Harper & Row, 1989, c1988. 226 p. 0060972459
Set between 1912 and 1924, this novel depicts the destruction and dissulution of the traditional Ojibwa Indians through disease, education, religion, and a pernicious treaty and land allocation scheme. Erdrich uses alternating narrators, the traditional Nanapush and the increasingly masochistic Christian convert Pauline, to contrast the Ojibwa's Earth centered philosophy and the settlers' notion of progress. HR.

1995. Hale, Janet Campbell*. *The jailing of Cecelia Capture.* NY: Random House, 1985. 201 p. 0394543270
AR: *Bloodlines : odyssey of a native daughter.*
As an Indian law student spends her 30th birthday in the drunk tank, her thoughts return to her native roots in Idaho and her heart begins to reconnect with Earth-centered values.

1996. Harrison, Jim. *Dalva.* NY: Dutton, 1988. 324 p. 052524624X
Harrison combines the story of a conservation-oriented rancher who loves Indian culture and the journal of her ancestor, a botanist and minister, to relate the history of the genocide of the American Indian and the parallel devastation of the West.

1997. Highwater, Jamake*. *Dark legend.* NY: Grove, 1994. 245 p. 0802114776
An adaptation of the *Nibelungenlied* set in pre-Columbian tropical America. Both earthly and heavenly beings are infected by degrading greed and lust leading to the rape of the land. The dwarf's theft of gold from the water maidens seems symbolic of civilization's scorn toward feminine, life-giving nature. For sophisticated adults.

1998. Hillerman, Tony. *Sacred clowns.* NY: HarperCollins, 1993. 305 p. 006016767X
Navajo police agents follow a serpentine trail through various clans seek-

ing the thread that links two brutal murders, a missing teenager, a band of lobbyists trying to put a toxic dump site on Pueblo land, and an invaluable momento given to the tribe by Abraham Lincoln. This is the most overtly ecofictional Hillerman novel, but all are HR AOT. #

1999. Hogan, Linda*. *Mean spirit*. NY: Atheneum Collier, 1990. 374 p. 0689121016
AR: Hogan's poetry.
Considers the exploitation of Indian people and land during the Oklahoma oil boom of the 1920s.

2000. Houston, James A. *Spirit wrestler*. West Seneca, NY: Ulverscroft, 1982, c1980. 479 p. : map. 0708907687
An ailing shaman is brought to the white authorities. He relates how Morgan, a white man obsessed with eskimo life, has interfered with his health, powers, and wife. An exciting mix of legend, action, and the conflict between two very different philosophies of man's place in nature.

2001. ___. *The white dawn : an Eskimo saga*. San Diego: Harcourt Brace Jovanovich, 1989, c1971. 275 p. : drawings 015696256X
During a springtime whaling expedition in the Arctic, a small boat full of whalers loses the main ship and lands in Eskimo country. They are taken in by the Eskimos and well treated, learning the Inuit way of living in harmony with nature, and even mating with the women. When the rigors of winter arrive, though, the white men resort to lust, greed and pride, inevitably leading to conflict. Mature themes.

2002. King, Thomas*. *Green grass, running water*. Boston: Houghton Mifflin, 1993. 360 p. 0395623049
AR: *Medicine River*.
How does the government honor treaties that will last "as long as the grass grows and the waters run?" By damming the river and flooding the grass. Four ageless traditional Indians escape from a mental hospital to recruit modern Indians, Sun Dancers, and Coyote to stop the development and restore harmony to the land. This raucous comedy is HR AOT. #

2003. Kinsella, W. P. *Dance me outside*. Boston: Godine, 1993, c1977. 158 p. 0879239824
Contains short stories about life among the Indians of Alberta. Features an inherent love of and connection to nature.

2004. ___. *The moccasin telegraph and other Indian tales*. Boston: Godine, 1993, c1983. 192 p. 0879239816
More Earth-oriented short stories about the Indians of Alberta.

2005. La Farge, Oliver. *Laughing boy.* Cutchogue, NY: Buccaneer, 1981, c1929. 182 p. 0899663672
Portrays a young man and a culture caught between traditional, land-centered ways and modern civilization.

2006. Lesley, Craig. *River song.* NY: Laurel, 1990, c1989. 307 p. 0395430836
AR: *Winterkill.*
A Native American father and son who live by the Columbia River become involved in a murderous dispute over native fishing rights and the survival of their traditional culture. This thriller is HR.

2007. Lucas, Janice. *The long sun.* NY: Soho, 1994. 280 p. 1569470138
In the early 1700s a family is trapped by hostile Apalachee Indians, rescued by Tuscarora Indians, and learns to live joyously among them for six years. Then one of their children is trapped by settlers and they are forced to decide where their loyalties lie. Recommended despite tinges of the noble savage stereotype.

2008. Manfred, Frederick Feikema. *King of spades.* Lincoln: Univ. of Nebraska, 1983, c1966. 304 p. 0803281218
AR: *Green Earth.*
By 1876 most of the Indians have been moved to reservations and an Indian-raised white woman is left to witness the end of a mystical union of land and people as miners, ranchers, and other settlers squabble over and desecrate the prairies .

2009. ___. *Lord Grizzly.* Lincoln: Univ. of Nebraska, 1983, c1954. 281 p. 0803281188
AR the Buckskin man tales: *Conquering horse, Scarlet plume, King of spades, Riders of judgment.*
A trapper badly mauled by a bear and abandoned by his companions, crawls 200 miles to safety with revenge on his mind. A vision of the bear and the influence of his Indian wife transform his hatred into forgiveness and he attains a greater unity with God and nature as a result. HR.

2010. ___. *The manly-hearted woman.* Lincoln: Univ. of Nebraska, 1985, c1975. 198 p. 0803281277 (pbk) 0803230923
Manly Heart is a Dakotah woman who disguises herself as a man to become a shaman (who ceremonially disguises "himself" as a woman) and who falls in love with the leader of a war party.

2011. ___. *Scarlet plume.* Lincoln: Univ. of Nebraska, 1983, c1964. 378 p. 080328120X
In a modern version of *Romeo and Juliet* a young Native American boy and

white girl fall in love despite the growing tension between the two cutures which culminated in the Sioux Uprising of 1862 and the hanging of 25 Indians on Christmas day.

2012. Mathews, John Joseph*. *Sundown*. Intro. by Virginia Mathews. Norman: Univ. of Oklahoma, 1988, c1938. 312 p. 0806121602
A mixed-blood Osage returns home in the 1920s to find a land and its people undergoing rapid transition. The effects of the discovery of oil and cultural efforts to assimilate the Osages are explored.

2013. Mayhar, Ardath. *People of the mesa*. NY: Diamond, 1992. 297 p. 155773674X
The Anasazi Indians are living a life rich in ritual and in harmony with nature until they are threatened by a marauding tribe. A young man who can commune with the Earth spirits is the only hope they have to survive, but can chants and prayers stop arrows and spears?#

2014. McNichols, Charles Longstreth. *Crazy weather*. Lincoln: Univ. of Nebraska, 1994, c1944. 195 p. 0803282192
During four days of "glory-hunting" with an Indian comrade a white teenager realizes that he must choose between two cultures.

2015. McNickle, D'Arcy*. *Wind from an enemy sky*. Albuquerque: Univ. of New Mexico, 1988, c1978. 265 p. 0826311008
AR: *Surrounded* .
Two Indian brothers take different paths in life, Henry as a rancher, Bull as a traditional tribal leader. Bull becomes involved in a deadly dispute with the government and contractors over a dam which violates a holy place and limits the Indians' water supply.

2016. Momaday, N. Scott*. *House made of dawn*. NY: Perennial, 1989, c1968. 212 p. 0060916338
Pulitzer Prize.
Young Abel is caught between two worlds: that of his Native-American father rooted in the land and its teachings and that of 20th-century society. HR AMT. #

2017. Mourning Dove*. *Coyote stories*. Lincoln: Univ. of Nebraska, 1990. 246 p. 0803281692
AR: *Cogewea*.
Mourning Dove, one of the first Native American novelists, recorded her people's legends in a form which she considered both respectful to the legends and accessible to white audiences. She uses Coyote and other legendary animals to offer observations on living in a balance with nature and

dealing with your own personal ogres. #

2018. Owens, Louis*. *The sharpest sight*. Norman: Univ. of Oklahoma, 1992.
263 p. 0806124040
Mixed blood Choctaw Cole McCurtain and Mexican American Mundo
Morales search a Mississippi swamp to find and bury the bones of Cole's
murdered older brother.

2019. ___. *Wolfsong*. Albuquerque, NM: West End, 1991. 249 p. 093112266X
AR: *The bone game.*
When a Native American returns to his home in the Cascade Mountains he
reconnects with the spiritual power of his ancestral roots as a result of his
shock over the devastation caused by logging. He must decide whether to
participate in his uncle's guerrilla war against mining.

2020. Shea, Robert. *Shaman*. NY: Ballantine, 1991. 519 p. 0345360486
Depicts the history, times, and surrender of the famous chief Black Hawk to
white soldiers. This event marked the end of an Earth-centered Indian way
of life in the area.

2021. Silko, Leslie Marmon*. *Almanac of the dead*. NY: Penguin, 1992. 792 p.
: map. 0140173196
An ambitious, multilevel novel which tells the tale of the Americas from
the native perspective. Starting in contemporary Tucson it moves through
time in the Americas, Africa, emerges in "the fifth world" and concludes
with prophecies on holistic healing, eco-warriors, and the triumph of native
peoples and the Earth itself. HR.

2022. ___. *Ceremony*. NY: Penguin, 1986, c1977. 262 p. 0140086838
AR: Silko's poetry.
A Native American attempts to cure his illness and his tribe's malaise
through living a ceremony that returns him to harmony with the land, its
creatures and fellow humans. HR.

2023. ___. *Storyteller*. NY: Arcade, 1989, c1981. 288 p. 155970005X
Stories from different Native American tribes. .

2024. ___. *Yellow woman*. Ed. by Melody Graulich. New Brunswick, NJ: Rutgers Univ., 1993. 235 p. 0813520045 0813520053 (pbk)
This legend about a woman abducted by a mountain spirit explores the
relatedness of sexuality, the land, personal identity, and storytelling. Contains an introduction by Graulich, Silko's version of the story, a interview of
Silko by Kim Barnes, and eight essays by various authors. HR, especially
for storytellers.

2025. Smith, Patrick D. *Forever island; and, Allapattah*. Englewood, FL: Pineapple, 1987. 386 p. : ill. 0910923426
In *Forever island* an 86 year old Seminole realizes that developers are posioning the marsh to kill wildlife, so he shoots an old alligator to spare it a slow, agonizing death and is arrested for poaching. His attorney attempts to shift the blame to the developers. In *Allapattah*, a guide abandons two poachers who have shot one of only four remaining crocodiles, thus beginning a series of acts meant to preserve the Everglades and traditional Seminole culture. AMT. #

2026. Speare, Elizabeth George. *The sign of the beaver*. Santa Barbara: ABC-Clio, 1988, c1983. 146 p. 1557360375
Left alone to guard the family's wilderness home in 18th-century Maine, a boy is hard-pressed to survive until local Indians teach him their skills. #

2027. Standing Bear, Luther*. *Stories of the Sioux*. Lincoln: Univ. of Nebraska, 1988, c1934. 79 p. : ill. 0803291876 (pbk) 080324194
Contains 20 short stories focusing on the physical and spiritual animal helpers of the Sioux (buffalo, dog, horse, eagle, deer, hawk, and wolf); the wisdom of medicine men; and the resourcefulness of individuals. #

2028. Storm, Hyemeyohsts*. *Seven arrows*. NY: Ballantine, 1985, c1972. 374 p. : ill. 0345329015
Storm depicts massive changes in the lives of northern Plains Indians in the second half of the 19th century through the stories of over a dozen characters. He combines this outward action with legends and vision quest imagery to explore the themes of change, renewal, the circle of life, androgyny, survival, respect for the land, and understanding. Controversial due to charges of cultural appropriation and historical inaccuracy.

2029. Vizenor, Gerald Robert*. *Darkness in Saint Louis Bearheart Bearheart : the heirship chronicles*. Minneapolis: Univ. of Minnesota, 1990, c1978. 254 p. 0816618518 0816618526 (pbk)
A young AIM activist befriends an old man who has written a novel set after the ecological and economic collapse of industrial America. In the man's novel, a group of Indian and non-Indian people make a quest to Pueblo Bonito, but are ravaged by environmental hazards and evil spirits. HR, but not for genteel readers due to scenes of graphic violence.

2030. Vollmann, William T. *Fathers and crows*. NY: Viking, 1992. 990 p. 0670843334
Historian Francis Parkman, Ignatius of Loyalo, Blessed Catherine Tekakwith, Samuel Champlain, and the Wyandot and Huron Indians are featured in this continuation of the saga of the confrontation between conquest ori-

ented European cultures and nature oriented Indians. HR.

2031. ___. *The ice-shirt*. NY: Viking, 1990. 415 p. : ill. 0670832391
Sequels: *Fathers and crows, The rifles*.
The Vikings use ruthless aggression, shape changing, and axes in their conquest of Greenland and early efforts to supplant the natives. Vollman
employs Icelandic eddas, Inuit legends, and compelling drama to create a
parallel between the fall of Eden and the despoiling of the new world. HR.

2032. ___. *The rifles*. NY: Viking, 1994. 411 p. 0670848565
A fictionalized account of the 1845-48 expedition by Franklin to find the
NW Passage is combined with the present day story of the narrator who
nearly dies in an abandoned arctic weather station in his attempt to experience the hardships encountered by Franklin. Meanwhile, back in the 19th
century, the Europeans give the Indians rifles, which they use to slaughter
wildlife to the point of extinction, just like the white men. HR.

2033. Waters, Frank*. *Frank Waters, a retrospective antholology*. Ed. by Charles
L. Adams Athens, OH: Swallow, 1985. 218 p. 0804008744 0804008752 (pbk)
Contains excerpts from 17 of Waters' fiction and nonfiction books, demonstrating that similar themes such as valuing our place on the land and a
coalescence of Indian and Western thinking dominate both genres. An
excellent sampler for those unfamiliar with Waters. HR.

2034. ___. *The lizard woman*. Austin: Thorp Springs, 1984, c1930. 136 p.
0914476998
Original title: *Fever pitch*.
Two young American engineers and a Mestizo bar-girl suffer both physical
hardship and psychic peril as they guard a gold strike in a desolate desert
in Baja California.

2035. ___. *The man who killed the deer*. NY: Pocket, 1984, c1942. 266 p.
0671555022
A young man returns from boarding school where he was taught to reject
traditional ways. When he kills a deer out of season he transgresses both
the white man's regulations and the Indian practice of asking the deer's
permission and thanking it. Ironically, his trial becomes part of the Indian
struggle to reclaim sacred land from the government. HR AOT. #

2036. ___. *Pike's Peak : a mining saga*. Athens, OH: Swallow, 1987, c1971. 743
p. 0804009007
A revision and combination of three books: *The wild Earth's nobility, Below
grass roots*, and *Dust within the rock*. A multigenerational family mining saga
which deals with the themes of love vs. overexploitation of the land, the

clash of values between Indians and settlers over it, and the alienation caused by losing a personal connection with it.

2037. ___. *The woman at Otowi Crossing*. Athens, OH: Swallow, 1988. 314 p. 0804008930
A novel based on the life of Edith Warner who ran a tearoom and formed a link between the nuclear scientists working at Los Alamos and the local natives. When she is stricken with cancer she becomes more Indianized, extolling the unity of all living things.

2038. Weaver, Will. *Red earth, white earth*. NY: Pocket, 1987, c1986. 402 p. 0671619888
Guy Pehrsson returns from Silicon Valley to his boyhood farm home in Minnesota where bankers threaten to repossess the farm and local Indians have restored their claim to the land to stop its possible destruction by developers.

2039. Welch, James*. *Fool's crow*. NY: Penguin, 1987. 391 p. : ill., map. 0140089373
Set in Montana around 1870, this novel presents a vivid, non-romanticized depiction of the nomadic life of the Blackfoot Indians, the centrality of the black horn (bison) and other animals to their material and spiritual lives, the many ways that white expansionism threatened their existence, and the lively debate about how to deal with that threat. This powerful page-turner is HR.

2040. ___. *Winter in the blood*. NY: Penguin, 1986, c1974. 176 p. 0140086447
AR:*The Indian lawyer*, *The Death of Jim Loney*, *Killing Custer*.
This story of life on and around Montana's Fort Belknap Indian Reservation paints a fairly positive picture of re-emerging native values on the reservation contrasted to the alcoholism, brutality, and hopelessness of the towns around it. HR.

2041. West, Paul. *The place in flowers where pollen rests*. NY: Collier, 1989, c1988. 490 p. 002038260X
Pushcart Prize.
A young Hopi man returns home from the horror of Viet Nam and the sleaze of Los Angeles. With the help of a Kachina carver he attempts to restore the harmonious balance of nature and self. Due to its unusual syntax and rambling sentences, this is a fairly tough read for all but sophisticated adults, but a rewarding one.

Books for young adults:

2042. Armer, Laura Adams. *Waterless Mountian.* NY: Knopf, 1994, c1931. 222 p. 067984502X
AR: Frances Gilmor's *Windsinger.*
Newbery Award, 1931.
A young Navajo named Younger Brother learns to be a Singer ("medicine man") and the practical knowledge of how to make a living while living harmoniously with the Earth. Emphasizes his personal connection to animals, the elements, and the spirit world. Contains depictions of Navajo rituals. Occasionally overromanticized, but still HR. #

2043. Austin, Mary Hunter. *The basket woman : a book of fanciful tales for children.* NY: AMS, 1969, c1904. 220 p. 0404004296
Contains 14 stories, most based on Native American legends. HR. #

2044. Cooper, James Fenimore. *The last of the Mohicans.* Adapted by Les Martin. NY: Random House, 1993. 111 p. : ill. 067994706X 0679847065
A simplified retelling of the story about the exploits of a young white man and his Mohican friends during the French and Indian War. Contrasts between the Indians' and settlers' views about the land and nature are presented fairly well, but the Mohicans are sometimes "noble savages." HR GR 5-8..

2045. Dorris, Michael*. *Morning Girl.* NY: Hyperion, 1992. 74 p. : ill. 1562822845 1562822853
Morning Girl and her younger brother Star Boy take turns describing their life in pre-Columbian America. In her last entry she witnesses the arrival of the first Europeans and realizes that her nature-centered world is imperiled. HR primarily for GR 4-8, but also for adults. #

2046. George, Jean Craighead. *Julie.* NY: HarperCollins, 1994. 240 p. : ill. 0060235284 0060235292
Publishers Weekly Best Book, 1994.
When Julie returns to her father's Eskimo village, she struggles to find a way to save her beloved wolves in a changing Arctic world and falls in love with a young Russian man. HR GR 5-9.

2047. ___. *Julie of the wolves.* Santa Barbara: ABC-Clio, 1987, c1972. 184 p. : ill. 1557360537
Newbery Award.
Sequel: *Julie.*
While running away from home and an unwanted marriage, a 13-year-old Eskimo girl becomes lost on the North Slope of Alaska and is befriended by a wolf pack. HR GR 5-9.

2048. ___. *The talking earth*. NY: Harper & Row, 1987, c1983. 151 p.
0064402126
AR Spanish translation: *La tierra que hable*.
Billie Wind ventures out alone into the Florida Everglades to test the legends of her Indian ancestors and learns the importance of listening to the earth's vital messages. Somewhat overdidactic, but still recommended for GR 5-9.

2049. ___. *Water sky*. NY: Harper & Row, 1989, c1987. 212 p. 0064402029
A boy who goes to Barrow, Alaska, to live with friends of his father for awhile learns the importance of whaling to the Eskimo culture. GR 5-9.

2050. Hobbs, Will. *Beardance*. NY: Atheneum, 1993. 197 p. 0689318677
While accompanying an elderly rancher on a trip into the San Juan Mountains, a Ute Indian boy tries to help two orphaned grizzly cubs survive the winter while performing his spirit mission. HR GR 5-9.

2051. ___. *Bearstone*. NY: Atheneum, 1989. 154 p. 0689314965
Sequel: *Beardance*.
A troubled Indian boy goes to live with an elderly rancher whose caring ways help him become a man. They share a love and respect for nature, particularly bears. GR 5-9.

2052. Holling, Holling Clancy. *Paddle-to-the-Sea*. NY: Houghton Mifflin, 1986, c1942. 61 p. 0395150825
Caldecott Honor Book, 1942.
A young boy carves an Indian figure and sends him off in a small canoe on a long, adventurous journey through the Great Lakes to the sea. Provides a geographic, historical, and naturalistic account of the area. HR GR 4-8.

2053. Houston, James A. *Drifting snow : an Arctic search*. NY: Puffin, 1994, c1992. 150 p. : ill. 0140365303
AR: *Ice swords*, and for elementary level *Akavak*, *Ghost paddle*, *Kiviok's magic journey*, and *Wolf run*.
Having been taken from her Arctic home when a tiny child, a teenager returns to look for her parents and learn once again about her Eskimo culture.

2054. ___. *The falcon bow*. NY: Puffin, 1992, c1986. 94 p. : ill. 0140360786
Kungo, a young Inuit, seeks to prevent a bloody feud when his people and a rival tribe find themselves in competition for dwindling food supplies.

2055. ___. *Frozen fire : a tale of courage*. NY: Aladdin, 1992. 149 p. : ill.
0689716125
AR: *Black Diamonds*.

Determined to find his father who has been lost in a storm, a young white boy and his Eskimo friend brave wind storms, starvation, wild animals, and wild men during their search in the Canadian Arctic.

2056. ___. *River runners : a tale of hardship and bravery*. NY: Puffin, 1992. 142 p. 014036093X
Two young boys, who have been sent into the Canadian interior to set up a fur-collecting station, are befriended by a Naskapi Indian family. GR 5-9..

2057. Irwin, Hadley. *We are Mesquakie, we are one*. Old Westbury, NY: Feminist, 1980. 115 p. 0912670851 0912670894 (pbk)
A young Mesquakie girl grows up at a time when her people are forced to move from their home in Iowa to a reservation in Kansas and encouraged to adopt white culture. She sees how disconnection from the land has demoralized her people and forms a plan to return home to buy the land back. GR 5-9.

2058. L'Engle, Madeleine. *Dragons in the waters*. NY: Dell, 1982, c1976. 293 p. 0440917190
A boy's trip to Venezuela culminates in murder and the discovery of an Indian tribe dating from the time of Simon Bolivar. When he decides to stay with the tribe he is warned that he will revert to savagery, but he points out that they live in harmony with nature and have a saner attitude to life and death while civilization is ravaging the environment. HR. #

2059. Lopez, Barry Holstun. *Crow and Weasel*. NY: HarperPerennial, 1993, c1990. 79 p. : col. ill. 0066975288
Two beings with both animal and human characteristics engage in a spirit quest which teaches them the value of unity with nature and how the earth will provide for them if they care for it. Written at an elementary level, but HR all ages. #

2060. Markoosie*. *Harpoon of the hunter*. Montreal: McGill-Queen's Univ., 1970. 81 p. : ill. 0773501029
After tracking down a wounded polar bear a young boy must make his way home alone through hazardous conditions after his friends are killed.

2061. Mayne, William. *Drift*. NY: Dell, 1990, c1985. 166 p. 0440403812
Lost in the snowy forest, Rafe Considine is taken prisoner by two Indian women who teach him to live off the land and to communicate with bears and other animals. He then finds himself adrift when an ice flow breaks away from the shore. #

2062. McNickle, D'Arcy*. *Runner in the sun : a story of Indian maize*. Albu-

querque: Univ. of New Mexico, 1987, c1982. 249 p. : ill. 0826309747
A teenage boy travels to Mexico to find a hardy strain of corn which will
survive drought and save his people from famine. Describes the ecological
beliefs and lifestyles of the Pueblo Indians, the development of maize, and
its importance in tribal life and trade. HR. #

2063. Mowat, Farley. *The snow walker.* NY: Bantam, 1984, c1977. 209 p.
0776422098
This collection of stories about Inuit life, mythology, attitudes toward
nature, and mistreatment by white culture is sometimes brilliant, but
sometimes suffers from the "noble savage" stereotype. AOT, but only if read
with a dismissal of this stereotype. #

2064. Naylor, Phyllis Reynolds. *To walk the sky path.* NY: Dell, 1992, c1973.
144 p. 0440406366
Ten-year-old Billie, a Seminole Indian, is caught between the cultures when
his family moves away from the natural setting of the Florida Everglades
and nearer the white man's civilization. GR 4-8.

2065. O'Dell, Scott. *Island of the Blue Dolphins.* Boston: Houghton Mifflin,
1990. 181 p. : ill. 0395536804
Newberry Award, 1990.
Left alone on a beautiful but isolated island off the coast of California, a
young Indian girl spends eighteen years, not merely surviving through her
enormous courage and self-reliance, but also finding a measure of happiness in her solitary life. HR. #

2066. Paulsen, Gary. *Canyons.* NY: Dell, 1991, c1990. 184 p. 0440210232
Finding a skull on a camping trip in the canyons outside El Paso, Texas,
Brennan becomes involved with the fate of a young Apache Indian who
lived in the late 1800s. GR 4-8.

2067. ___. *Dogsong.* NY: Puffin, 1987, c1985. 177 p. 0140322353
A 14-year-old Eskimo boy who feels assailed by the modernity of his life
takes a 1400 mile journey by dog sled across ice, tundra, and mountains
seeking his own "song" of himself. GR 5-9.

2068. ___. *Mr. Tucket.* NY: Delacorte, 1994. 166 p. 0385311699
In 1848, while on a wagon train headed for Oregon, a 14-year-old is kidnapped by Pawnees and then falls in with a one-armed trapper who teaches him how to live in the wild. GR 5-9.

2069. ___. *The night the white deer died.* NY: Dell, 1991. 105 p. 0440210925
A teenage girl and an old Indian are brought together by the same haunt-

ing dream. GR 4-8.

2070. Richter, Conrad. *The light in the forest*. NY: Bantam, 1990, c1953. 117 p.
0553268783
AR: *A country of strangers.*
A four-year-old is captured by the Delaware Indians and adopted by one of
the tribal leaders. After 11 years, he is forced to return to his original home
and parents. His love for and loyalty to his Indian parents, his cousin Half
Arrow, and nature lead him to reject white civilization. HR GR5-adult. #

2071. Root, Phyllis. *The listening silence*. NY: HarperCollins, 1992. 106 p. : ill.
0060250925 0060250933
A young Indian girl who can enter the minds and sensations of animals
and other people triumphs over her fears and proves herself worthy to be
the mystical healer of her village. GR 4-8.

2072. Sandoz, Mari. *The horsecatcher*. Lincoln: Univ. of Nebraska, 1986,
c1957. 192 p. 0803241666 0803291604 (pbk)
Unable to kill, a young Cheyenne is scorned by his tribe when he chooses
to become a horse catcher rather than a warrior. Emphasizes the respect of
the Cheyenne for elders, animals, and nature. GR 5-8.

2073. ___. *The story catcher*. Lincoln: Univ. of Nebraska, 1986, c1963. 175 p.
0803291639
A young Sioux warrior earns the right to be called historian for his tribe
after numerous adventures and trials which test his ability to tell the story
of his people with truth and courage. GR 7-10.

2074. Sharpe, Susan. *Spirit quest*. NY: Bradbury, 1991. 122 p. 0027823555
An 11-year-old vacationing on a reservation off the coast of Washington
becomes friends with a young Quileute Indian who is preparing for his spirit quest. GR 4-8.

2075. Underwood, Paula. *Who speaks for Wolf : a native American learning story*.
San Anselmo, CA: Tribe of Two, 1991. 51 p. : ill. 1879678012
An Indian tribe learns an important lesson after it ignores a hunter's warning and settles in the heart of a great community of wolves. This read-
aloud book is recommended for elementary schools and families.

14. Frontier and Pioneer Life

"Return with us now to the glorious days of yesteryear . . . " when men were men and justice was rendered with lead, when the lowly buffalo was replaced with the noble cow, when redskins indiscriminately slaughtered settlers, when missionaries desired only to spread the word of God, when women and children lived only to serve their brave menfolk that the evil wilderness might be tamed for Progress in the holy cause of Manifest Destiny. That's only a slight exaggeration of the image of the old West as portrayed in much popular fiction, film, television, music, and art.

Documents from that era, historical research, and some fiction depicts a different frontier, one where relations between white and red were complex and the savagery was mostly against the latter, where different settlers disagreed about how the land should be used, where women often supplied voices of reason, where nature was as much an inspiration as a force to be tamed, and where the hats of Wyatt Earp, Buffallo Bill, and Custer were charcoal at best, not lilly white. That is the frontier you will find in this chapter and the one on Native Americans.

The resettlement of North America was a multi-stage process dramatized in the work of Cooper, Fergusson, Guthrie, London, Manfred, Richter, and Rolvaag. Although Cooper's Native Americans are sometimes portrayed as noble savages, the ones in Guthrie, Manfred, and Richter are as real and complex as those portrayed by Welch or Erdrich. Criticism of the Manifest Destiny perspective and the consequent rape of the land can be found in novels by Austin, Cather, Cooper, Garland, Guthrie, London, Manfred, Richter, Stegner, Waters, and especially in Norris' *Octopus*, an important precursor of the modern Environmental Action novel. A realistic portrayal of whites and Indians, men and women, can be found in novels by Austin, Cather, Jackson, Strobridge, Stegner, and even in the sometimes overly sentimental Wilder. The anthology *Westward the Women* (1737) and the literary study *Prairie Women* (2548) are also recommended.

Books for adults:

2076. Austin, Mary Hunter. *Isidro*. Upper Saddle River, NJ: Literature House, 1970, c1904. 424 p. 0839800703
After being released when wrongly accused of murder, the son of a hacien-

da owner is nearly killed in a forest fire set by soldiers to kill Indians. In addition to offering an exciting adventure, Austin contrasts the missionaries' exploitative attitudes toward Indians and the land, with the Indians' more naturalistic orientation. AOT#

2077. ___. *Western trails : a collection of short stories.* Ed. by Melody Graulich. Reno: Univ. of Nevada, 1987. 309 p. 087417127X
AR:*The ford, Starry adventure, One smoke stories.* HR AOT#

2078. Berry, Don. *Moontrap.* Eugene, OR: Comstock, 1991, c1962. 346 p. 089174007
The story of Johnson Monday, a trapper in Oregon around 1850 who lives much like the Indians and is oppressed by the coming of settlers who bring the tedium of farming, the oppression of government, and the mind control of missionaries. Monday is forced to make a choice between the settlers of his own race and the people of the race he has grown to love. AMT. #

2079. ___. *To build a ship.* Eugene, OR: Comstock, 1977, c1963. 224 p. 0891740295
A group of explorers and settlers isolated on the Oregon Coast and overwhelmed by wilderness must use native materials and their own wits to build a schooner to escape.

2080. ___. *Trask.* Eugene, OR: Comstock, 1990, c1960. 373 p. 0891740015
Set in coastal Oregon in 1848. A young trapper-farmer encounters both natural hardships and the threat of death at the hands of Indians who realize that he represents a society which will eventually destroy them. Trask is befriended by a Clatsop Indian, adapts their customs, and makes a vision quest. Berry's first novel is his best and is HR AMT. #

2081. Cooper, James Fenimore. *The leatherstocking tales.* NY: Literary Classics of the US, 1985, c1823-41. 2 v. 0940450208 (v. 1) 0940450216 (v. 2)
Contents: *The pioneers, The last of the Mohicans, The prairie, The pathfinder, The deerslayer.* A sweeping chronicle of the settlement of upstate New York and the northern plains. Demonstrates changes in both landscape and consciousness from Indian days, to early exploration and fur trapping, to settlement. Some depictions of Native Americans are romanticized, others more complex. HR AOT. #

2082. Dillard, Annie. *The living.* NY: HarperCollins, 1992. 397 p. 0060168706
Around the turn of the century, settlers strive to make a new life among nature and the Indians in an abundant but sometimes challenging setting.

2083. Fergusson, Harvey. *Wolf song.* Lincoln: Univ. of Nebraska, 1981, c1927. 206 p. 0803268556

Sequels: In those days, The blood of the conquerers.
An evocation of nature and the "mountain man" era in the Sangre de Chris-to Mountains. AR.

2084. Fulton, Len. The grassman. Berkeley: Thorp Springs, 1974. 275 p.
0914476262 0914476270 (pbk)
Set in Wyoming during the final days of the open ranges, this novel depicts conflict over water and other resources and the impact of frontier depriva-tions on cowboys and settlers.

2085. Garland, Hamlin. Main-travelled roads. NY: Cutchogue, 1991, c1891.
299 p. 0899665551
A collection of gritty short stories that counter the romantic stereotype of pioneer life found in late 19th century fiction. The natural rigors are diffi-cult enough, but Garland presents the existing economic system as the greatest hardship to be overcome.

2086. Gloss, Molly. The jump-off creek. Boston: Houghton Mifflin, 1989. 186
p. 0395510864
AR: Outside the gates.
In 1895 a woman seeks freedom and renewal on an Oregon homestead. A vivid depiction of how pioneers dependend on each other and nature to survive.

2087. Grainger, Martin Allerdale. Woodsmen of the West. Seattle: Fjord, 1988,
c1908. 199 p. 0940242346 (pbk) 0940242354
A thinly veiled autobiographical novel about an upper class Englishman working as a logger in British Columbia. Includes excellent descriptions of the lush virgin forests, the exhausting and dangerous work, the rough and tumble life of logging camps, and the methods used to destroy the forests.

2088. Guthrie, A. B. The big sky. Boston: Houghton Mifflin, 1992, c1947. 386
p. 0395611539
Sequel: The way west.
Fearing he has accidently killed his father, a young man flees to become a "mountain man" in Montana, develops a life in harmony with nature and the Indians, but is tempted to an act of jealous rage. HR AOT. #

2089. ___. Fair land, fair land. Boston: Houghton Mifflin, 1982. 262 p.
0395325110
Sequels: The big sky, The way west.
Regretting that he had helped guide the wagon train, Summers heads back to Montana to live among the Indians, only to be encroached upon by other settlers. The trilogy thus completes the saga of the passage of a way of life from Indian times to homesteading. AOT. #

2090. ___. *Mountain medicine*. Eugene, OR: Comstock, 1991, c1960. 152 p.
0891740570
Original title: *The big it.*
A collection of short stories and other selections, mostly about life in the
west, some with conservationist themes.

2091. ___. *The way West*. Boston: Houghton Mifflin, 1993, c1949. 340 p.
0395656621
Summers guides a wagon train to Oregon before realizing that the goals and
ideals of the settlers who wish to tame the land are incompatible with those of
the Indians and trappers who had lived upon and valued it just as it was. AOT. #

2092. Haig-Brown, Roderick Langmere. *Timber*. Corvallis: Oregon State
Univ., 1993. 410 p. 0870715143 0870715151
AR: *Panther.*
A novel about logging, friendship, and unionization set in the forests of the
Pacific Northwest in the 1930s.

2093. Hausman, Gerald, and Zelazny, Roger. *Wilderness*. NY: Forge, 1994.
302 p. 0312856547
Intersperses the stories of John Colter fleeing Blackfeet Indians over 150
miles in and around what is now Yellowstone National Park, and Hugh
Glass, crawling 100 miles after being mauled by a bear. Their intimate
knowledge of wilderness and resourcefulness allow them to survive.

2094. Hoagland, Edward. *Seven rivers west*. NY: Penguin, 1987, c1986. 319 p.
: map. 0140102760
Cecil Roop and a troupe of human, equine, and canine companions search
the remote Rockies first for a trainable grizzly bear, then for a Bigfoot. Combines old West action, vivid depictions of the prairies and mountains, and
unique characters. AOT, particularly dog lovers. #

2095. Jackson, Helen Hunt. *Ramona*. NY: New American Library, 1988,
c1884. 362 p. 0451522087
A depiction of the San Fernando Valley in Indian-Mexican hacienda times
which contains vivid descriptions and appreciation of the natural landscape.

2096. London, Jack. *The collected Jack London : thirty-six stories, four complete
novels, a memoir*. NY: Barnes & Noble, 1991. 1061 p. 0880295961
This excellent collection for teens and adults is a good choice for individuals and small libraries with limited budgets. HR. #

2097. ___. *In a far country : Jack London's tale of the West*. NY: Jameson, 1987.
320 p. 0915463369

Two mismatched gold prospectors are brought together in a wilderness cabin. Lacking the will and prudence to join forces to survive, they fall prey to both their inner dreads and the overpowering forces of nature. #

2098. ___. *The sea-wolf*. NY: Tor, 1993, c1904. 305 p. 0812522761
When a brutal sea captain rescues the only survivors of a shipwreck and refuses to return them to land, a battle of wills ensues wherein civilized city dwellers are forced to muster the courage to confront both natural dangers and the captain's cruelty. AOT. #

2099. ___. *To build a fire and other stories*. Des Moines, Iowa: Perfection Form, 1991. 115 p. 1563120054
The title story tells a tale of the human confrontation with the beauties and dangers of the Alaskan wilderness. #

2100. ___. *The valley of the moon*. Middleton , CA: Rejl, 1988, c1913. 192 p. 0961418117
Two wanderers in turn of the century northern California reject the pressures of the city and urban values in favor of the freedom of the road and a simpler lifestyle. AOT. #

2101. Manfred, Frederick Feikema. *The golden bowl*. Brookings: S.D. Humanities Foundation, 1992, c1944. 237 p. 0963215701
AR: *The chokecherry tree, Eden Prairie, The man who looked like the Prince of Wales, Milk of wolves, This is the year*.
In Manfred's first novel a dust bowl victim decides that nature is an evil force and farming a futile activity. He leaves the land, but returns to discover that it is bountiful if one is willing to work naturally with it rather than fight against natural forces.

2102. ___. *Of lizards and angels : a saga of Siouxland*. Norman: Univ. of Oklahoma, 1992. 617 p. 0806124172
The tale of three generations of a family living in the prairies between 1880 and 1960. The 'old lizard,' primal human instinct, is merged with the redeeming value of compassion. HR.

2103. Norris, Frank. *The octopus : a story of California*. NY: Penguin, 1986. 656 p. 0140390405
When railroad companies try to make further encroachments on rich agricultural land in the 1880s, they face a variety of forms of resistance from the farmers. This precursor of ecofiction based on the Mussell Shoals Massacre has influenced many contemporary authors and activists. HR.

2104. Richter, Conrad. *The fields*. Athens: Swallow, 1991. 161 p. 0821409794

Sayward Wheeler raises her own family against a backdrop of wilderness and forest being tamed by settlers, the development of community, and the movement of adventurous souls further westward. AMT. #

2105. ___. *The rawhide knot and other stories.* Lincoln: Univ. of Nebraska, 1985, c1935. 205 p. 0803289162
AR: *The mountain on the desert.*
These stories illustrate the relationships between settlers, Indians, and the land itself on the prairies. #

2106. ___. *The sea of grass.* Athens: Swallow, 1992, c1937. 149 p. 0821410261
Portrays a 25 year struggle in New Mexico between ranchers who wished to preserve the natural open range for grazing and farmers who wished to plow it despite lacking irrigation water. AOT. #

2107. ___. *The town.* Athens: Swallow, 1991, c1950. 300 p. 0821409808
Sequel: *The fields.*
Pulitzer Prize, 1951.
Sayward Wheeler's tenth child rebels from her simple, nature loving, pioneer values by becoming editor of a newspaper and attacking his older brother, now the governor of Ohio. Thus the rapid change of wilderness to pioneer farm communities to civilization is completed. HR AOT. #

2108. ___. *The trees.* Athens: Swallow, 1991, c1940. 167 p. 0821409786
Sequels: *The town*, *The fields.*
In the late 18th century, the Luckett family moves west to the then deep woods of the Ohio territory, where young Sayward survives her mother's death, her father's failures, and the harsh natural conditions to raise her younger siblings. This very realistic depiction of pioneer life is HR AOT. #

2109. Rolvaag, O. E. *Giants in the earth : a saga of the prairie.* NY: Harper-Collins, 1991, c1927. 453 p. 0060830476
Translation of I *de dage.*
Sequels: *Peder victorius*, *Their father's gold.*
When Per and Beret Hansa immigrate to South Dakota in the early 1870s Per finds satisfaction in taming the prairie while Beret's longing for her native valleys and dread of the open plains drive her slowly insane. Ironically, Beret's sanity is restored by a minister while Per dies in a blizzard trying to locate a minister for a dying friend. HR AMT. #

2110. Sanders, Scott R. *Wilderness plots : tales about the settlement of the American land.* Columbus: OSU, 1988, c1983. 117 p. : ill. 0814204724
Contains 50 brief fictional sketches of people facing the wilderness during the exploration and settlement of the Ohio River Valley. #

2111. Stegner, Wallace Earle. *Angle of repose*. NY: Penguin, 1992, c1971. 569 p. 014016930X
Pulitzer Prize, 1971.
Angle of repose, *The spectator bird*, and *Crossing to safety* were published in one volume by the Quality Paperback Book Club, 1993.
A dual story wherein an aging, disabled historian attempts to retain his independence while writing a book about his grandmother. Based on Mary Hallock Foote, she is a cultured easterner who moves west with her husband to work on a variety of mining and irrigation projects. She struggles with increasingly rougher conditions, keeping her family solvent through nature writing and drawings. HR.

2112. ___. *The Big Rock Candy Mountain*. NY: Penguin, 1991, c1938. 563 p. 0140139397
In the early 20th century, Bo Mason tries to make his way in the west, struggling against the forces of nature and culture alike. Strongly evokes western America in a time of transition away from the frontier.

2113. Strobridge, Idah M. *Sagebrush trilogy : Idah Meacham Strobridge and her works*. Reno: Univ. of Nevada, 1990, c1904-09. 1 v. (various pagings) 0874171644
Contents: In *miners' mirage-land*, *The loom of the desert*, *The land of purple shadows*.

Books for young adults:

2114. MacLachlan, Patricia. *Sarah, plain and tall*. NY: Harper & Row, 1987, c1985. 58 p. 0064402053
A mail-order bride moves from her beloved home in coastal Maine to the plains of Kansas. She learns to understand and love her new environment and family despite hardships. HR GR 4-8.

2115. McClung, Robert M. *Hugh Glass, mountain man*. NY: Morrow Junior, 1990. 166 p. 0688080944
Another ed. entitled *Hugh Glass, mountain man: left for dead* publ. by Beech Tree Books, 1993.
A fictionalized biography of the legendary hero of the Old West, who as a fur trapper in 1823 survived an attack by a grizzly bear.

2116. Wilder, Laura Ingalls, and Lane, Rose Wilder. A *little house sampler*. NY: Perennial, 1989, c1988. 243 p. 0060972408
Articles and fictional works by Laura Ingalls Wilder and her daughter creating a chronological account of their lives and depicting the settlement of the northern prairies. Good family reading. #

15. The New West

Frontier times were more complex than often portrayed in popular culture, but were fairly simple compared to the many conflicts and contradictions of the American West today. In addition to the issues depicted in frontier literature, the new Westerner might ponder such issues as atomic testing, animal rights, endangered species, logging, racism and sexism, sexual orientation, the sanctuary movement, and the family. If you think things are the same as ever out at the ranch, Henderson, Nelson, Roderus, and Schaefer have some surprises for you.

Perhaps the most pervasive theme in this chapter is the relation between people and the land. This is a central concern of Adamson, Agee, Barber, Broder, Canary, Duncan, Eastlake, Ford, Gish, Kingsolver, Lord, Maclean, McGuane, Nelson, O'Brien, Schaefer, and Stegner. See "Land Use" in the subject index for related works in other chapters.

A few authors incorporate metafictional writing styles wherein time, place, perspective, and character are either unconventional or constantly shifting. *The Meadow* itself is Galvin's protagonist. Garber's *The Historian* and his feral cousin, Simms, and Romtvedt's Don Galeano in *Crossing Wyoming* travel through time and space at will. Eastlake employs the stream of consciousness of a man slowly sinking in quicksand. The results are both challenging and entertaining. Similar techniques are employed by such contemporary Native American authors as Dorris, Erdrich, King, and Welch.

Relevant anthologies and studies include *The Interior Country* (1717), *In the Dreamlight* (1724), *The Last Best Place* (1727), *Writers of the Purple Sage* (1731), *New Writers of the Purple Sage* (1730), *Southwest Stories* (1734), *Inheriting the Land* (1747), and *Where the Bluebird Sings to the Lemonade Springs* (2597)

Books for adults:

2117. Adamson, Yvonne. *Bridey's mountain.* NY: Delacorte, 1993. 613 p. 0385308507
A third generation Coloradan saves her mountain land from developers when she wins $25 million in the state lottery, but becomes caught up in her own ambitions and those of others. The modern protagonist and her life are less compelling than those of her mother or grandmother in this uneven multigenerational novel.

2118. Agee, Jonis. *Strange angels.* NY: Ticknor & Fields, 1993. 405 p. 039560835X
When a rancher dies, his three children of different mothers squabble over the estate. Depicts how business, government, and personal greed are degrading both the natural and social life of family farms and Indian reservations. Occasionally stereotypical and predictable, but generally well written.

2119. Barber, Phyllis. *And the desert shall blossom.* Salt Lake City: Signature, 1993. 281 p. 1560850361
When a Mormon family moves to Nevada to help build Hoover Dam they are confronted with questions about the relationships between the individual, family, God, and nature.

2120. Broder, Bill. *The sacred hoop : a cycle of earth tales.* SF: Sierra Club, 1992. 238 p. : ill. 0871565838
Portrays the larger picture of humankind's changing relationship with the Earth, the Gods, and the universe. HR AOT. #

2121. Canary, Brenda Brown. *Home to the mountain.* NY: Avon, 1982, c1975. 217 p. 0380611279
When a young man returns to his mother's home in the Oklahoma hills to discover the truths surrounding her death he meets a young Indian widow. They develop a powerful bond between one another and the mountain.

2122. Duncan, David James. *The river why.* SF: Sierra Club, 1983. 294 p. 0871563215
A fly fishing fanatic moves to an Oregon river where he hooks some big questions about nature, conservation, and spirituality. Despite some moments of spiritual-literary excess this is recommended for adults, especially anglers.

2123. Eastlake, William. *Portrait of an artist with twenty-six horses.* LA: Seven Wolves, 1991, c1963. 221 p. 0962738743
Caught up to his neck in quicksand, a rancher ruminates on his life and its meaning, nature, white and Indian cosmology, and related ideas. This literary experiment which strings independent stories along the lines of a possibly dying man's thoughts is sometimes jumbled but is HR for sophisticated adult readers.

2124. Ford, Richard. *Rock Springs.* NY: Atlantic Monthly, 1987. 235 p. 0871131595
Contains ten stories set mainly in Montana. The writing is quite lyrical, but Wallace Stegner takes issue with Ford's poetic license in inaccurately describing real places.

2125. ___. *Wildlife*. NY: Atlantic Monthly, 1990. 177 p. 0871133482
A 16 year old Joe Brinson and his family move to Great Falls in 1960 and their lives change in unexpected ways. Includes vivid descriptions of the Montana wilderness and forest fires.

2126. Fromm, Pete. *The tall uncut*. Santa Barbara: Daniel, 1992. 163 p. 0936784954
Winner of *Sierra's* fourth annual Nature Writing Contest.
AR: *King of the mountain* .
Contains 17 short stories, most set in the wild, dealing with nature, hunting, fishing, and human relationships.

2127. Galvin, James. *The meadow*. NY: Holt, 1992. 230 p. 0805016848
A multigenerational novel about the joys and struggles of a family on a small, remote ranch in the Rockies. It vividly depicts how the actual main character, the meadow itself, changed over 100 years. HR. #

2128. Garber, Eugene K. *The historian : six fantasies of the American experience*. Minneapolis: Milkweed, 1993. 233 p. 0915943573
In a metafictional series of stories spanning the 19th century, the historian's intellectual pursuit of his muse, Clio, runs counterpoint to his cousin Simms' outdoor adventures and metamorphoses. Is Simms actually part bear? AOT. #

2129. Gish, Robert. *First horses : stories of the new West*. Reno: Univ. of Nevada, 1993. 134 p. 0874172101 087417211X (pbk)
AR: *When Coyote Howls*
Gish explores relations between Anglos, Chicanos, and Indians just after WW II. Although not overtly ecofictional, these lively stories brilliantly evoke the landscapes and mindscapes of the Southwest.

2130. Henderson, William Haywood. *Native*. NY: Dutton, 1993. 250 p. 0525935746
A ranch foreman risks abuse by and estrangement from his fellow macho ranchers when he falls in love with a young hired hand and meets an Indian who is trying to restore the mystical, androgynous berdache tradition.

2131. Kingsolver, Barbara. *Animal dreams*. NY: HarperCollins, 1990. 342 p. 006016350X
Edward Abbey Award for Ecofiction, 1991.
AR: *Homeland*.
A young woman returns to Arizona searching for the truth of her past. She meets an old Apache boyfriend who introduces her to Indian culture and also becomes involved in the Sacnctuary movement through her sister. HR AMT. #

2132. ___. *The bean trees*. NY: Harper Perennial, 1992, c1988. 232 p. 0060915544
Sequel: *Pigs in Heaven*.
Twentyish Taylor leaves Kentucky for a new life in the Southwest and finds an abused Cherokee baby named Turtle in her car. In Tuscon she works at a tire store which is a front for the sanctuary movement for Central American political refugees. She finds new meaning in life through her difficult but rewarding relationship with Turtle, her work in the sanctuary movement, and the beauty and power of the desert and mountains. HR AMT. #

2133. ___. *Pigs in heaven*. NY: HarperCollins, 1993. 343 p. 0060168013
On a trip to Hoover Dam Taylor and Turtle save a life, but when Cherokee lawyer Annawake Fourkiller sees them on "Oprah" she is sure that the Indian child Turtle has been illegally adopted. Taylor and Turtle go on the lam and end up in Heaven, Oklahoma where they learn to value family, community, and the continuity provided by living close to the land. HR AMT. #

2134. Kittredge, William, and Carver, Raymond. *We are not in this together*. Port Townsend, WA: Graywolf, 1984. 128 p. 0915308436
Contains eight short stories set in the small towns, ranches, and wilderness in the West by two masters of the genre. HR.

2135. Lopez, Barry Holstun. *Desert notes : reflections in the eye of a raven ; River notes : the dance of herons*. NY: Avon, 1990. 138 p. 038071109
Sequel: *Field notes*.
Desert notes is a highly personal and poetic evocation of the desert, *River notes* of rivers. Lopez is very open in his appreciation and effectively lays a conservation message between the lines. HR AOT. #

2136. ___. *Field notes : the grace note of the canyon wren*. NY: Knopf, 1994. 176 p. 0679434534
Contains twelve very vivid stories set primarily in various back country and wilderness locales. *Pearyland* is particularly notable. HR AOT. #

2137. ___. *Winter count*. NY: Avon, 1993, c1981. 112 p. : ill. 0380719371
These brief stories featuring animals and the western landscape resemble natural history sketches more closely than traditional short stories. Not Lopez's best work, but still recommended AOT. #

2138. Lord, Nancy. *Survival*. Minneapolis: Coffee House, 1991. 161 p. 0918273846
Stories set in Alaska which evoke the wilderness, the sometimes beneficent and sometimes hostile forces of nature, and the struggles of the often unusual people who are drawn to live there and who are inevitably

changed by those forces.

2139. Maclean, Norman. *A river runs through it and other stories*. NY: Pocket, 1992, c1976. 256 p. 0226500551
AR: *Trout and grayling, an angler's natural history, Young men and fire*.
A tough but poignant tale about a father and two sons who were drawn together by their almost mystical love of fly fishing, but torn apart by the recklessness of one son. HR.

2140. Matlin, David. *How the night is divided*. Kingston, NY: McPherson, 1993. 201 p. 092970133X
A story set on the edge of the Mojave Desert in the late 1940s and early 50s featuring immigrant Jewish farmers and an Earth-wise Kiowa foreman. Depicts the destruction of nature via B-52 bomber practice, radiation from Nevada atomic bomb tests, the logging of old growth redwoods, and the extinction of the California grizzly bear.

2141. McGuane, Thomas. *Nothing but blue skies*. NY: Vintage, 1994. 349 p. 0679747788
Land and livestock speculation, fly fishing, Montana rivers, and male camaraderie are just a few of the many forms of western culture discussed.

2142. Nelson, Kent. *All around me peaceful*. NY: Delta, 1989. 397 p. 0440550068
AR: *Cold wind river*.
Neil Shanks' investigations into his family's questionable lumber profits are interrupted when he joins a friend to find a friend lost in the wilderness. What will happen on his return?

2143. ___. *Language in the blood*. Salt Lake City: P. Smith, 1991. 251 p. 0879053941
Edward Abbey Award for Ecofiction, 1992.
An ornithologist becomes involved with smuggling Guatemalan refugees across the Sonoran desert where he learns about nature, love, determination, and loyalty. HR.

2144. ___. *The middle of nowhere*. Salt Lake City: P. Smith, 1991. 197 p. 0879053984
Contains 13 excellent stories, approximately half of which have natural settings and themes. The title story is a powerful coming-of-age tale set in the desert outside Tuscon. "The spirits of animals" is about a Native American woman disrupting an antelope hunt. AOT. #

2145. O'Brien, Dan. *In the center of the nation*. NY: Atlantic Monthly, 1991. 374

p. 0871134411
Bankers and developers play divide, payoff or foreclosure, and conquer
against family ranches and farms in South Dakota, but are surprised by the
gritty resistance of the residents. HR AMT. #

2146. Ravvin, Norman. *Cafe des Westens.* Red Deer, Alta.: Red Deer College,
1991. 216 p. 0889950792
As Calgary becomes a slick oil boomtown, the Cafe becomes the haunt of
both dwellers of a bygone era and younger people repulsed by the destruction of the city's old landmarks and neighborhoods. A tale of the struggle
between developers and traditionalists, each seeking a very different version of Utopia.

2147. Roderus, Frank. *Mustang war.* NY: Doubleday, 1991. 203 p.
0385418450
When the Beale brothers overgraze their own land, they start slaughtering
mustangs on public land to provide more forage for their cattle, but are
caught in the act by two environmental activists. Bloody and gritty.

2148. Romtvedt, David. *Crossing Wyoming.* Fredonia, NY: White Pine, 1992.
263 p. 1877727237
In this metafictional romp a historian time travels around Wyoming and to
Uruguay, Buffalo, the Dakotas, England, Guatemala, and various other
places between the years 1493 and 2079. Romtvedt's cautionary message is
couched in hilarious satire, and is HR for adventurous adult readers.

2149. Smith, Mitchell. *Due north.* NY: Simon & Schuster, 1992. 333 p.
0671738771
After a young woman's husband is eaten by a grizzly before her eyes she
summons up the courage to continue living in Alaska's Brooks Range as a
trapper. She visits her mother and older sister in Seattle, realizes she does
not belong in the city, and returns to Alaska for more adventure and self-
healing.

2150. Stegner, Wallace Earle. *All the little live things.* NY: Penguin, 1991,
c1967. 345 p. 0140154418
Sequel: *The spectator bird.*
When two eastern retirees move to California in the 1960s their peace and
quiet are threatened by a "hippie developer." Due to a scarcity of hippie
developers, this book should be taken with a pinch of organic sea salt.

2151. ___. *Collected stories of Wallace Stegner.* NY: Penguin, 1991. 525 p.
0140147748
Some of these stories have formed parts of his novels, others have not.

About half deal with nature or the west.

2152. ____. *The spectator bird*. NY: Penguin, 1990, c1976. 214 p. 0140139400
National Book Award, 1976.
Two senior citizens continue their search for their roots, their place in society, and their place in nature. HR.

2153. Sutherland, William L. *News from Fort God*. Minneapolis: Mid-List, 1993. 212 p. 0922811172
Jack is the minister of a church which worships Ronda, a radio-nature deity who demands vegetable sacrifices, and the evil waitress Lorraine, who serves only cold scrambled eggs. His wife Ellen is a teacher, nurse, and "soul traveler" whose out of body trips are symbolic of the women who have followed their men to Alaska and now long to escape.

16. Country Life

These books are about living in the country, the clash of rural and urban values, and rural land development issues in the 20th century. For books about country life in earlier times see the Frontier chapter.

Writers whose books focus on rural-vs.-urban values or development issues include Barrett, Berry, Choyce, Clark, Dykeman, Greber, Greenberg, Harrison, Haslam, Hoffman, Johnson, Moore, and Smiley. Related books in other chapters include those by Faulkner (1760-61), Hawthorne (1765), Lawrence (1771-73), Steinbeck (1787-89) Bodett (2495), Douglas (2504), Garland (2510), Adamson (2117), Agee (2118), McGuane (2141), Nelson (2142-44), O'Brien (2145), and Stegner (2150-52)

Psychological aspects of rural life are considered by Choyce, Craven, Dodge, Hale, Haun, Powys, Rawlings, West, Williams, and Wright.

The highlights of the young adult section include the popular authors Paulsen and George, Anaya's *Bless Me, Ultima,* and the classic novels of the Limberlost Swamp by Stratton-Porter. Books about land use and development conflicts by Herzig (2550), Hurmence (2551), and West (2580) can be found in the teen Environmental Action section.

Books for adults:

2154. Barrett, Andrea. *The forms of water.* NY: Pocket, 1993. 292 p. 067179521X
Octagenarian Brendan Auberon convinces his nephew Henry (who once sold family land to a developer) to "borrow" a van so he can see his land again before he dies. Henry's sister and her ex-husband assume Henry has kidnapped Brendan and plans to sell the rest of the land, so they pursue. Eventually, they develop an appreciation of each other and Henry learns to value both his family and the land.

2155. Bass, Rick. *Platte River.* Boston: Houghton Mifflin, 1994. 145 p. 0395680808
Contains three novellas. In *Mahatma Joe,* an aging evangelist in the wilds of Montana and a young woman raise food for the missions. In *Field events,* two athletic brothers encounter a giant semi-wild man and train him to be a discus champ. In *Platte River,* an injured former football player visits an ex-

teammate to deliver a speech to a class and to go steelhead fishing while his lover back home is packing to leave. AOT. #

2156. ___. *The watch.* NY: Washington Square, 1989. 190 p. 0671692224
Short stories about nature and various other subjects.

2157. Berry, Wendell. *The Port William Fellowship series.* NY or SF: Pantheon or North Point, 1983-94. 6 v.
Contents: *A place on Earth* (0865470448) — *Nathan Coulter* (0865471843) — *The wild birds* (0865472173) — *Remembering* (0865473315) — *Fidelity* (0679748318) — *Watch with me* (0679434690).
Novels and stories set primarily in rural Kentucky, focusing on the respectful relation of people to the land and each other, or in Berry's words: "the deep marriage between the land and its people."

2158. Choyce, Lesley. *The second season of Jonas MacPherson.* Saskatoon: Thistledown, 1993, c1989. 156 p. 1895449057
Dartmouth Book Award.
An old man living a simple, rural life on the eastern shore of Newfoundland reminisces about nature, the last Micmac Indians, and the perils and joys of commercial fishing; helps rescue a beached whale; and struggles to convince his success driven son and his new daughter-in-law that he can continue to live independently on the land. HR for adults, particularly seniors.

2159. Clark, Walter Van Tilburg. *City of trembling leaves.* Reno: Univ. of Nevada, 1991, c1945. 690 p. 0874171806
Contrasts the glitz and glitter of Reno and the people attracted to it with the harsh but beautiful deserts and mountains surrounding it and the down to Earth backcountry inhabitants. By opposing nature, the Renoites lose touch with all that is sacred and become neurotic. A flawed but fascinating novel.

2160. Craven, Margaret. *Walk gently this good Earth.* NY: Dell, 1981, c1977. 192 p. 0440394848
A Montana ranch family attempts to maintain its traditional values and love of nature in the face of development and changing societal norms. Describes mountain climbing, exploring glaciers, and coming of age on the ranch and in the wild. AMT. #

2161. Dodge, Jim. *Fup.* Berkeley: City Miner, 1986, c1983. 94 p. : ill. 0933944047
An aging, rural man who fears death is given a new understanding of and lease on life through caring for a duck with a broken wing ("all Fup duck.") This delightful novella which combines humor and spiritual insight is HR. #

2162. Dorner, Marjorie. *Winter roads, summer fields*. Minneapolis: Milkweed, 1992. 203 p. : ill. 0915943867
Contains 12 stories about farm life set in the Midwest between 1935 and 1991 that reveal that some things about country life have changed, but other traditions persevere.

2163. Dykeman, Wilma. *Return the innocent earth*. Newport, TN: Wakestone, 1994, c1973. 444 p. 0030666406
AR: *The tall woman, The far family*.
A southern Appalachian family progresses from farming to operating a successful canning company, but divisions arise between Jon, who wants to preserve the Earth and their traditional ways, and his cousin Stull, who cares only for the bottom line.

2164. Escamill, Edna. *Daughter of the mountain : un cuento*. SF: Aunt Lute, 1991. 208 p. 0933216823 (pbk) 0933216831
Text in both English and Spanish. A grandmother teaches a granddaughter to overcome her conflicts by forming her own ties to the land as a way to survive. This spiritual evocation of the desert is told in a series of traditional fables (cuentos) and is HR AOT. #

2165. Fulton, Len. *Dark other Adam dreaming*. Paradise, CA: Dustbooks, 1975. 214 p. 0913218480 0913218499
Set in the isolated Vermont dairyland in 1948, this novel tells the story of a farm boy's struggle with his peers, the wild land , and with ancient archetypes and modern family ghosts.

2166. Giono, Jean. *The man who planted trees*. Toronto: CBC, 1989. 51 p. 0887943624
AR: *Joy of man's desiring*.
A lone Frenchman living in a remote section of the Pyrenees begins planting acorns in the deforested landscape, and over the course of thirty years the land is restored. This inspirational example that one person really can make a difference is HR. #

2167. Glisson, J. T. *The creek*. Gainesville: Univ. of Florida, 1993. 267 p. 0813011841 081301185X (pbk)
A personal account of Glisson's childhood in the backwoods of Florida recounting the natural setting, the unusual inhabitants, country living, and a "warm, high strung, and eccentric" author named Marjorie Kinnan Rawlings.

2168. Greber, Judith. *Mendocino*. NY: Crown, 1988. 356 p. 051756761X
A multigenerational saga of a Russian family that moves to the Mendocino, initially struggles with both prejudice and nature, establishes a close and

positive relationship to the land, and eventually, in 1973, must find a family that will protect the family home from encroachment by resort developers.

2169. Greenberg, Joanne. *No reck'ning made*. NY: Holt, 1993. 296 p. 0805025790
A teacher and mother living in a small mountain has built a life around the values of education, family, honor, and love of the land. In the 1960s, social change and land development come to the Rockies, threatening to despoil the land and undermine her values.

2170. Hale, Robert D. *The elm at the edge of the earth*. NY: Norton, 1990. 351 p. 0393028615
A shy eight year old boy is sent to a pastoral County Home when his mother becomes ill.

2171. Harrison, Jim. *Farmer*. NY: Dell, 1989. 160 p. 0385282281
Joseph is tempted by the sea, the pursuit of wealth, and other women, but he realizes that his life as a farmer, the land itself, and his ongoing relationships give his life meaning and value that his pipedreams could never replace.

2172. ___. *Julip*. Boston: Houghton Mifflin, 1994. 275 p. 0395488850
Contents: *Julip*, *The seven-ounce man*, *The beige dolorosa*.

2173. Haslam, Gerald W. *Condor dreams & other fictions*. Reno: Univ. of Nevada, 1994. 203 p. 0874172276 0874172322 (pbk)
Contains 25 short stories set mostly in the multicultural, blue collar Central Valley of California. The title story is a tale of the connections between a father, his son, and the endangered California condor. HR AOT. #

2174. ___. *That constant coyote : California stories*. Reno: Univ. of Nevada, 1990. 197 p. 0874171601 087417161X (pbk)
Short stories which depict working class life in the San Joaquin Valley and chronicle the evolution of the valley from wilderness, to farm and oil country, to suburban sprawl. Haslam is the author or editor of many works of fiction and nonfiction. This is probably his most overtly ecofictional. All are HR AMT. #

2175. Haun, Mildred, and Gower, Herschel. *The hawk's done gone : and other stories*. Nashville: Vanderbilt Univ., 1984, c1968. 356 p. 0826512135
Contains 23 stories set in the Smokey Mountains from 1860 to 1940.

2176. Hoagland, Edward, and Ehrlich, Gretel. *City tales by Edward Hoagland and Wyoming stories by Gretel Ehrlich*. Santa Barbara: Capra, 1986. 86, 76 p. 0884962431
This back-to-back chapbook contains three stories by Hoagland; and three

by Ehrlich. A fine contrast of settings and styles by two excellent writers. #

2177. Hoffman, William. *The land that drank the rain*. Baton Rouge: LSU, 1982. 245 p. 0807110043
A realtor abandons his marriage and business to seek redemption living self-sufficiently in the devastated coal country of Appalachia. His efforts to grow vegetables from the abused land parallel his own healing process, but encounters with the local populace, his former business partner, and ex-wife force him to confront the outside world. For mature adults only.

2178. Johnson, Josephine Winslow. *The inland island*. Columbus: OSU, 1987, c1969. 159 p. 0814204503
This lyrical and powerful book contrasts the natural world on an abandoned Ohio farm with the horrors of industrialism as revealed in the carnage and environmental destruction of the war in Viet Nam. Not as good as her classic, *Now in November*, but still HR.

2179. ___. *Now in November*. NY: Feminist, 1991, c1934. 275 p. 1558610332 1558610359 (pbk)
Pulitzer Prize, 1934.
The daughter of a Midwestern farmer tells the tale of a horrific decade when her family faced drought, debt, and a sense of imminent disaster. Reminiscent of Willa Cather. HR.

2180. Moore, Ruth. *Spoonhandle*. Nobleboro, ME: Blackberry, 1986, c1946. 377 p. 0942396499
In a small, depressed Maine fishing village Pete and his sister Agnes are wiling to violate land and sea for money, but traditional fisherfolk Willie and Hod live and work according to a higher morality.

2181. Paulsen, Gary. *Clabbered dirt, sweet grass*. Ill. by Ruth Wright Paulsen. NY: Harcourt Brace Jovanovich, 1992. 120 p. : ill. 0151181012
A semi-autobiographical work considering the passing of the seasons and life on a traditional American family farm. Written for adults, but also recommended for older teens. #

2182. Powys, John Cowper. A *Glastonbury romance*. Woodstock, NY: Overlook, 1987, c1933. 1120 p. 0879512822 0879513063 (pbk)
This pastoral novel set in England in the 1930s combines the legend of the Grail, human romance, and the romantic qualities of nature.

2183. Rawlings, Marjorie Kinnan. *Cross Creek*. NY: Collier, 1987, c1942. 368 p. 0020238207
An unsuccessful reporter and author moves to the wilds of the Everglades.

The winged, finned, four legged and two legged denizens of the swamp provide her with great material for best sellers and a new outlook on life. HR.

2184. ___. *Short stories*. Ed. by Roger L. Tarr. Gainesville: Univ. of Florida, 1994. 376 p. 081301252X 0813012538 (pbk)
Contains 23 stories, most set in the Everglades or elsewhere in Florida, many featuring Rawling's wisecracking literary alter-ego Quincey Dover. HR.

2185. Smiley, Jane. A *thousand acres*. NY: Ballantine, 1992, c1991. 371 p. 0449907481
After his family has farmed the same land for four generations, a farmer suddenly retires and turns it over to his three daughters. Their efforts to expand the farm put it into debt for the first time ever, launching a series of events that could destroy the family and the farm. Strong parallels to *King Lear*.

2186. Webb, Mary Gladys Meredith. *Gone to earth*. North Pomfret, VT: Trafalgar Square, 1993, c1917. 288 p. 0860681432
AR: *Golden arrow*.
An unschooled young Welsh woman who loves nature and animals is torn between her love for a rather effete minister and a down-to-Earth country squire. Somewhat reminiscent of Hardy and Lawrence, but not as powerful.

2187. West, John Foster. *The summer people*. Boone, NC: Appalachian Consortium, 1989. 243 p. 0913239593
After spending a summer alone in her family's summer home in North Carolina recovering from her husband's death in Vietnam, Anna DeVoss must decide whether to return to the city or live a quieter life in the Appalachians.

2188. Williams, Philip Lee. *The heart of a distant forest*. Atlanta: Peachtree, 1991. 221 p. 1561450588
A dispirited, retired historian returns to his native Georgia to die, but finds a new understanding of and lease on life through nature and in the renewal of an old friendship. Recommended for adults, particularly seniors.

2189. Wright, Harold Bell. *The shepherd of the hills*. Gretna, LA: Pelican, 1992, c1907. 304 p. 0882898841
A sophisticated stranger fleeing grief and disappointment travels deep into the Ozarks to become a shepherd. The locals fail to understand him at first, but they bond with one another and the Earth over time.

2190. ___. *The winning of Barbara Worth*. Holtville, CA: Quellen, 1983, c1911. 512 p. 0961847301
Primarily a rather mawkish love story, this also describes the development of the California citrus industry and the creation of the Salton Sea.

Books for young adults:

2191. Anaya, Rudolfo A. *Bless me, Ultima*. NY: Warner, 1994, c1972. 272 p. :
ill. 0446600253
An alienated New Mexico teenager seeks answers to his questions about
life from Ultima, a magical healer who draws her power from the natural-
spiritual world. This classic Chicano coming of age novel is HR. #

2192. George, Jean Craighead. *My side of the mountain*. NY: Puffin, 1991,
c1959. 177 p. : ill. 0140348107
Newberry Award, 1960.
A young boy relates his adventures during the year he spends living alone
in the Catskill Mountains including his struggle for survival, his dependence on nature, his animal friends, and his ultimate realization that he
needs human companionship. HR GR 5-10.

2193. ___. *On the far side of the mountain*. NY: Dutton, 1990. 170 p.
0525445633
Sequel: *My side of the mountain*.
Sam's peaceful existence in his wilderness home is disrupted when his sister
runs away and his pet falcon is confiscated by a conservation officer. GR 5-10.

2194. Hesse, Karen. *Phoenix rising*. NY: Holt, 1994. 182 p. 0805031081
Nyle learns about relationships, caring, and death when Ezra, who was
exposed to a radiation leak from a nuclear plant, comes to stay at her
grandmother's Vermont farm. Contains vivid scenes of nature and country
life in New England. GR 6-9.

2195. Luger, Harriett Mandelay. *The elephant tree*. NY: Dell, 1986, c1978. 112
p. 0440923948
AR: *Chasing trouble*.
Lost together in the California desert, two tough kids from hostile gangs
are forced to acknowledge their mutual dependence.

2196. Morey, Walt. *Canyon winter*. NY: Dutton, 1994, c1972. 202 p.
0140368566
Stranded for six months in the Rocky Mountains following an airplane
crash, a 15-year-old boy is taken in by an old hermit who teaches him the
ways of the wilderness. GR 5-10. #

2197. Murphy, Robert William. *The pond*. NY: Dutton, 1964, 1980 printing.
254 p. : ill.
AR: *A certain island*, *The golden eagle*, *The mountain lion*, *The peregrine falcon*, *The
phantom setter*, *The stream*, *Wild geese calling*.
Junior Animal Book Award, 1985.

A 14-year-old's hiking and hunting trip at his father's cabin in the Virginia woods brings him greater maturity and a respect for all living things. Recommended more for adults than modern teens due to its nostalgic tone. #

2198. Paulsen, Gary. *The cookcamp*. NY: Orchard, 1991. 115 p. 0531059278 0531085279
During WW II a boy is sent to live with his grandma, a cook in a camp for workers building a road through the wilderness. GR 5-9.

2199. ___. *The foxman*. NY: Scholastic, 1991, c1977. 119 p. 0590440977
A town boy sent to live on a remote wilderness farm forms a friendship with an elderly, disfigured man who teaches him many things.

2200. ___. *Harris and me : a summer remembered*. San Diego: Harcourt Brace, 1993. 157 p. 0152928774
Sent to live with relatives on their farm because of his unhappy home life, an eleven-year-old city boy meets his distant cousin Harris and is given an introduction to a whole new world.

2201. ___. *Hatchet*. NY: Viking Penguin, 1988, c1987. 195 p. 014032724X
Sequel: *The river*.
After a plane crash a 13-year-old spends 54 days in the wilderness, learning to survive with only the aid of a hatchet given him by his mother. He also learns how to survive his parents' divorce.

2202. ___. *The haymeadow*. NY: Delacorte, 1992. 195 p. : ill. 0385306210
A 14-year-old John gains self-reliance during the summer he spends up in the Wyoming mountains tending his father's herd of sheep.

2203. ___. *The island*. NY: Dell, 1990, c1988. 202 p. : ill. 0440206324
A 15-year-old discovers himself and the wonders of nature when he leaves home to live on an island in northern Wisconsin.

2204. ___. *The river*. NY: Delacorte, 1991. 132 p. 0385303882
Because of his success surviving alone in the wilderness and the profound changes he has experienced, Brian is asked to undergo a similar ordeal to help scientists learn more about the psychology of survival. Recommended for all teens, including reluctant readers.

2205. ___. *Tracker*. NY: Scholastic, 1990, c1984. 90 p. 0590440985
A 13-year-old must track a deer in the Minnesota woods for his family's winter meat. Doing so, he finds himself drawn to the doe who leads him. He begins to hate his role as hunter.

2206. ___. *The voyage of the Frog*. NY: Orchard, 1989. 141 p. 0531058050 0531084051
When David goes out on his sailboat to scatter his recently deceased uncle's ashes to the wind, he is caught in a fierce storm and must survive many days on his own as he works out his feelings about life and his uncle.

2207. ___. *The winter room*. NY: Dell, 1991, c1989. 103 p. 0440404541
A young boy growing up on a northern Minnesota farm describes the scenes around him and recounts his old Norwegian uncle's tales of an almost mythological logging past.

2208. Schaefer, Jack. *Old Ramon*. NY: Walker, 1993. 102 p. : ill. 0802774032
Newbery Honor Book.
A wise old shepherd teaches a young boy lessons about survival, bravery, wisdom, and friendship as he shows him how to care for a flock of sheep in the harsh Mojave Desert. HR.

2209. Stratton-Porter, Gene. *Freckles*. NY: Puffin, 1992, c1904. 267 p. 0140351442
Orphaned and maimed, Freckles' bitterness about his fate is lessened when he is hired to guard a stretch of lumber in the wild Limberlost and, after meeting the beautiful "Swamp Angel," he determines to find out about his past. All of Stratton-Porter's books are HR for all ages and family reading. #

2210. ___. *A girl of the Limberlost*. NY: Puffin, 1992, c1909. 413 p. 0140351434
Deeply wounded by her embittered mother's lack of sympathy for her aspirations, Elnora finds comfort in the nearby Limberlost Swamp, whose beauty and rich abundance provide her with the means to better her life. HR. #

2211. ___. *The keeper of the bees*. Bloomington: Indiana Univ., 1991, c1921. 515 p. 025335496X 025320691X (pbk)
A wounded WW I soldier takes over a bee garden in a then-rural valley near Los Angeles, and with the assistance of Little Scout saves it from being claimed and ruined by the former owner's daughter. Contains a great deal of bee and nature lore. #

2212. ___. *Laddie : a true blue story*. Wheaton, IL: Tyndale House, 1991, c1913. 420 p. 0253331137 0253204585 (pbk)
A romanticized autobiographical novel about Gene and her brother Laddie growing up in the the Limberlost Swamp country. Possibly her best work. HR. #

17. Ecofeminist Fiction

Ecofeminist perspectives can be found in the work of a wide variety of creative novelists. Several anthologies are available, the best being *Sisters of the Earth* (1713) a superb collection of fiction, poetry, and essays. *Wild Women* (1744), *Westward the Women* (1737), *Hear the Silence* (1751) and *Woman and Nature* (2582)are also recommended.

Precursors of ecofeminist fiction include Austin, Cather, Gilman, and Jewett. They argued that the land had an inherent value, and should be worked *with* (as opposed to worked over) as naturally as possible. Native American ecofeminists include Erdrich (1992-4), Silko (2021-24), Cameron (1984), and Pijoan (1998) Also see *Spider Woman's Granddaughters* (1712) and *A Gathering of Spirit* (1718)

Ecofeminist fiction experienced an incredible flowering between 1974 and 1982 as the feminist and environmental movements grew and coalesced. Major novels published during those years, in chronological order, include Laurence's *The Diviners*, Charnas' *Walk to the End of the World*, Russ' *The Female Man*, Alther's *Kinflicks*, Piercy's *Woman on the Edge of Time*, McIntyre's *Dreamsnake*, Charnas' *Motherlines*, Robinson's *Housekeeping*, Wilhelm's *Juniper Time*, Hoffman's *Angel Island*, Griffin's *Woman and Nature*, Atwood's *Surfacing*, and Walker's *The Color Purple*. Other recommended ecofeminist novelists include Le Guin, Houston, Bush, Coleman, Elgin, Davenport, and Astley.

You don't have to pass a physical to write an ecofeminist novel. London's *A Daughter of the Snows* and *Smoke Belew*, Kesey's *Sometimes a Great Notion*, Watson's *Book of the River*, Manfred's *Manly-hearted Woman*, Eastlake's *Dancers in the Scalp House*, Waters' *The Woman at Otowi Crossing*, and even *Dalva* by the normally macho Harrison include ecofeminist characters or themes. Ecofeminists have made major contributions to the traditionally male genres of science fiction and mystery writing. See the introduction to those chapters for details. Female authors such as Auel, Harrison, Pierce, Shuler, and Thomas dominate contemporary Prehistory fiction, typically laying an ecofeminist message between the lines.

Although over a third of the books in this chapter are also appropriate for young adults, few if any ecofeminist novels have been written specifically for teens. Austin's *Basket Woman* (2043), Dickinson's *A Bone from a Dry Sea*, (1795) and Cooper's *Earthchange* (2545) may come the closest. Fantasy authors Bell (2431-36), Board (2438), Furlong (2444-45), Le Guin

(2450), and Morgan (2455) offer feminist visions of various alternative times and places which are recommended for both teens and adults.

Books for adults:

2213. Alcala, Kathleen. *Mrs. Vargas and the dead naturalist*. Corvallis, OR: Calyx, 1992. 170 p. 0934971269 0934971250 (pbk)
Although not actually about naturalists, this book depicts the influence of the natural environment and traditional culture on Mexican women and Chicanas.

2214. Alther, Lisa. *Kinflicks*. NY: Dutton, 1977, c1975. 528 p. 0394498364
When a woman's husband suspects her of adultery he throws her out of the house and she returns home to her dying mother. A series of flashbacks follows her fluctuating path from being a high school cheerleader, to an antiwar-lesbian-organic gardening feminist, and back to a more conventional but finally-scorned wife. This early ecofeminist novel received strongly mixed reviews.

2215. Arnow, Harriette Louisa Simpson. *The dollmaker*. Lexington: Univ. of Kentucky, 1985, c1954. 549 p. 0813115442
A family moves from a Kentucky farm to a defense plant in Detroit during WW II. Contrasts the two divergent lifestyles and considers the spiritual pain of abandoning the land for the city.

2216. Arthur, Elizabeth. *Beyond the mountain*. St. Paul, MN: Graywolf, 1993, c1983. 224 p. 1555971717
Although haunted by a tragic avalanche, Artemis Philips joins a women's climbing expedition to Nepal where she must face the ghosts of her past and the perils of the Himalayas. Will disaster strike again?

2217. Atwood, Margaret Eleanor. *The handmaid's tale*. NY: Fawcett, 1986. 443 p. 0449448290
In the future Republic of Gilead ecological disasters have ravaged the land and only a few women, called Handmaids, can still bear children. One Handmaid is sent to the home of the commander to bear his child but revolutionaries plan to use her to assassinate him.

2218. ___. *Surfacing*. NY: Ballantine, 1987, c1972. 231 p. 0449213757
A young artist and three of her friends who want to "play back to nature" search for her father in the Canadian backwoods where she experiences a deepening sense of foreboding about her father and her past. This landmark work of ecofeminist fiction is HR.

2219. Austin, Mary Hunter. *Cactus thorn*. Reno: Univ. of Nevada, 1988. 122 p. 0874171350
Desert denizen Dulcie Adelaid inspires New Yorker Grant Arliss to campaign for conservation, but later kills him. This early ecofeminist novel went unpublished for 60 years due to Dulcie's lack of remorse for the killing.

2220. ___. *Stories from the country of Lost borders*. Ed. by Marjorie Pryse. New Brunswick: Rutgers Univ., 1987. 267 p. 0813512182 (pbk) 0813512174
A reprint of *Lost borders* and *The land of little rain* in one vol. Two collections of short stories set in frontier times in the high desert of California. Austin depicts strong female characters who value the Earth for its own sake and are skeptical about development of the desert. HR.

2221. Brossard, Nicole. *Mauve desert*. Toronto: Coach House, 1990. 202 p. 0889103895
A woman experiences heightened consciousness and a depper meaning of life in the desert. Recommended only for sophisticated adults due to complex, postmodern style.

2222. Bryant, Dorothy. *The kin of Ata are waiting for you*. NY: Moon, 1976, c1971. 220 p. 0349407296 0394732928 (pbk)
Original title: *The comforter*.
Depicts an ecofeminist utopia where people and nature are respected.

2223. Bush, Catherine. *Minus time*. NY: Hyperion, 1993. 341 p. 1562828819
A young woman who aspires to be an astronaut realizes that the conflicts within her family and the degradation of the environment are inextricably connected. HR. #

2224. Cameron, Anne*. *Tales of the Cairds*. Madeira Park, BC: Harbour, 1989. 191 p. 155017004X
Contains nine Celtic matriarchal creation tales about Earth-centered women whose beliefs, religion, and lives are dedicated to nurturing people and the environment. Implicitly presents parallels between Celtic and Native American mythology and values.

2225. Cather, Willa. *Later novels*. NY: Library of America, 1990. 988 p. 0940450526
Contents: A *lost lady* — *The professor's house* — *Death comes for the archbishop* — *Shadows on the rock* — *Lucy Gayheart* — *Sapphira and the slave girl*. Except for *The professor's house*, these novels are somewhat less ecofictional and powerful than her earlier work, and are recommended mainly for true Cather devotees.

2226. ___. *My Antonia*. NY: Penguin, 1994, c1915. 304 p. 0140187642
Two friends grow up on the prairie together, but later their lives contrast sharply as he moves east to seek his fortune while she stays on in Nebraska to raise a family. A beautiful evocation of the land and people as they shaped one another. HR. #

2227. ___. *O pioneers!* NY: Dover, 1993, c1913. 128 p. 0486277852
The children of settlers on the Nebraska prairies engage in conflict over Alexandra's love for the land and use of natural farming practices and her brothers' short-term profit orientation. HR. #

2228. ___. *The professor's house*. NY: Vintage, 1990, c1925. 283 p. 0679731806
In this book Cather explicitly presents ecological values which are implicit in much of her work. #

2229. ___. *The song of the lark*. NY: Bantam, 1991, c1915. 369 p. 0553213911
A talented young Swedish woman escapes provincial Colorado society to become a great opera singer in Chicago, but later returns to the west to gain a deeper sense of her own nature in a more natural setting. The strongest theme here is of the woman artist's struggle to be accepted. #

2230. ___. *Three novels and selected stories*. NY: Barnes & Noble, 1993. 660 p. 1566191084
Contents: *O pioneers!* — *The song of the lark* — *Alexander's bridge* — *The profile* — *The willing muse* — *Eleanor's house* — *On the gull's house* — *Flavia and her artists* — *The marriage of Phaedra* — *The garden lodge*. This anthology is HR. #

2231. Charnas, Suzy McKee. *Motherlines*. NY: Berkley, 1979, c1978. 246 p. 0425041573
One of *Science fiction's 100 best novels*.
Alldera finds the Motherline Tribe of escaped slaves who are more nature oriented than the men of the Holdfast, but are equally hierarchical and repressive. Alldera and Daya leave Motherline, but return to be partly accepted. This is both a fine adventure and a warning to other feminists not to repeat the errors of patriarchy. HR AOT. #

2232. ___. *Walk to the end of the world*. NY: Ballantine, 1977, c1974. 214 p. 0345256611
Sequels: *Motherlines*, *The furies*.
The survivors of a nuclear holocaust form a repressive colony named Holdfast in which women have the status of slaves or even domestic animals. Alderra escapes Holdfast in a desperate attempt to find the independent Motherline tribe. HR AOT. #

2233. Clark, Eleanor. *Camping out*. NY: Putnam, 1986. 223 p. 0399131221
AR: *Gloria mundi*.
Two women are caught making love on the beach of a remote Vermont lake by a drifter who may be an escaped murderer. He damages their canoe to bar escape, holds them captive, and rapes one of them, who enjoys the multiple orgasms. They escape and meet a deputy sheriff, but do not mention their former captor.

2234. Coleman, Jane Candia. *Stories from Mesa country*. Athens: Swallow, 1991. 142 p. 080400949X
A collection of short stories, most set in the rural Southwest.

2235. Cunningham, Elizabeth. *The return of the goddess : a divine comedy*. Barrytown, NY: Station Hill, 1992. 384 p. 088268115X
When developers threaten to fell an ancient grove of mystic trees with reputed healing power they find opposition from four disparate people.

2236. ___. *The wild mother*. Barrytown, NY: Station Hill, 1993. 355 p. 0882681478
A modern Adam ventures into the Empty Lands where the immortal descendents of Lillith live in harmony with nature. He captures a wife who bears a daughter and a son before escaping. Adam eventually uses their now ten year old daughter Ionia as bait. Lillith is captured, and Ionia must chose between living in a wild or civilized state.

2237. Douglas, Carole Nelson. *Cup of clay*. NY: T. Doherty, 1991. 329 p. 0312851464
Sequel: *Seed upon the wind*.
Reporter Alison Carver begins a quest to find a magical cup in the "otherworld" of Veil. Darker forces emerge after she heals some ruined farmland, forcing her and former rival Rowan to seek "the Heart of the Earth" to heal the planet. This novel has its moments, but is uneven and recommended only for dedicated ecofeminist or fantasy readers.

2238. ___. *Seed upon the wind*. NY: T. Doherty, 1992. 319 p. 0312851472
When Carver returns the magical cup to Earth she learns that a blight is occuring in Veil. She returns so she and Rowan can make another quest to heal the planet.

2239. Elgin, Suzette Haden. *Native tongue III : earthsong*. NY: DAW, 1994. 255 p. 0886775922
AR: *Native tongue, Native tongue II : the Judas rose*.
After the extraterrestrials leave Earth, economic and environmental collapse ensue, a female linguist presents a startling plan for the survival of

the human race and the restoration of environmental quality.

2240. Galford, Ellen. *The fires of Bride*. Ithaca, NY: Firebrand, 1988. 229 p. 0932379427 0932379419 (pbk)
Maria Melleny finds her ancestral roots in the Celtic triple goddess Bride and travels to the Outer Hebrides where ancient seaweed women, archaeological clues, and natural phenomena guide her on.

2241. Gearhart, Sally Miller. *The wanderground : stories of the hill women*. Boston: Alyson, 1984, c1979. 196 p. 0932870554
The hill women escape the environmental ruin and violent sexism of Dangerland to establish a lesbian ecofeminist society. They use telepathy and mystical natural powers in their constant struggle for survival. An otherwise powerful work is marred by pervasive reverse sexism: all the women are wonderful, and all the men utterly evil.

2242. Gilman, Charlotte Perkins. *Herland and selected stories*. NY: Signet, 1992. 349 p. 0451525620
Herland is an entirely female utopia. Gilman explores and criticizes existing roles of men and women and promotes the establishment of a more nurturing, female-oriented attitude toward the Earth. This early feminist novel is a strong precursor of current ecofeminist writing. AMT. #

2243. Gomez-Vega, Ibis. *Send my roots rain*. SF: Aunt Lute, 1991. 204 p. 1879960044 (pbk) 1879960052
Carole leaves her home in Brooklyn to seek out her roots in Pozo Seco, Texas. The desert offers challenges to her survival as well as personal insights and healing.

2244. Griffin, Susan. *Woman and nature : the roaring inside her*. NY: Collins, 1979, c1978. 263 p. 0060907444
AR: *Made from this Earth*.
A literary dialogue of sorts wherein Griffin traces patriarchical judgments about nature and about women; the consequent separations of body, thought, emotion, and nature; separation of feminist from patriarchal thought; and creation of an ecofeminist vision.

2245. Hoffman, Alice. *Angel Landing*. NY: Berkley, 1993, c1980. 296 p. 0425139522
An explosion at a nuclear power plant construction site unites a formerly half-hearted social worker and "the bomber" who uses her concern to break a pattern of violence without losing his sense of commitment. A septuagenarian vegetarian environmental activist character adds humor and depth to the story.

2246. Houston, Pam. *Cowboys are my weakness*. NY: Norton, 1992. 171 p. 0393030776
Stories about smart women who love wild places and are susceptible to falling in love with wild men who are hard to pin down to a commitment. An ironic feminist picaresque.

2247. Jewett, Sarah Orne. *Best stories of Sarah Orne Jewett*. Augusta, ME: Tapley, 1988. 282 p. : ill. 0912769335
A collection of stories written between 1874 and 1900, most set in rural or coastal Maine. The obvious appreciation of nature and subtle portrayal of feminist ideas and ideals prefigure the later rise of ecofeminism. HR AOT. #

2248. ___. *Country by-ways*. Irvine, CA: Reprint Services, 1988, c1881. 249 p. 0781213045
These stories about nature and country life in late 19th century New England are not as strong as *The white heron* or or Jewett's other best work, but are still recommended for AOT. #

2249. ___. *The country of the pointed firs*. Boston: Godine, 1991. 197 p. 0879238941
Life in a decling Maine sea coast town. HR AOT. #

2250. ___. *Deephaven*. Portsmouth, NH: Randall, 1993, c1877. 310 p. 0963611100
Two young Boston women vacation in the decaying port town of Deephaven and discover the contrasts between the lives of the seafarers and farmers. She effectively underplays her plea for enduring, nature centered values. HR AOT. #

2251. ___. *Novels and stories*. NY: Library of America, 1994. 937 p. 0940450747
Her four most famous novels and a variety of stories published in one volume. HR AOT. #

2252. ___. *A white heron and other stories*. Irvine, CA: Reprint Services, 1988, c1914. 254 p. 0781213088
The famous title story tells the tale of a shy young girl living with her grandmother in turn of the century Maine. She meets a hunter who enlists her aid in finding a great white heron and must choose between their friendship and saving the bird. HR. #

2253. Kauffman, Janet. *The body in four parts*. Saint Paul: Graywolf, 1993. 128 p. 1555971792
Recommended only for those who enjoy complex, postmodern literary

style in and of itself, regardless of the quality of the writing.

2254. Kesey, Ken. *Sometimes a great notion*. NY: Penguin, 1988, c1964. 628 p. 0140045295
A logging family bucks a close-knit timber community to deliver a shipment of logs in defiance of a strike. Includes vivid descriptions of the rigors of logging and the harsh Oregon coastal climate and landscape. Kesey prefigures the ecofeminist novel with the strong, nature loving wife, Viv, who finally asserts her independence from a patriarchal clan. HR AMT. #

2255. Laurence, Margaret. *The diviners*. Chicago: Univ. of Chicago, 1993, c1974. 389 p. 0226469352
A Scottish settler struggles as a single mother in southern Ontario, overcoming hardships through her love of wilderness, her daughter, and her writing. Includes accounts of the settlers' relations with the Indians and the observations of an ecophilosophical water witcher.

2256. Le Guin, Ursula K. *Always coming home*. NY: Bantam, 1987 c1985. 523 p. : ill., maps. 0553262807
Published with a cassette, *Songs of the Kesh*, by Harper & Row, 1985.
American Book Award nominee.
Le Guin interweaves story, fable, poem, music, and artwork to depict the society of the Kesh, a group of peaceful, low-tech, nature loving feminists on the Northwest coast. HR.

2257. ___. *The eye of the heron*. NY: Harper, 1991, c1978. 198 p. 0061001384
Luz leads the peaceful, nature loving residents of Shantytown away from the exploitative city people into the wilderness. They attempt to build a new and harmonious life, but regaining simplicity isn't all that simple. Le Guin explores the complex connection between nature and human nature. AOT. #

2258. ___. *Searoad : chronicles of Klatsand*. NY: HarperCollins, 1991. 193 p. 0060167408
A primitive village on the Oregon coast slowly becomes an overdeveloped resort area, partly due the town's businesswomen. Though the women are depicted as resourceful and compassionate, their "success" has a negative consequence. Le Guin seems to question whether the portrayal of women by other ecofeminists is overly romanticized.

2259. Le Guin, Ursula K., and Sanders, Scott R. *The visionary : the life story of Flicker of the Serpentine of Telina-Na by Ursula Le Guin and Wonders hidden: Audubon's early years by Scott Sanders*. Santa Barbara: Capra, 1984. 43, 79 p. 0884962199
Texts bound back to back in one volume. Le Guin tells a tale of a future

nonindustrial, ecologically-based society of only 100 million people. (A prequel to her novel, *Always coming home*) Sanders recounts a fictionalized version of Audubon's adventures in nature from ages six to eighteen. #

2260. London, Jack. *A daughter of the snows.* Oakland: Star Rover House, 1987, c1902. 333 p. 093245836X
A feminist heroine experiences adventures and hardships in the Klondike where she is pitted against crude conditions and crude men. This fine first novel foreshadows the rise of ecofeminism. AOT. #

2261. ___. *Smoke Bellew.* NY: Dover, 1992, c1912. 223 p. 0486273644
A citified young man heads for the Klondike gold fields where he learns to deal with the challenges of the Arctic from a rugged outdoorsman and an equally rugged and courgeous woman. Like *Daughter of the snows*, this novel prefigures ecofeminism. AOT. #

2262. McIntyre, Vonda N. *Dream snake.* NY: Bantam, 1994, c1978. 312 p. 0553296590
AR: *Starfarers, Transition, The exile waiting, Metaphase, Nautilus* .
Nebula Award, 1978, Hugo Award, 1979.
A healer undertakes a perilous jouney to replace her dreamsnake, the animal used for insight and healing. HR AOT. #

2263. Piercy, Marge. *He, she, and it.* NY: Knopf, 1991. 446 p. 0679404082
In the mid 21st century mega-corporations have plundered Earth's resources, ravaged the environment, and sharply curtailed human freedoms. When Shira Shipman is divorced and the corporation takes her son away, she escapes to a Jewish free town where she meets a "man" who is both promising and ominous. HR.

2264. ___. *Woman on the edge of time.* NY: Fawcett Crest, 1986, c1976. 381 p. 0449210820
One of *Science fiction's 100 best novels.*
After Consuelo Ramos is subjected to neurological experiments she foresees two possible futures: a hierarchical, militaristic one wherein both human freedom and the environment are degraded, or an androgynous, nature loving society using appropriate technology. This classic ecofeminist novel is HR AOT. #

2265. Robinson, Marilynne. *Housekeeping.* NY: Bantam, 1989, c1980. 187 p. 0553346636
A teenage girl whose parents have died and her suicidal aunt draw reconciliation, love, and emotional healing from each other and living among the natural wonders of Idaho. This classic ecofeminist text is HR AMT. #

2266. Russ, Joanna. *The female man.* Boston: Beacon, 1986, c1975. 213 p. 0807063134
One of *Science fiction's* 100 *best novels.*
AR: *Picnic on Paradise.*
Four genetically identical women from different planets have been encultur-ated into four very different personalities. Their discussions and adventures deal primarily with feminism and secondarily with the environment. HR for adults despite its cryptic style and occasional oversimplified posturing.

2267. Sarton, May. *The house by the sea : a journal.* NY: Norton, 1977. 287 p. 0393075184
Noted poet Sarton relates how her move to a remote house on the Maine seacoast rejuvenated her inner life and writing by providing a new set of natural stimuli and solitude. She describes how the sea, the fields and woods, the plants and animals fueled her spiritual development and cre-ative urges.

2268. ___. *Plant dreaming deep.* NY: Norton, 1983, c1968. 189 p. 0393301087
Sarton tells the story of what compelled her to seek roots in a remote vil-lage, finding her house, gardening, animals, birds, and the violent changes of the climate.

2269. Schulman, Audrey. *The cage.* Chapel Hill, NC: Algonquin, 1994. 228 p. 1565120353 1565120590
The story of the only woman on a job to shoot pictures of polar bears. "The cage" is literally a device to protect photographers from the bears, but it also represents the learned helplessness which limits women's opportuni-ties and accomplishments. HR.

2270. Slonczewski, Joan. *Daughter of Elysium.* NY: Morrow, 1993. 521 p. 0380972220 0688125093
When the Windclans settle in the ocean world of Shora they find it difficult to adapt to the ways of the Elysians and become involved in dangerous intrigues with the Sharers, the Elysians, and inhabitants of neighboring planets.

2271. ___. *A door into ocean.* NY: Avon, 1987, c1986. 406 p. 0380701502
John W. Campbell Memorial Award.
Sequel: *Daughter of Elysium.*
Describes the ocean world of Shora where a clan of nature oriented, paci-fist women known as the Sharers have learned to accept the ways of the Elysians, a high-tech society of seemingly ageless beings. HR.

2272. Sojourner, Mary. *Sisters of the dream.* Flagstaff: Northland, 1989. 363 p. 0873584864

AR: "Belly," winner of *Sierra's* 1993 Nature Writing Contest.
The story of a modern Chicago woman's move to the Southwest, her friendships with other women, their efforts to make sense of their lives, their
struggles against sexism and environmental hazards, and her connection
to the natural and spirit world through dreams of a twelfth century Hopi.
Recommended for most adults, especially those interested in the Hopi.

2273. Starhawk. *The fifth sacred thing.* NY: Bantam, 1993. 486 p. 0553095226
In mid 21st century California the northern neopagan, bisexual Witches
resist the efforts of the southern Stewards to steal their water. Describes life
in an ecofeminist Utopia. By using the term "Stewards" Starhawk implies that
the Christian stewardship or "conservation for use" are excuses for pillaging
resources which need to be supplanted by an ecofeminist ethic.

2274. Stegner, Lynn. *Undertow.* Dallas: Baskerville, 1993. 367 p. 1880909022
A marine biologist, the victim of an abusive childhood, struggles to find
meaning and love in a hopeless relationship with a married man and in her
work at a sea-lion sanctuary. Received mixed reviews.

2275. Tepper, Sheri S. *The gate to Women's Country.* NY: Foundation, 1988.
278 p. 0385247095
Three centuries after a nuclear holocaust, men and women have been separated into two colonies, the former a military camp, the latter a center of
government, commerce, law, and culture. At fifteen boys must choose
between military life or being eunuch "servitors." The use of Greek plays is
intriguing, but the book lacks Tepper's usual subtlety.

2276. Vonarburg, Elisabeth. *In the mothers' land.* NY: Bantam, 1992. 487 p.
055329962X
After man creates the Decline through toxic waste and pollution, Elli, the
creator, radically limits the number of males who can be born and survive
the Malady. When a female named Lisbei is afflicted with the Malady she is
forced to confront the edict of Elli.

2277. ___. *The silent city.* NY: Bantam, 1992, c1988. 261 p. 0553297899
AR: *The Maerlande chronicles.*
When the Outside is ravaged by war, the people closed away in the City
genetically engineer a girl, Elisa, with powers of transformation and rejuve-
nation. Elisa eventually must chose between saving the City which created
her or the whole Outside world.

2278. Walker, Alice. *The color purple.* NY: Harcourt Brace Jovanovich, 1992,
c1982. 290 p. 0151191549
AR: *Possessing the secret of joy.*

In the early 20th century a young Black woman who has been abused and has two children by her "father" has her life transformed by a blues singer who loves her and teaches her self respect. Eventually she is reunited with her sister and children in Africa. Love and nature are contrasted with hatred and artificial social relationships. HR AMT. #

2279. ___. *The temple of my familiar.* NY: Pocket, 1990, c1989. 417 p. 0671683993
Moves between centuries and continents to explore relations between the sexes, races, and various species. Recommended for adults, particularly New Age readers.

2280. Watson, Ian. *The book of the river.* NY: DAW, 1986, c1984. 256 p. 0886771056
Sequels: *The book of the stars, The book of being.*
In Yaleen's world, only women can sail on the black current at all, and nobody can cross it or know what may lie on the other side. When the current mysteriously recedes, all manner of other divisions also disappear.

2281. Watson, Sheila. *The double hook.* Toronto: McClelland & Stewart, 1989, c1966. 130 p. 0771099983
After domineering Mrs. Potter is killed by her son, her neighbors see her ghostly image fishing in her favorite stream. Later, when a barren young woman commits suicide by burning the Potter house down, Mrs. Potter's ghost is released to become a spirit of natural rebirth. An early, but provocative variation on the ecofeminist theme of woman as nature and nature as woman.

2282. Wilhelm, Kate. *Juniper time.* NY: Harper & Row, 1979. 280 p. 0060146575
One of *Science fiction's* 100 *best novels,* American Book Award nominee, 1980. After drought devastates the west, the government resettles people in concentration camps called Newtowns. A linguist escapes to the Pacific Northwest where she survives with the help of Indians. HR AOT. #

2283. Winterson, Jeanette. *Sexing the cherry.* NY: Vintage, 1991, c1989. 167 p. 0679733167
A shape-changer and her son, Jordan, live by the 'stinking Thames' in 17th century London. Jordan is reincarnated in 1990 to fall in love with an environmental activist and join her in her efforts to fight water pollution. AMT. #

18. Mystery and Detective Stories

The first significant modern mystery with an environmental theme was written by an acknowledged master of the genre, John D. MacDonald. He wrote three novels about the sleazy side of land development: A *Flash of Green* in 1962, *Condominium* in 1977, and *Barrier Island*, his final book, in 1987. MacDonald was very influential, and by the time of his demise a wide variety of writers were transforming the traditional whodunit to the subgenre of "whodunit to the environment?" Nearly two thirds of the books in this chapter were published between 1992 and 1994.

Development deals, high-tech secrecy, corporate-political corruption, endangered species and animal rights controversies are all inherently controversial and loaded with subterfuge: perfect mystery material. Preservation of endangered species, animal rights, hunting, poaching, or fishing are the subject of novels by Barr, Chabonneau, Craig, D'Apuget, Eulo and Mack, Gill, Goddard, Hiassen, Landers, McQuillan, Moore, Quinn, Straley, Van Gieson, and Wallace. Land development or corporate pollution and corruption are considered by Blake, Philip, Jim Hall, Patricia Hall, Healy, Langton, Leon, Maron, McCabe, Russell, Simonson, Stabenow, Stephenson, Straley, Van Gieson, Wambaugh and Wren.

Another trend is the prevalence of female authors and protagonists, with over half of these books featuring one or both. The majority have an ecofeminist subtext: working Mother Nature's beat can be mighty confusing and dangerous, the perfect job for a smart and sensitive but tough and decisive woman.

Young adult readers had such a heroine for over a quarter-century: Miss Pickerall, the creation of Ellen MacGregor and, later, Dora Pantell. Unfortunately, they are all out of print now. The prolific Jean Craighead George, author of many juvenile nonfiction and environmental action books, has written four books which she refers to as "ecological mysteries."

Books for adults:

2284. Barr, Nevada. *Track of the cat.* NY: Putnam, 1993. 238 p. 0399138242
When a West Texas park ranger is mysteriously killed by a normally reclusive mountain lion and several innocent lions are killed in retaliation, ranger Anna Pigeon suspects that two-legged predators are the real killers. Although the presentation of strong female rangers and a coroner are posi-

tive, the depiction of all males as testosterone driven killers weakens the book's credibility.

2285. Blake, Jennifer. *Shameless*. NY: Fawcett Columbine, 1994. 352 p. 0449906183
A "nearly divorced" woman has an affair with a man who has returned to Louisiana to take over his late father's lumber mill. He wants to sell to a British firm which will infuse new money into the depressed economy while speeding environmental degradation. This leads to conflict between them, which escalates as people start disappearing and a dead body turns up.

2286. Charbonneau, Louis. *The ice : a novel of Antarctica*. NY: D.I. Fine, 1991. 336 p. 1556111770
Sequel: *White harvest*.
When a joint American/Soviet research team discovers dying, oil-soaked birds in a part of Antarctica where there was no reported drilling, they must race against both the encroaching winter and a corporate "scientific" expedition which is concealing its true activity.

2287. ___. *White harvest*. NY: D.I. Fine, 1994. 288 p. 1556113625
A marine biologist confronts an ivory poaching operation in Alaska.

2288. Craig, Philip R. *Off season : a Martha's Vineyard mystery*. NY: Scribner, 1994. 240 p. : map. 0684196174
A conflict between hunters and animal rights activists leads to the death of a recluse who was killed by his own bow and arrow. Fisherman/gourmet chef J. W. Jackson's investigations turn up connections to politicians and to organized crime, not to mention some great scallops.

2289. D'Alpuget, Blanche. *White eye*. NY: Simon & Schuster, 1994. 256 p. 0671620053
The investigation of the death of an Australian animal rights activist leads to a research scientist who is working on two vaccines to fight the "white eye" infection. One protects animals, while the other, which sterilizes them, could also be used for "politically painless mass sterilization" of humans. HR.

2290. Doss, James D. *The shaman sings*. NY: St. Martin's, 1994. 230 p. 0312105479
Publishers Weekly Best Book, 1994.
When a grad student makes a breakthrough in room-temperature super-conductivity, she is murdered and the police chase a janitor who has been framed. A local police chief, a reporter, and a Ute shaman track the real killer. HR.

2291. Eulo, Ken, and Mauck, Joe. *Claw.* NY: Simon & Schuster, 1994. 319 p. 0671799630
After Rajah the tiger shreds one of his keepers, an anthropologist named Shindler discovers an unusual microbe. This discovery is suppressed by Shindler's "suicide." When Rajah escapes into the wilderness, the police must pursue not only the killer cat, but the scientists behind the debacle.

2292. Gill, Bartholomew. *Death on a cold, wild river : a Peter McGarr mystery.* NY: Morrow, 1993. 251 p. 0688128815
A detective discovers that the death of a world-class fly fisher was no accident, but who killed her: the "flyfishing cowboy," the poacher she reported, a professional and personal rival, or someone yet to be discovered?

2293. Goddard, Kenneth W. *Prey.* NY: TOR, 1992. 335 p. 031285112X
When a Fish and Game cop infiltrates a gang of big game poachers he discovers that it is part of a secret cadre of financiers, industrialists, and criminals sworn to destroy the environmental movement. When he and his girlfriend are discovered, the government abandons them and they are forced to flee into the wilderness.

2294. Hall, Jim. *Mean high tide.* NY: Delacourte, 1994. 371 p. 0385307985
AR: *Gone wild.*
After the mysterious underwater death of his girlfriend, an investigator joins forces with a former cop and becomes caught up in a deadly web of violence, obsession, and ecological vengeance.

2295. Hall, Patricia. *The poison pool.* NY: St. Martin's, 1993. 250 p. 0312098944
This British whodunit set in Yorkshire features a town choked by grit and coal dust, a failing economy, moral corruption, and steadily worsening environmental and social conditions.

2296. Healy, J. F. *Foursome : a John Cuddy novel.* NY: Pocket, 1993. 344 p. 0671795562
When three of four friends are killed by a crossbow, a detective is called upon to prove the innocence of the fourth friend's husband. Because the foursome were considered intruders who built ostentatious homes and disturbed wildlife, there is no shortage of suspects, including Shotgun Ma Judson, a double-amputee environmental activist, and various sleazeball relatives of the Boston foursome.

2297. Hiaasen, Carl. *Double whammy.* NY: Putnam's, 1987. 320 p. 039913297X
A journalist, an environmentalist, and a mysterious swamp dweller join

forces to investigate and fight the related ills of cheating in professional bass fishing, political corruption, the development of a giant condo complex on a toxic waste site, and the devastation of the Everglades by developers.

2298. ___. *Native tongue*. NY: Fawcett Crest, 1992, c1991. 407 p. 0449221180
When the precious blue tongued mango voles at North Key Largo's Amazing Kingdom of Thrills are stolen, the trail leads former journalist Joe Winder to sleazy real estate developer Francis X. Kingsbury and more trouble than he ever imagined.

2299. ___. *Skin tight*. NY: Putnam, 1989. 319 p. 0399134891
Sleazy plastic surgeon Rudy Graveline (who epitomizes the corruption and greed that underlie the devastation of Florida's environment) will go to any lengths, including murder, to protect his status and wealth. An environmental journalist emerges from retirement and attempts to use both words and radical action to confound Graveline and his cohorts.

2300. ___. *Strip tease*. NY: Knopf, 1993. 320 p. 1568950497
A fictional exploration of extortion and the connections between organized crime and politics. Although not as explicitly ecofictional as Hiaasen's other books, it clearly implies that the illegal and elected mobs both profit from development at any environmental cost.

2301. Hoyt, Richard. *Bigfoot*. NY: TOR, 1993. 224 p. 0312852789
Detective John Denson and his possibly shamanistic partner Willie Prettybird is hired by a Russian scientist to find Bigfoot, but they face literally murderous competition from land developers, the British Museum, local bigfoot aficionados, a TV star, and mountaineeers.

2302. Landers, Gunnard. *The deer killers*. NY: Walker, 1990. 209 p. 0802711340
Sequel:*The violators*. AR:*The hunting shack*, *Rites of passage*.
Special agent Reed Erickson of the U.S. Fish and Wildlife service goes undercover to combat poaching and the destruction of the forest for commercial gain. #

2303. ___. *The violators*. NY: Walker, 1991. 198 p. 0802711790
Prequel: *The deer killers*.
Reed Erickson travels to Yellowstone to arrest poachers and stop the slaughter of endangered species. #

2304. Langton, Jane. *God in Concord*. NY: Viking, 1992. 338 p. : ill. 0670842605
AR: *National enemy* and other Homer Kelly books.

On the land around Walden Pond, preservationists, ex-cop Homer Kelly and the elderly residents of a trailer park who are dying at an inexplicably fast rate are pitted against developers and a corrupt planning commission. A lively romp.

2305. Leon, Donna. *Death in a strange country*. NY: HarperCollins, 1993. 290 p. 0060170085
A detective story set in Italy featuring the Mafia, corruption, military coverups, and illegal hazardous waste disposal.

2306. MacDonald, John D. *Barrier Island*. NY: Ballantine, 1987, c1986. 259 p. 0449131793
A mystery set in the proposed Gulf Islands National Seashore, featuring a conservation-oriented real estate man , a sleazy developer, a crooked judge, a gutless assistant US attorney, and a noble National Park ranger. HR.

2307. ___. *Condominium*. NY: Ballantine, 1985, c1977. 478 p. 0449207374
After an unscrupulous developer builds ecologically unsound, physically unsafe condominiums in the Florida Keys, a hurricane leads to disaster. A little long-winded, but still HR.

2308. ___. *A flash of green*. NY: Fawcett, 1984, c1962. 336 p. 0449126927
Developers take over Palm City Florida and use bribery, corruption, blackmail, and attempts to induce fear of nature (snakes, 'gators, etc.) in the general public so they can develop rare swamp lands. This ecofictional thriller is a classic of the genre and is HR.

2309. Maron, Margaret. *Shooting at loons*. NY: Mysterious, 1994. 229 p. 0892964472
During a fishing trip off the North Carolina coast Judge Deborah Knott discovers the body of an old angler who had opposed a land development scheme.

2310. Matthiessen, Peter. *Killing Mister Watson*. NY: Vintage, 1991. 372 p. 0679734085
When Mr. Watson moves to the Everglades in the late 19th century, his neighbors see him as a wealthy and powerful man who will bring development and wealth to the area. Later, rumors that he may be a killer lead them to conspire to kill him. Includes very vivid descriptions of the Everglades landscape and animals which can be benificent or deadly.

2311. McCabe, Peter. *Wasteland*. NY: Scribner, 1994. 260 p. 0684196816
A contractor/bribery artist who is working on a waste dump for a corrupt land developer, attacks a geologist for making a pass at his wife. He

becomes the prime suspect in the geologist's murder the next day. Can he figure out who set him up before going to trial?

2312. McQuillan, Karin. *Cheetah chase*. NY: Ballantine, 1994. 320 p. 0345381831
Prequel: *Elephant's graveyard*.
Safari operator/private investigator Jazz Jasper joins forces with a widowed animal rights activist and a local detective to track down the killers of an investigative reporter who may also be killing the cheetah population. HR.

2313. ___. *Deadly safari*. NY: St. Martin's, 1990. 293 p. 0312038089
Novice safari leader Jasper encounters strange phenomena such as zebras who protect their own from lions and safari members who do not protect their own from mysterious murders.

2314. ___. *Elephants' graveyard*. NY: Ballantine, 1993. 260 p. 0345381823
Sequel: *Cheetah chase*.
When an anti-poaching activist is found buried by a wildlife watering hole, Jasper assumes the path will lead directly to poachers, but she uncovers scientific and familial rivalries, mating baboons, red herrings, and pitfalls aplenty while trying to track the murderers down.

2315. Moore, Barbara. *The wolf whispered death*. NY: Dell, 1988, c1986. 215 p. 0440201179
A new Mexico rancher is killed, but are the tracks around his body those of a feral dog, wolf, or the white werewolf of Indian lore? A veterinarian and his doberman track the animal into Navajo country, but who is really the hunter and who the prey? AMT. #

2316. Poyer, David. *The dead of winter*. NY: T. Doherty, 1988. 336 p. 0812507886
Under the cover of deer hunting, Paul Michaelson pursues the men who killed his son, but the expert hunter he pursues is after him, too.

2317. ___. *Louisiana blue*. NY: St. Martin's, 1994. 304 p. 0312104944
AR: *Bahama blue*.
Two divers encounter not only the natural perils of the deep, but the questionable environmental and criminal activities of oil companies when they take a job working on an underground oil pipeline. Includes vivid depictions of deep-sea diving.

2318. Quinn, Elizabeth. *Murder most grizzly : a Lauren Maxwell mystery*. NY: Pocket, 1993. 211 p. 0671749900
When an investigator for the Wild America Society investigates the death

of a biologist she discovers that he may not have been mauled by the suspected grizzly bear after all. Who did kill him, and why?

2319. Russell, Alan. *The forest prime evil.* NY: Walker, 1992. 207 p. 0802732046
A murder investigation involving a protracted civil disobedience campaign.

2320. Sherwood, John. *Creeping Jenny.* NY: Scribner's, 1993. 256 p.
0684196131
When a nursery worker is abducted, a horticulturist/detective follows the
trail to an abandoned building where she discovers that militant environmentalists are planning to disrupt flower shows. An interesting idea, but
this clumsy novel suffers from stereotypes, obvious red herrings and a lack
of true mystery or suspense. Not recommended.

2321. Simonson, Sheila. *Mudlark.* NY: St. Martin's, 1993. 240 p. 031209874X
After a "California carpetbagger" receives a bag of dead seagulls on her
doorstep, the promoter of a proposed resort in an environmentally sensitive area is found dead, and the murder suspect's house is firebombed, a
private investigator finds herself in the midst of controversy and danger.

2322. Stabenow, Dana. A *cold day for murder.* NY: Berkley, 1992. 199 p.
042513301X
Edgar Award, 1993.
After a congressman's son and an investigator sent out after him disappear
in the Alaskan wilderness, Kate Shugak is called upon. Conflicts between
Aleut and white values come into play. Eventually, two drunken pipeline
workers use a stolen excavator in a grim but hilarious battle with a state
trooper's helicopter. HR.

2323. ___. A *cold-blooded business.* NY: Berkley, 1994. 231 p. 042514173X
Kate Shugak searches for drug dealers and murderers in Alaskan oil camps
and the wilderness.

2324. ___. *Dead in the water.* NY: Berkley, 1993. 217 p. 042513749X
Shugak tries to discover what happened to two men who mysteriously dis-
appeared from a fishing boat. Kate, who has rebelled against her bossy
Aleut grandmother, learns to value her native culture through her acquaintance with two basketmakers. The New Age movement is satirized in the
character Andy.

2325. Stephenson, Neal. *Zodiac : the eco-thriller.* NY: Atlantic Monthly, 1988.
283 p. 0871131811
Environmental thriller featuring a New Age Sam Spade and evil polluters in
Boston Harbor.

2326. Straley, John. *The curious eat themselves*. NY: Soho, 1993. 264 p. 093914994X
When detective Cecil Younger investigates the sexual assualt of a female cook on a remote mining site he discovers that she was preparing to blow the whistle on the mine's environmental hazards and then finds her body floating in the Ketchikan estuary. Straley probes corruption, collusion, and the environmental degradation of Alaska with power and grace. HR AMT. #

2327. ___. *The woman who married a bear*. NY: Soho, 1992. 225 p. 0939149648
While investigating the "closed" murder of an Indian hunting guide, investigator Cecil Younger is hunted by someone protecting the killer. Real life investigator Straley incorporates elemental forces, wilderness adventure, vivid descriptions of encounters with bears, and a Tlingit legend to great effect in this thriller. HR AMT. #

2328. Strieber, Whitley. *The wolfen*. NY: Avon, 1988, c1978. 275 p. 0380704404
A murderer with unusual strength and speed, incredible vision, and the apparent ability to transform or disappear at will is loose in New York City. Who or what could it be? Recommended for adventurous AOT, but not for the squeamish. #

2329. Taylor, Elizabeth Atwood. *The Northwest murders*. NY: St. Martin's, 1992. 278 p. 031207753X
A private investigator retreats to a remote cabin to recover from chronic fatigue syndrome, but when two backpackers are attacked and scalped she is employed to find evidence to defend the accused Karuk Indian and to determine the motives that will lead to the real murderer.

2330. Van Gieson, Judith. *The lies that bind*. NY: HarperCollins, 1993. 259 p. 0060177055
Environmental lawyer/investigator Neil Hamel gets involved in a case involving racism, murder, the drug trade, international assassinations, and devastating development around Albuquerque and Phoenix. A good read, but not as overtly ecofictional or exciting as some of her other books.

2331. ___. *North of the border*. NY: Pocket, 1993, c1988. 170 p. 0671769677
Called upon to investigate threatening notes sent to her ex-lover after he adopts a Mexican baby, Neil Hamel must deal with smuggling rings, sleazy political machinations, and an attempt to make a windfall profit by turning an old gold mine into a nuclear waste dump.

2332. ___. *The other side of death*. NY: HarperCollins, 1991. 216 p. 0060165812

When the death of a vocal opponent of the proposed "world's tallest, largest, ugliest building" is ruled a suicide, Neil Hamel decides to sniff out the real rats behind it.

2333. ___. *Raptor.* NY: Harper & Row, 1990. 246 p. 0060161671
When a poacher of the rare arctic falcon is killed, Neil Hamel defends the conservationist charged with the murder. Alone in a rugged wilderness, Neil herself becomes the target of human predators. Probably her best book to date. HR.

2334. ___. *The wolf path.* NY: HarperCollins, 1992. 232 p. 0060168048
Another heady fictional brew featuring Neil Hamel, this time involving the reintroduction of wolves to the New Mexico, the murder of a federal official, and a radical environmentalist as the fall guy. HR.

2335. **Wallace, David Rains.** *The turquoise dragon.* SF: Sierra Club, 1985. 230 p. 0871568195
AR: *The wilder shore.*
Forester George Kilgore learns that a murdered biologist had discovered the breeding grounds of one of the world's rarest animals. Kilgore finds himself plunged into a wilderness intrigue involving strange detectives, cocaine smugglers, and self described "redneck loggers." Naturalist Wallace slips vivid and accurate portrayals of the wilderness into this gripping mystery. HR AOT. #

2336. ___. *The vermilion parrot.* SF: Sierra Club, 1991. 217 p. 0871566303
Kilgore becomes the manager of a land preserve where he discovers that rare California Condors are being removed from the area so it can be logged. He then encounters a feathered, winged, talking dinosaur which has been hidden in a bunker by the feds. Kilgore, the dinosaur, and a female Soviet spy conspire to confront the condor poachers in Mexico. Rains couches his environmental message in a delightful farce which is HR AMT. #

2337. **Wallingford, Lee.** *Clear-cut murder.* NY: Walker, 1993. 219 p. 0802732313
Prequel: *Cold tracks.*
Carver and Trask return when a confrontation between a logging company and radical environmentalists results in the murder of a would-be peacemaker.

2338. ___. *Cold tracks.* NY: Walker, 1991. 204 p. 0902757839
Sequel: *Clear-cut murder.*
When national forest cop Frank Carver and ranger Ginny Trask investigate the murder of an illegal alien working on a Christmas tree plantation, they also discover a "suicide" that was really a murder, blackmail, and other unsavory activities.

2339. Wambaugh, Joseph. *Finnegan's week.* NY: Morrow, 1993. 348 p. 0688128017
A San Diego police detective joins forces with two female cops to chase a killer with a 55 gallon drum of toxic chemicals.

2340. White, Randy Wayne. *The man who invented Florida.* NY: St. Martin's, 1993. 294 p. 0312098669
AR:*The heat islands, Sanibel flat.*
Aging Tucker Gatrell claims he has found the fountain of youth in a sulfur spring on land he once owned, but when marine biologist/P.I. Doc Ford discovers that it made Tucker's horse grow new genitals and revitalized Tucker's best friend, Ford suspects it is a scheme by Tucker to get his land back from developers. Dark humor.

2341. Williams, Walter Jon. *Days of atonement.* NY: Tor, 1992. 437 p. 0812501802
A copper mine closes in early 21st century Atocha, New Mexico, and unemployed people accept the arrival of the super-secret Advanced Technologies Labs. When strange events start taking place and a man drops dead in the police chief's office, he is forced to deal with the hazardous activity at ATL.

2342. Wren, M. K. *Wake up, darlin' Corey.* NY: Ballantine, 1990, c1984. 218 p. 0345350715 (pbk)
When a family double-crosses an environmental group by selling their unspoiled Oregon coast property to a resort developer, their environmentalist daughter dies in a mysterious car "accident." A bookstore owner/P.I. searches for her missing diary, which he believes will reveal the killer.

Books for young adults:

2343. Byars, Betsy Cromer. *McMummy.* NY: Viking, 1993. 150 p. 0670849952
Looking after an eccentric scientist's greenhouse doesn't seem any stranger than the other odd jobs taken by Mozie and his partner Battie until they discover a large, mummy-shaped pod on one of the plants.

2344. George, Jean Craighead. *The fire bug connection.* NY: HarperCollins, 1993. 148 p. 0060214902 0060214910
AR: *Hook a fish, catch a mountain.*
A 12-year-old receives European fire bugs for her birthday. When they fail to metamorphose, growing grossly large and exploding instead, she uses scientific reasoning to determine the cause of their strange death.

2345. ___. *The missing 'gator of Gumbo Limbo.* NY: HarperCollins, 1992. 148 p.

006020396X 0060203978
Sixth-grader Liza, one of five homeless people living in an unspoiled forest
in southern Florida, searches for a missing alligator destined for official
extermination and studies the delicate ecological balance keeping her outdoor home beautiful. GR 4-9.

2346. ___. *Who really killed Cock Robin?* NY: HarperCollins, 1991, c1971. 160
p. : ill. 0060219807 0060219815
An eighth-grader follows a trail of environmental clues to try and figure out
what ecological imbalances might have caused the death of the town's
best-known robin. GR 4-9.

2347. Guiberson, Brenda Z. *Turtle People*. NY: Atheneum, 1990. 104 p.
068931647X
Depressed by the continuing absence of his beloved father and the distant
behavior of his mother, an 11-year-old retreats to the natural setting of a
remote island near his Washington State home where he finds a great
archeological mystery.

2348. Sanders, Scott R. *Bad man ballad*. NY: Bradbury, 1986. 241 p.
0027782301
A 17-year-old and a lawyer in early 19th century Ohio set out to find a murderer who might be a "Bigfoot."

2349. Warner, Gertrude Chandler. *Bus station mystery*. Chicago: Whitman,
1974. 127 p. : ill. 0807509760
The Alden children, isolated in a storm at a small bus station, are led into
a mystery centering on a polluted river. GR 3-7.

19. Science Fiction and Fantasy

Wells' *The Island of Doctor Moreau*, published in 1896, might be considered the first science ecofiction novel, but Doyle of Sherlock Holmes fame also created a more obscure sleuth named Professor Challenger at around the same time.

The perspective and values of pulp "horse operas" and "space operas" are similar: nature (wilderness or space) is a malevolent force to be conquered by macho men with powerful weapons for the sake of progress and profit. Red blood once flowed freely over the imaginary purple planets or equally fantastic purple sage, but that's changing. Asimov, Mr. Robot himself, explored the Gaia hypothesis in *Foundation's Edge* and *Foundation and Earth*. Even Pournelle and Niven, the literary equivalents of creepy 12 year olds with atomic slingshots, cranked out *The Legacy of Hereot* .

The "new wave" of science fiction which emerged during the 1960s and continues to the present, pays less attention to weaponry and more to social and environmental issues. The title of Stadler's 1971 anthology *Ecofiction* (now out of print) may be the first written use of that term. Notable authors demonstrating strong environmental values include Aldiss, Ballard, Bear, Brin, Brunner, Butler, Harrison, Herbert, Le Guin, Lem, Lessing, McIntyre, Robinson, Slonczewski, Wilhelm, and Vance.

While "space operas" tend to be penned by men, women have held up more than half of the sky in ecologically-oriented science fiction. Many ecofeminist authors have effectively employed fantasy and science fiction to explore sexism and the environmental crisis, typically offering positive alternatives to both. While Butler and Lessing's work can be found here, other books by Le Guin, Slonczewski, McIntyre, and Wilhelm, and the novels of Bush, Charnas, Douglas, Piercy, and Russ are listed in the Ecofeminist fiction chapter. Relevant anthologies include *Future Primitive* (1738) and *Isaac Asimov's Earth.* (1720)

There is a strong connection between science fiction and disaster novels. Notable sci-fi authors in the Environmental Crisis chapter include Herzog, Kilian, Norwood, Stewart, Strieber and Kunetka, and Turner.

A commonly accepted distinction between science fiction and fantasy is that the former is based upon what may be scientifically plausible, while fantasy is a purely imaginative realm. Since the best contemporary science fiction and fantasy is both genre- and gender-busting it may be more accurate to use the newer term "speculative fiction." You certainly don't need to be a sci-fi fan to appreciate the excellent work of Ballard, Brin, Le Guin, Piercy, Stewart, Turner, Vonnegut, or Wilhelm.

The dominance of science fiction over fantasy in the adult section is reversed among the young adult books, where less than a quarter of the titles are sci-fi and the rest fantasy. The magical animals, Arthurian, and Druidic characters often found in these novels are popular with self-selected readers of all ages. Many "hard" science fiction books for teens are formulaic, plot-heavy space operas with little reflection on the environment or anything else. On the contrary, the "what if" nature of fantasy provides an excellent vehicle for the consideration of environmental and social issues.

Books for adults:

2350. Adams, Douglas. *The hitchhiker's quartet.* NY: Harmony, 1986. 624 p. 0517564254
Contents: *The hitchhiker's guide to the galaxy* — *The restaurant at the end of the universe* — *Life, the universe and everything* — *So long, and thanks for all the fish* — *Young Zaphod plays it safe.* Arthur Dent learns that bulldozers are about to level his house for a highway project. His secretly alien neighbor, Ford Prefect, then tells him that Earth is about to be destroyed to build an instellar highway. Dent and Prefect hitchhike around the galaxy trying to discover where the dolphins went and why Earth was selected for destruction. After an hilarious quest they return to Earth, which was not demolished after all. HR. #

2351. Aldiss, Brian Wilson. *Helliconia trilogy.* NY: Collier, 1992-93. 3 v. Contents: v.1 *Helliconia spring* (0020160909) — v.2 *Helliconia summer* (0020160909) — v.3 *Helliconia winter* (0020160925)
AR: *Earthworks, Greybeard, Hothouse, Neanderthal planet.*
Presents life on Helliconia, a planet with a seasonal cycle of 2,500 Earth years. It suffers extreme heat in summer; extreme cold, darkness, and barbarism in winter; and plague in both winter and spring. Earth has been monitoring Hellaconia by satellite, and soon finds itself facing similar challenges due to the nuclear winter. HR.

2352. Asimov, Isaac. *Foundation's edge.* NY: Ballantine, 1983. 426 p. 0345308980
AR sequels: *Foundation, Foundation and empire, Second foundation.*
Representatives of the technological First Foundation and the mentalic Second Foundation arrive on Gaia, a planet where all organisms are separate individuals sharing the overall consciousness of the planet itself.

2353. ___. *Foundation and earth.* Garden City, NY: Doubleday, 1986. 356 p. 0385233124

Asimov contrasts the violence, pettyness and arrogance of other worlds agains the peaceful higher consciousness of Gaia. Robots established Gaia as the first step of creating a galaxy-wide conscious being to protect humans from themselves.

2354. Baker, Will. *Shadow hunter.* NY: Pocket, 1993. 373 p. 0671790463
In the early 22nd century the northern hemisphere is high tech and prosperous, but the southern is an irradiated desert. When the son of an official is attacked by a wild animal and then kidnapped on a hunting trip, a conservative faction arises which seeks to kill the primitive "ginks" and all other disorderly "primitive wildlife."

2355. Ballard, J. G. *Concrete island.* New York: Farrar Straus Giroux, 1994, c1974. 126 p. 0374524130
When an injured motorist is stranded on a traffic island in London he is overwhelmed by the industrial waste surrounding him. Like Robinson Crusoe he adapts to his "island," but adaptation to a dehumanizing environment leads to insanity. An allegory for the degradation of the soul in industrialized society.

2356. ___. *The crystal world.* NY: Farrar Straus Giroux, 1988, c1966. 216 p. 0374520968
One of *Science fiction's best 100 novels.*
A group of men lost in the West African jungle become entranced by a fascinating crystalized forest. Is it just a strange phenomenon or is nature petrifying itself? HR.

2357. ___. *The day of creation.* NY: Collier, 1989, c1987. 253 p. 0020415141
Dr. Mallory moves to Central Africa, but his medical work is stymied by drought and guerilla warfare. He dreams of bringing water to the region, and one day a new river emerges from beneath a shifted tree trunk. Mallory believes he has caused the river, tries to trace it to its source, and becomes involved in personal and political intrigue.

2358. ___. *The drowned world.* NY: Carroll & Graf, 1987, c1962. 175 p. 0881843245
One of *Science fiction's 100 best novels.*
When solar radiation melts the ice caps, the seas rise, coastal areas in low plains flood, the temperature climbs, and humans live mostly in polar regions. A doctor returns to the swamp which was once London on a scientific expedition. His dreams take his mind on a "night journey" leading him deeper into the drowned world. This powerful forewarning of global warming is HR AOT. #

2359. ___. *High-rise.* NY: Carroll & Graf, 1988, c1975. 169 p. 0881844004

One of *Science fiction's* 100 *best novels.*
Early 21st century high-rise apartment living is completely automated, divorced from nature, and supposedly utopian, but they forgot to program the theoretically obsolete dark side into the computer. Violence, backlash, perversity, and mass destruction ensue, as their technology can't fool Mother Nature or human nature. HR for non-squeemish adults.

2360. Barron, T. A. *The Ancient One.* NY: Philomel, 1992. 367 p. 0399218998
Prequel: *Heartlight.*
While helping her Great Aunt Melanie try to protect an Oregon redwood forest from loggers, thirteen-year-old Kate goes back five centuries through a time tunnel to face the evil creature Gashra, who is bent on destroying the same forest. AOT. #

2361. ___. *Heartlight.* NY: Philomel, 1990. 272 p. 0399221808
Sequel:*The ancient one.*
Kate and her grandfather use one of his inventions to travel faster than the speed of light on a mission to save the sun from a premature death. AOT. #

2362. Bear, Greg. *Anvil of stars.* NY: Warner, 1992. 434 p. 0446516015
Prequel: *Forge of God.*
After Earth has been destroyed by space probes, surviving Earthlings make a pact with benefactor machines to destroy the eco-killing probes before other planets are lost. They create a giant spaceship from the rubble left from Earth to find and destroy the probes. AMT. #

2363. ___. *The forge of God.* NY: TOR, 1987. 474 p. 0312930216
Sequel: *Anvil of stars.* AR: *Eon, Eternity.*
In 1996 when the sixth moon of Jupiter mysteriously disappears, a 500 foot cinder cone appears in Death Valley as does a perfect imitation of Ayers Rock in Australia. AOT. #

2364. Benford, Gregory. *Timescape.* NY: Bantam, 1992, c1980. 499 p. 0553297090
One of *Science fiction's* 100 *best novels.*
In 1998 the oceans are dying from pollution, threatening the survival of almost all living things. Scientists send a message to the year 1963, hoping that action will be taken then to save the Earth. Contrasts the optimism and naive faith in technology of the early sixties with worsening current and future conditions. HR.

2365. Bishop, Michael. *Unicorn mountain.* NY: Bantam, 1989, c1988. 418 p. 0553279041
A rancher, an AIDS sufferer, and two Ute Indians have their lives trans-

formed by a band of unicorns, but then must save them from a mysterious disease.

2366. Bova, Ben. *Empire builders*. NY: TOR, 1993. 383 p. 0312851049
Sequel NOT recommend: *Privateers*. AR: *Colony, Kinsman, Millenium*.
Scientists determine that the world has less than ten years before a global warming catastrophe. Dan Randolph takes action to stop it, but is betrayed and sent to the moon. He then attempts to forge alliances to return to Earth and save it. AOT. #

2367. Brin, David. *Earth*. NY: Bantam, 1990. 601 p. 0553057782
In the mid 21st century humanity is struggling to reconcile high technology, materialism, resource depletion, and pollution when black holes developing within the Earth shift the focus from conservation to a desperate struggle for planetary survival. HR.

2368. ___. *The postman*. NY: Bantam, 1986. 321 p. 0553257048
In the near future after nuclear war and its consequences have devastated the Earth and dispirited its survivors, an itinerant storyteller dons a postal uniform to spread his message of hope and renewal. Reaffirms the importance of myth and shared values and of respect for one another and the Earth. HR.

2369. ___. *Startide rising*. NY: Bantam, 1988, c1983. 462 p. 055327418X
Hugo & Nebula Awards, 1983.
Sequel: *The uplift war*.
Both books publ. in one vol. as *Earthclan*, Doubleday, 1987. The advanced Progenitors have "uplifted" all the other intelligent species in the galaxy, who are all thus indebted to them, except for Earth where human intelligence has arisen on its own and the humans have "uplifted" dolphins and chimpanzees. HR. #

2370. ___. *The uplift war*. West Bloomfield, Mich.: Phantasia, 1987. 506 p. 0932096441
Prequel: *Startide rising*.
A small colony of humans and chimps battles the Progenitors over a secret experiment designed to produce the next uplift. HR. #

2371. Brunner, John. *The sheep look up*. NY: Ballantine, 1981, c1972. 461 p. 0060105585
Prequel: *Stand on Zanzibar*.
After all life in the Mediterranean has been polluted out of existence, western American waterways are flooded with poisonous gas, and the entire biosphere is threatened with extinction, humanity must come up with a

solution to restore the damaged environment. HR.

2372. ___. *Stand on Zanzibar*. NY: Ballantine, 1987, c1968. 505 p. 0345347870
Sequel:*The sheep look up*.
One of *Science fiction's* 100 *best novels*, Hugo Award Winner, 1969.
In the 21st century, seven billion people experience a ravaged environment, oppression, violence, and the hysteria of extreme crowding stress, but don't realize that they are all slipping into madness. A powerful outcry against violence, overpopulation, drugs, greed, and stupidity. HR.

2373. Butler, Octavia E. *Xenogenesis trilogy*. NY: Warner, 1988-90. 3 v. Contents: v.1 *Dawn* (0445207795) — v.2 *Adulthood rites* (0445209038) — v.3 *Imago* (0445209771)
AR: *Wild seed*.
A poisoned Earth is saved by biotechnologists from Oankali, but humans are reduced to mere breeding stock for their alien masters. v.2: Human rebels bicker among themselves and begin descending into savagery. Will humanity bring itself to the brink of destruction once again? v.3: The humans who have escaped their Oankali masters are forced deeper into the jungle where they must settle their differences to maintain any hope of re-establishing a humane society.

2374. Chant, Joy. *The grey mane of morning*. NY: Bantam, 1980, c1977. 332 p. : ill. 0553012002
Sequel: *Red moon and black mountain*.
Chant continues the story begun in the earlier novel, but with more complex characters, greater emphasis on mythic figures, an even more complex view of the environment, and lush, naturalistic sexuality. HR AMT. #

2375. ___. *Red Moon and black mountain : the end of the House of Kendreth*. NY: Bantam, 1983, c1981. 255 p. 0553233114
Prequel: *The grey mane of morning*.
Three children are drawn to a starlit land and into a struggle between the hunters known as the Hurnai and the Vandarei star warriors. The attitude toward nature is one that rejects the oversimplified pastoral approach in favor of one emphasizing multiplicity, diversity, and complexity. HR AMT. #

2376. Crace, Jim. *Continent*. NY: Harper & Row, 1986. 138 p. 0060157240
David Higham and Whitbread Awards, *Guardian* Prize for Fiction.
Seven linked stories about an imaginary seventh continent, describing its flora and fauna, tribes and communities, lore and superstition, sexuality, conflicts and predicaments, and changes over time. The people are Earth oriented and have a strong collective unconscious. HR.

2377. Dick, Philip K. *Blade runner : Do androids dream of electric sheep?* NY: Ballantine, 1987, c1968. 216 p. 0345350472
Original title: *Do androids dream of electric sheep?* One of *Science fiction's100 best novels*. After nuclear war and subsequent radiation have destroyed almost all animal life on Earth, artificial animals become status symbols and highly humanized androids escape back to Earth. A viscious android bounty hunter falls in love with a female android. The book is much deeper than the film, questioning what constitutes humanity, and offering a cautionary note about nuclear war and environmental degradation. HR.

2378. DiSilvestro, Roger L. *Living with the reptiles.* NY: D. I. Fine, 1990. 284 p. 1556111746
Time travelers visit a pre-industrialized earth. HR AMT. #

2379. Doyle, Arthur Conan. *The complete Professor Challenger stories.* Ware, GB: Wordsworth, 1989. 577 p. 1853269271
Contains five science fiction stories by the creator of Sherlock Holmes. The novella "The poison belt" which tells of the Earth's passage through a belt of toxic interstellar gas has a subtle environmental moral. "When the world screamed" is particularly powerful. HR for mystery or sci-fi readers.

2380. Farmer, Philip Jose. *The lavalite world.* NY: Lightyear, 1993, c1977. 282 p. 0899684017
AR: The *Riverworld* series.
When Anana and the trickster Kickaha arrive in Los Angeles the overpopulation and pollution make them wish they were back on the world of Tiers. The living mountains, man-eating trees, and other unusual characters make for exciting reading. HR.

2381. Foster, Chris. *Winds across the sky : a love story.* Lower Lake, CA: Aslan, 1992. 118 p. : ill. 0944031439
A reporter, an actress, a humpback whale, and a redwood overcome emotional pain and the human-imposed separation between living things. They learn to psychically communicate with one another for mutual survival. This ecological love story is a bit cloying, but will appeal to romance readers and some New Age fans. #

2382. Frank, Pat. *Alas, Babylon.* NY: HarperCollins, 1993, c1959. 312 p. 0060812540
One of *Science fiction's 100 best novels*.
AR: *Mr. Adam, Forbidden area*.
At first this appears to be a standard nuclear holocaust novel, with civil defense totally ineffective against mass destruction, radiation, and social collapse. It takes an optimistic turn as a group of survivalists emerge to

form a semi-idyllic pastoral society. Mainly of interest as a naive, survivalist fantasy.

2383. Grant, Richard. *Rumors of spring.* NY: Bantam, 1987. 439 p.
0553051903 0553343696 (pbk)
Earth's final forest starts growing uncontrollably and the First Biotic Crusade begins.

2384. Harrison, Harry. *Make room! Make room!* NY: Bantam, 1994, c1966. 288 p. : ill. 0553564587
AR: *One step from Earth.* AR teens: *The California iceberg.*
In the early 21st century there are 40 million people in New York and environmental destruction is so advanced that there are few natural sources of food. All they have to eat is a strange green substance called Soylent. What might it be made of? AMT. #

2385. ___. *West of Eden trilogy.* NY: Bantam, 1984-88. 3 v. Contents: v.1 *West of Eden* (0553050656) — v.2 *Winter in Eden* (0553051636) — v.3 *Return to Eden* (0553053159).
As a new ice age emerges, Kerrick rallies to unite the human tribes against encroachments by the warlike Yilane, an advanced reptilian species. v.2: As the glaciers spread south, the Yilane resort to biological and genetic warfare to capture teritory held by humans. v.3: Kerrick and the other humans find temporary safe haven and harmony on an island, but must inevitably face Yilane hostility.

2386. Herbert, Frank. *Dune.* NY: Ace, 1990, c1965. 535 p. 0441172717
One of *Science fiction's* 100 best novels.
AR:*Children of Dune, God emperor of Dune, Dune chapterhouse, Dune messiah, Heretics of Dune, The notebooks of Frank Herbert's Dune, Soul catcher.*
The natives of a drought-stricken planet struggle to overthrow robber Baron Barkonnen to end their oppression and gain control of a mind altering drug produced by giant sand worms. Herbert presents a theme of a degraded world thrown out of ecological balance throughout the series. HR AOT. #

2387. Herbert, Frank, and Ransom, Bill. *The ascension factor.* NY: Ace, 1989. 381 p. 0441031277
Prequels: *The Lazarus effect, The Jesus incident.*
As the land of planet Pandora continues to be engulfed by the sea, the remaining humans attempt to restore land and build settlements. To survive, they must defeat the forces of the evil Director, but resistance fighters make a last-ditch effort to employ the sentient kelp to aid them in their cause.

2388. ___. *The Jesus incident.* NY: Ace, 1987, c1979. 416 p. 0441385397

Sequels:*The ascension factor, The Lazarus effect.*
Like the world of Dune, the planet Pandora is a desert demanding feats of great adaptation to the environment. In this case it involves mating a human being and a sentient kelp to produce the hybrid female Vata.

2389. ___. *The Lazarus effect.* NY: Ace, 1987, c1983. 393 p. 0441475213
The authors continue to use the planet Pandora and the hybrid Vata to explore the themes of survival, adaptation, and genetic engineering.

2390. Kirby, T. J. *Dangerous nature.* NY: Zebra, 1993. 351 p. 0821740342
A disabled scientist seems to be on the verge of a breakthrough on limb regeneration when animal activists break into his lab, release the animals, and steal the serum. What will happen if the experimental drug is released into the environment?

2391. Knight, Damon. *Why do birds : a comic novel of the destruction of the human race.* NY: TOR, 1992. 272 p. 031285174X
Ed Stone says he's been in suspended animation since the 1930s when he was kidnapped by aliens. They've returned to Earth to convince its leaders to build a massive vault to place the entire human population in suspended animation to survive the planet's impending ecological doom. #

2392. Kube-McDowell, Michael P. *The quiet pools.* NY: Ace, 1990. 371 p. 0441699111
The rapidly deteriorating environment of 21st century Earth speeds the effort to colonize other planets. One subplot features environmental activists using computers and other high-tech devices in their struggle against corporate polluters.

2393. Le Guin, Ursula K. *The word for world is forest.* NY: Ace, 1989, c1972. 169 p. 0441909159
Hugo Award, 1972.
AR: *The dispossessed* .
A group of colonists who are determined to exterminate the green skinned forest people known as the Antsheans find that their aggression and weapons may not be sufficient to overcome the Antsheans' natural lore and determination to survive. AOT. #

2394. Lem, Stanislaw. *The Futurological Congress : from the memoirs of Ijon Tichy.* San Diego: Harcourt Brace Jovanovich, 1985, c1974. 149 p. 0156340402
Translation of *Ze wspomnien Ijona Tichego; kongres futurologiczny.*
When a conference on overpopulation is attacked by revolutionaries, a cosmonaut is shot but frozen in liquid nitrogen until 2039. He is revived in a world where "psychem" drugs blind people to overpopulation, environmental

decay, the extinction of all non-human animals, and major food shortages.

2395. ___. *Solaris*. San Diego: Harcourt Brace Jovanovich, 1987, c1970. 204 p. 0156837501
When Kris Kelvin arrives on the planet Solaris he discovers strange Phi-creatures, including a replica of the wife he had driven to suicide. He discovers that the ocean itself is a vast living intelligence whose only means of communicating with humans is to create replicas of their memories. Lem's use of the ocean here, and sentient "inanimate" objects ranging to full planets in other novels foreshadows the Gaia principle.

2396. Lessing, Doris May. *Canopus in Argos : archives*. NY: Vintage, 1992. 1228 p. 0679741844
Also published separately in 5 volumes. An ambitious saga which chronicles millions of years of evolution including ecological cataclysms, re-evaluations of male-female relationships and the human relation to the Earth, new ways of thinking and feeling, the creation of social and ecological utopias, and the hazards of misleading rhetoric and false sentiment. Recommended for patient, sophisticated adults.

2397. Martin, George R. R. *Tuf voyaging*. NY: Baen, 1986. 374 p. 0671559850
Space trader Haviland Tuf recovers an abandoned seedship stocked to provide new foodstocks for the planet S'uthlam. He warns them to control their population. The S'uthlamese prepare for expansionistic wars to provide even more food instead, so Tuf introduces a contraceptive crop. This critique of overpopulation, reliance on science to solve social problems, and the inherent dangers of meddling with ecological balance is HR.

2398. May, Julian. *Intervention : a root tale to the galactic milieu and a vinculum between it and the saga of Pliocene exile*. Boston: Houghton Mifflin, 1987. 546 p. 0395437822
Human evolution seems to be reaching the next step when some people in various parts of the world are born with tremendous "metapsychic" powers Although the metapsychics are humankind's best hope against nuclear and environmental annihilation, the "normals" consider them a threat.

2399. McCaffrey, Anne, and Scarborough, Elizabeth Ann. *Powers that be*. NY: Ballantine, 1993. 311 p. 0345381734
AR sequel: *Power Lines*.
A major is forced into early retirement on the planet Petaybee. The sentient planet heals her, and when she learns that the mining company plans to evacuate all humans from Petaybee so they can pillage its mineral resources, she is forced to chose between Earth and her longtime employer or her new planet and its lifeforms. HR AOT. #

2400. McIntyre, Vonda N. *Star Trek* IV : *the voyage home*. NY: Pocket, 1986. 274 p. 0671632663
To save the Earth from destruction in the 23rd century, Admiral Kirk and his crew travel back in time to the 20th to recover a pair of humpback whales. Amusing adventure with a strong ecological message. This version is HR AOT. #

2401. McQuinn, Donald E. *Wanderer*. NY: Ballantine, 1993. 544 p. 0345378407
Sequel: *Warrior*.
500 years after a nuclear-biological war, the struggle between the Orthodox Church and the visionary Rose Priestess continue. This complex and grisly book is not McQuinn's best work, but will appeal to fans of apocalyptic literature or cryogenics.

2402. Moore, Ward. *Greener than you think*. NY: Lightyear, 1993, c1947. 322 p. 089968355X
When new lawn fertilizer brings dead Bermuda "devil grass" to life, it grows at amazing speed, takes over everything, and stymies every attempt to stop it.

2403. Morrow, James. *Towing Jehovah*. NY: Harcourt Brace, 1994. 371 p. 0151909199
An oil tanker captain, guilt-ridden by an earlier oil spill, accepts an assignment by the Angel Raphael to tow God's decaying, two-mile-long body from the tropical Atlantic to an arctic ice cave. Along the way he encounters militant atheists who want to destroy the body, crew members who want to eat it, and various adventures. Recommended for sophisticated adults.

2404. Myers, Edward. *Fire and ice*. NY: ROC, 1992. 432 p. 0451452119
The two outsiders fulfill their role in prophecy, with Forster becoming the man of darkness and Jesse the son of light. Jesse's retreat to the mountain triggers the action that might begin the end of civilization. #

2405. ___. *The mountain made of light*. NY: ROC, 1992. 420 p. 0451451368
Sequel: *Fire and ice*.
Anthropologist Jesse O'Keefe discovers a secret race descended from the Incas who fled conquest to save their mystical, Earth-centered culture. When another westerner arrives, they are both drawn into a prophecy and must make a quest to find the mountain made of light to save the people and themselves. #

2406. Niven, Larry; Pournelle, Jerry; and Barnes, Steven. *The legacy of Heorot*. NY: Simon & Schuster, 1987. 367 p. : maps. 0671640941
The National Geographic Society sends 200 men and women to Tau Ceti Four where they establish an idyllic settlement called Avalon. Their

attempt to exterminate the competitive, native species (Grendels) has unforeseen ecological consequences. AMT, but not for those offended by graphic sexuality. #

2407. Norton, Andre. *Breed to come*. NY: Ace, 1972. 288 p. 0441078982
When pollution drives all humans from Earth they leave an experimental virus behind that increases the intelligence in successive generations of the remaining animals. #

2408. ___. *Fur magic*. Hampton Falls, NH: D. M. Grant, 1992, c1968. 173 p. : ill. 0671299026
AR: *Lavender green magic, Octagon magic*.
A young boy takes the oath of magic and is transformed into an animal and transported into the world described in Indian legends. Animals are the only intelligent inhabitants and humans do not yet exist. #

2409. Robinson, Kim Stanley. *Escape from Kathmandu*. NY: Orb, 1994, c1989. 320 p. 0312890060
Two American adventurers in Nepal searching for Yeti (the so-called abominable snowmen) discover that one has been captured and attempt to free it from imprisonment. #

2410. ___. *Orange County trilogy*. NY: Berkley or TOR, 1984-1991. 3 v. Contents: v.1 *The wild shore* (0441888704) — v.2 *The Gold Coast* (0812552393) — v.3 *Pacific edge* (0815200563).
Sixty years after a nuclear war young Henry dreams of a democratic, ecological, high tech society and joins a resistance movement v.2: The resistance struggles against Orange County trends: an "autotopia" of malls, designer drugs, and star wars technology. HR AOT. #

2411. ___. *Green mars*. NY: Bantam, 1994. 535 p. 0553096400 0553373358 (pbk)
AR: *Red Mars, Blue Mars*.
Publishers Weekly Best Book, 1994.
The original colonists of Mars who opposed massive "terraforming" of the planet in *Red Mars* face new challenges from development-minded multinational corporations from Earth, leading to a Martian independence movement. HR AOT. #

2412. Silverberg, Robert. *Hot sky at midnight*. NY: Bantam, 1994. 327 p. 0553092480
AR: *The infinite web*.
Pollution, industrial espionage, ruthless megacorporations, genetic engineering, and orbital habitats all come into play in this sci-fi ecothriller.

2413. Simak, Clifford D. *City*. NY: Collier, 1992, c1952. 267 p. 0020253915
International Fantasy Award, 1953.
AR: *A choice of gods*.
A collection of tales supposedly written by dogs after they became the
dominant species, describing how cities became overpopulated; families
moved to estates; sentient dogs were bred to hunt people who had adapted primitive lifestyles; and, eventually, a peaceable kingdom with respect for
all lifeforms emerged..

2414. ___. *Way station*. NY: Collier, 1993, c1963. 210 p. 0020248717
One of *Science fiction's* 100 *best novels*, Hugo Award, 1963.
A tale featuring a provider of galactic traveler's aid and a deaf woman with
strong psychic-naturalistic communication and healing powers. This message of respect for different species and different kinds of people is HR.

2415. Slonczewski, Joan. *The wall around Eden*. NY: Avon, 1990, c1989. 288 p.
038071177X
AR: *Still forms on Foxfield*, a sci fi-fantasy novel about a Quaker settlement on
a remote planet.
Nuclear winter has exterminated most of the biosphere, but one small
town in rural Pennsylvania was saved by an alien force field. The residents
overcome their fear of the aliens and struggle to support themselves in a
primitive lifestyle, farming a small patch of ground.

2416. Smith, L. Neil. *Pallas*. NY: TOR, 1993. 447 p. 0312097050
A highly controlled communist collective and a violent, anarchical "Outside" subculture compete for control of Pallas, a terraformed asteroid.

2417. Tepper, Sheri S. *Grass*. NY: Bantam, 1990. 449 p. 0553285653
A husband-wife ambassadorial team sent to the planet Grass face puzzling
questions. Why the are residents apparently immune to a plague which is
destroying humanity elsewhere? Is it the result of a strange symbiosis of
the human settlers and the planet's native races? AOT. #

2418. Tolkien, J. R. R. *The lord of the rings*. Boston: Houghton Mifflin, 1991,
c1954-56. 1193 p. : ill., maps. 0395595118
Contents: *The fellowship of the ring* — *The two towers* — *The return of the king*. Read
Tolkien's *The hobbit* first. A multi-level fantasy saga pitting the evil Sauron
against the Captains of the West over the control of Middle Earth and of
nature itself. The humble hobbits (who live in harmony with nature and
desire no dominion over it) manage to survive, as more "heroic" figures on
the quest rise and fall around them. Familiarity with classical and
medieval mythology helpful, but not required. HR AMT. #

2419. Turtledove, Harry. *The case of the toxic spell dump.* Riverdale, NY: Baen, 1993. 367 p. 0671721968
An Environmental Perfection Agency employee is forced to find new resources within himself when he discovers an ancient Aztec god of death in the physical and spiritual murk of the Toxic Spell Dump.

2420. Vance, Jack. *The Cadwal chronicles.* NY: TOR, 1989-93. 3 v. Contents: v.1 *Araminta Station* (0812557093) — v.2 *Ecce and old Earth* (0812557018) — v.3 *Throy* (0312851332).
Araminta Station is a scientific outpost with a limited population studying and preserving the wilderness world of Cadwal. Police cadet Glawen Clattuo is drawn into a conflict with the Yips (humanoid settlers.)

2421. ___. *The dying earth.* NY: Baen, 1986, c1950. 220 p. 0671655647
Despite new technology which is magical and organic, the Earth is threatened by earlier technological excesses.

2422. Varley, John. *Millennium.* NY: Berkley, 1985, c1983. 249 p. 0441531830
99th century Earth is dying from the accumulated poisons of war and industrialism. A time travel team kidnaps disaster victims from the past in an attempt to rekindle human life on a distant planet, but crucial mistakes place the Earth and humanity's future in final jeopardy.

2423. ___. *The Ophiuchi hotline.* NY: Ace, 1993, c1977. 234 p. 0441634842
One of *Science fiction's* 100 *best novels.*
AR: *Steel beach,* and the Gaean trilogy:*Triton, Wizard, Demon.*
First-order intelligence space invaders land on Earth to save the second-order whales and dolphins from destructive third order humans by destroying human artifacts.

2424. Vonnegut, Kurt. *Galapagos.* NY: Dell, 1988, c1985. 295 p. 0440127793
AR: *Cat's cradle, Slaughterhouse five, Player piano.*
What begins as "the nature cruise of the century" alters the course of evolution as the new "humans" return to the land and sea.

2425. Wells, H. G. *The island of Doctor Moreau.* Athens: Univ. of Georgia, 1993, c1896. 239 p. 0820314110
The shipwrecked sailor Prendick is taken to a strange island where a doctor exiled from London is conducting experiments meant to turn animals into humans. The beings revert to their feral forms and kill Moreau and his assistant.

2426. Wilhelm, Kate. *Where late the sweet birds sang.* NY: Collier, 1990, c1976. 207 p. 0020264828

Hugo and Jupiter Awards, 1977.
After Earth is ravaged by nuclear warfare, radiation, and pestilence, a family tries to survive through bioengineering and cloning. When they discover they can replicate their own bodies, but not their humanity, they must decide whether to take their chances in what remains of the natural world. HR AOT. #

2427. Willis, Paul J. *No clock in the forest*. NY: Avon, 1993, c1991. 219 p. 0380720779
Sequel:*The stolen river*.
A mountaineer and two teenage hikers finds themselves in a parallel world when their climbing companion disappears into thin air on a mountaintop. They are called upon to help save the wilderness from an evil Princess and possible commercial development. AMT. #

2428. ___. *The stolen river : an alpine tale*. Wheaton, Ill: Crossway, 1992. 190 p. 0891076719
A research assistant encounters strange animals, a flowering ice ax, and other mysteries in the mountainous Three Queens Wilderness. Can he help marshall the Creator God to overpower the evil spirit El Ai? AMT. #

2429. Wolfe, Gene. *Endangered species*. NY: T. Doherty, 1989. 506 p. 0312931549
AR:*The book of the new sun*.
Includes 37 short science fiction stories on a variety of subjects including nature, and relations between men and women. AMT. #

Books for young adults:

2430. Barron, T. A. *The Merlin effect*. NY: Philomel, 1994. 261 p. 0399226893
When a 13-year-old joins the investigation of a strange whirlpool and possible sunken treasure, she becomes involved in a whale rescue and an age-old conflict between Merlin and the evil Nimue. An entertaining melding of mythology, environmental issues, and scientific research. HR GR 5-9.

2431. Bell, Clare. *Clan ground*. NY: Atheneum, 1984. 258 p. 0689503040
Sequels:*Ratha's creature, Ratha and thistlechaser*.
The stranger Orange-Eyes, recognizing that the one who controls fire can become absolute ruler, challenges Ratha's authority over a clan of intelligent wild cats living 25 million years ago. #

2432. ___. *The jaguar princess*. NY: TOR, 1993. 480 p. 0312856733
An Aztec slave girl with shape-changing and other powers is sought by the

Earth loving speaker-king of Tezcotzinco to prevent the onslaught of the imperialistic Tenochtitlan. Can Mixcatl change herself into a jaguar and save Tezotzinco?

2433. ___. *People of the sky.* NY: TOR, 1989. 344 p. 0812502612
Two centuries after they have emigrated to Oneway, a planet of plateaus and canyons like their native southwest, the Pueblo Indians have evolved their own high-tech society, but in Barranca Canyon the old ways persist, and people live in symbiotic harmony with the Aronans, a species of intelligent flying beasts. #

2434. ___. *Ratha and Thistle-Chaser.* NY: McElderry, 1990. 232 p. 0689504624
Crippled and tortured by paralyzing nightmares, a solitary cat finds a new life for herself with the strange tusked creatures of the seashore until her peaceful existence is disrupted by Ratha's scout, Thakur, who brings welcome companionship but also forces her to face her terrifying past. #

2435. ___. *Ratha's creature.* NY: McElderry, 1987, c1983. 259 p. 0689502621
PEN Los Angeles Award, ALA Best Books for Young Adults, and the International Reading Association's Children's Book Award.
25 million years ago a society of intelligent cats pushed close to extinction meets an enemy band of raiding predatory cats in a decisive battle which will determine the future for both. HR. #

2436. ___. *Tomorrow's sphinx.* NY: Dell, 1988, c1986. 292 p. 0440201241
Two unusual black cheetahs share a mental link, one cat coming from the past to reveal scenes from his life with the young pharaoh Tutankhamen, and one struggling to survive in a future world ravaged by ecological disaster. #

2437. Blackwood, Gary L. *The dying sun.* NY: Atheneum, 1989. 213 p. 0689314825
Set around 2050 during a new ice age, two young men leave the crime-ridden, warm South for a more primitive existence on a family farm.

2438. Board, Sherri L. *Ambrosia and the coral sun.* Newport Beach, CA: Tug, 1994. 249 p. 0963476777
While attempting to overthrow the evil ruler of her watery realm, a young Zel named Ambrosia teaches humans how to respect and care for the sea.

2439. Bosse, Malcolm J. *Captives of time.* NY: Delacorte, 1987. 268 p. 0385295839
AR: *Cave beyond time, The 79 squares, The barracuda gang.*
Orphaned by the brutal murder of their parents, Anne and her gentle but mute brother suffer great hardships as they travel across a dangerous,

pestilence-ridden Europe to their uncle, an armorer and clockmaker, and, after his death, to a distant city to deliver the commissioned plans of his precious clock. #

2440. Boston, L. M. *The Green Knowe series*. Ill. by Peter Boston. NY: P. Smith, 1984, c1955-1970. 5 v. : ill. Contents: *The children of Green Knowe* (0844662887) — *An enemy at Green Knowe* (084466152X) — *The river at Green Knowe* (0844661538) — *A stranger at Green Knowe* (0152817557) —*Treasure of Green Knowe* (0844662755)
AR: *The fossil snake.*
This ecologically-tinged series of fantasy novels takes place in various eras at the ancient English country manor of Green Knowe. HR GR 4-8.

2441. Burke, Terrill Miles. *Dolphin magic*. Fiddletown, CA: Alpha Dolphin, 1992-93. 2 v. 1880485656 (v. 1) 1880485699 (v. 2) 188048563X (set)
Book 1 subtitle: *The first encounter*, Book 2: *Adepts vs. inepts*. A New Age fantasy about human-dolphin relationships, telepathy within and between the species, and the fall of the ancient civilization of Lemuria. For young adults and New Age devotees. #

2442. Conly, Jane Leslie. *Rasco and the rats of* NIMH. NY: Harper & Row, 1986. 278 p. : ill. 0060213612 0060213620
Prequel: Mrs. *Frisby and the rats of* NIMH by Robert C. O'Brien.
A field mouse teams up with an adventurous young rat to try to prevent the destruction of a secret community of rats that can read and write from the rising waters caused by a dam near their home. GR 3-7.

2443. Cooper, Susan. *The dark is rising*. NY: Collier, 1986, c1973. 244 p. 0689710879
Newbery Honor Book, 1973.
On his 11th birthday Will Stanton discovers that he is the last of the Old Ones, destined to seek the six magical Signs that will enable them to triumph over the evil forces of the Dark. Cooper uses the land as both setting and character. HR. #

2444. Furlong, Monica. *Juniper*. NY: Knopf, 1991. 198 p. 0394832205
Prequel: *Wise child.*
While apprenticed to a nature loving witch, a young girl struggles to save her family from the evil machinations of her power-hungry aunt. #

2445. ___. *Wise Child*. NY: Knopf, 1987. 228 p. 0394891058 0394991052
Sequel: *Juniper.*
Abandoned by both her parents, 9-year-old Wise Child goes to live with Juniper, who has now become a Wiccan priestess and who begins to train

her in the ways of herbs and magic..

2446. Gurney, James. *Dinotopia : a land apart from time*. Atlanta: Turner, 1992. 159 p. : ill. 1878685236
In Dinotopia humans, dinosaurs, and other ancient and modern animals live together in a peaceful, nonsexist, innovative, creative, compassionate society. GR 3-8 and whimsical adults. #

2447. Hoover, H. M. *Orvis*. NY: Puffin, 1990, c1987. 186 p. 0140321136
On an Earth that has become an inhospitable wilderness from global warming, Toby and her friend Thaddeus find themselves lost in "the empty" with Orvis, an obsolete robot who is their only hope of protection and escape. GR 5-9.

2448. Hughes, Monica. *The crystal drop*. NY: Simon & Schuster, 1993. 212 p. 0671791958
AR: *Beyond the dark river.*
In the summer of 2011, the death of their mother sends Megan and Ian on a dangerous journey across a Canada ravaged by drought and the collapse of civilization. GR 5-9.

2449. Lawrence, Louise. *The warriors of Taan*. NY: Harper & Row, 1988. 249 p. 0060237376 0060237368
AR: *Star lord.*
Prince Khian needs the help of a woman, an Earthman, and the Stonewraiths to drive the despotic Earthling Outworlders from their home planet of Taan before the Outworlders completely plunder the resources and ruin the environment of their planet. #

2450. Le Guin, Ursula K. *Tehanu : the last book of Earthsea*. NY: Atheneum, 1990. 226 p. 0689315953
AHR prequels: A *wizard of Earthsea*, *The tombs of Atuan*, *The farthest shore.*
When Sparrowhawk, the Archmage of Earthsea, returns from the dark land stripped of his magic powers, he finds refuge with the aging widow Tenar and a crippled girl who possesses ancient Earth powers and an unknown destiny. HR. #

2451. L'Engle, Madeleine. *An acceptable time*. NY: Farrar Straus Giroux, 1989. 343 p. 0374300275
Polly's visit to her grandparents in Connecticut becomes an extraordinary experience as she travels back in time to play a crucial role in a prehistoric confrontation. The conflict involves a pair of Druids and the Earth-centered People of the Wind vs. the more rapacious People Across the Lake. #

2452. ___. *The arm of the starfish*. NY: Dell, 1980, c1965. 243 p. 0440901839
A marine biology student reporting to his summer job on the island of Gaea off Portugal finds himself in the middle of a power struggle between his boss and another group of Americans. At issue is a project with huge ecological and sociopolitical ramifications.

2453. Levy, Robert. *Clan of the shape-changers*. Boston: Houghton Mifflin, 1994. 192 p. 0395666120 0395666023
Having inherited the power to change into any kind of animal, a 16-year-old joins the fight against the greedy shaman Ometerer, who is attempting to steal this secret from her people and then destroy them. GR 5-9.

2454. Mayhar, Ardath. *A place of silver silence*. NY: Walker, 1988. 185 p. : ill. 0802768253
AR: *Medicine walk*.
Ten-year-old Andraia fights with the government not to destroy the planet where she is working alone, because of an intelligent life form she has discovered there. #

2455. Morgan, Robin. *The Mer-Child : a legend for children and other adults*. NY: Feminist, 1991. 64 p. : ill. 1558610537 1558610545
Relates the friendship between a little girl whose legs are paralyzed and a young boy whose mother is a mermaid and whose father is a human. Includes themes of friendship, loving nature, and accepting all people regardless of their race, age, or physical challenges. HR. #

2456. O'Brien, Robert C. *Mrs. Frisby and the rats of Nimh*. NY: Aladdin, 1986, c1971. 233 p. : ill. 0689710682
Sequel: *Rasco an the rats of NIMH* by Jane Leslie Conly.
Having no one to help her with her problems, a widowed mouse visits the rats whose former imprisonment in a laboratory made them wise and long lived. GR 3-7.

2457. Oppel, Kenneth. *Dead water zone*. Boston: Joy Street, 1993. 152 p. 0316651028
Muscular 16-year-old Paul tries to find his genetically stunted younger brother Sam in the polluted ruins of Watertown, where Sam is trying to cure himself with toxic "dead water" that alters the metabolism of those who drink it.

2458. Park, Ruth. *My sister Sif*. NY: Viking, 1991, c1986. 180 p. 0670839248
Fourteen-year-old Riko manages to get her delicate older sister Sif to their remote Pacific island home. An American scientist falls in love with Sif and discovers her connection with an underwater race, creating complications

in Riko's life. GR 5-9.

2459. Paxson, Diana L. *Master of earth and water*. NY: Morrow, 1993. 395 p. : ill. 0688125050
Sequel: *The shield between the worlds*.
In ancient Celtic Ireland young Demry learns the secrets of trees, stars, and water to battle beasts, monsters, and human enemies. #

2460. Phillips, Ann. *The Oak King and the Ash Queen*. NY: Oxford, 1984. 171 p. 0192714953
When Dan and Daisy wake up from a nap under an oak, they find themselves in a different world where they have been chosen King and Queen to witness the rites of the trees. This is both an exciting fantasy adventure story and an allegory about the importance of forest conservation. HR. #

2461. Reynolds, Susan Lynn. *Strandia*. NY: Farrar Straus Giroux, 1991. 277 p. 0374372748
Punished for fleeing her duties as a member of Strandia's privileged class, Sand is rescued by her dolphin friend M'ridan and learns of a disaster threatening her homeland. #

2462. Sanders, Scott R. *The engineer of beasts*. NY: Orchard, 1988. 258 p. 0531083837
Thirteen-year-old Mooch runs afoul of the repressive authorities controlling her floating domed city by helping an old engineer build realistic robot animals and by seeking her spiritual roots with the wild animals left on the outside. #

2463. Strieber, Whitley. *Wolf of shadows*. NY: Knopf, 1985. 105 p. 039487224X 0394972244
Notable Children's Trade Book in Social Studies, Outstanding Science Trade Book for Children, 1985.
In the terrible aftermath of a nuclear holocaust, a wolf and a human woman form a mysterious bond that brings each close to the spirits of the shattered earth. Told from the wolf's perspective. HR. #

20. Prehistory

An old adage contends that "You can't know where you're at unless you know where you've come from." The writers in this chapter speculate upon or idealize prehistoric life to show not just where we're from, but which values from those times might be reintroduced into a troubled modern world. The first modern novel to do so was probably Rosny's *La Guerre du Feu* (*The Quest for Fire*) which was published in 1920.

Contemporary prehistory fiction is dominated by five women: Auel, Harrison, Pierce, Shuler and Thomas. Mackey and Sarabande show promise of joining this popular group. All evoke a world featuring greater harmony between women, men, and animals. Hardship and conflict certainly exist, but can be overcome.

Ackroyd, Crace, and Wolverton present more complex, less plot-driven, and slightly more philosophical works. They are tougher reading but highly rewarding, especially Crace's *Gift of Stones* . If you want Fred and Wilma, rent the video.

The best book on prehistoric life for both adults and young adults may be Bruchac's brilliant *Dawn Land*. Simple in its characterizations and plot, but complex in its incorporation of Native American and classical Greek legends, it depicts an ecologically oriented way of life without overly romanticizing it. For other books by Bruchac see entries 1493-98.

Books for adults:

2464. Ackroyd, Peter. *First light*. NY: Grove Weidenfeld, 1989. 328 p.
0802111610
When a neolithic grave aligned to the stars is excavated, the ancient pattern of stars appears in the skies and all manner of unusual people are drawn to the secluded site.

2465. Allan, Margaret. *The mammoth stone*. NY: Signet, 1993. 391 p.
0451174976
A tale of the followers of the mammoth herds during the last ice age. An outcast learns the secrets of a talisman called the Mammoth stone and leads her people to a new way of life.

2466. Auel, Jean M. *Earth's children series*. NY: Bantam or Crown, 1982-1990. 4 v. 0553250426
Contents: v.1. *Clan of the Cave Bear* (0553250426) — v.2. *The valley of horses* (051754489X) — v.3. *The mammoth hunters* (0517556278) — v.4. *The plains of passage* (0517580497).
A Cro-Magnon orphan named Ayla is adopted by a wandering band of Neanderthals, but her intelligence and rapidly developing powers mystify and frighten her adoptive tribe. HR AOT. #

2467. Bruchac, Joseph*. *Dawn land*. Golden, CO: Fulcrum, 1993. 317 p. 155591134X
In this portrayal of prehistoric life among the Abnaki Indians, Young Hunter must make a quest to save his people from a race of cannibals. Bruchac skillfully depicts a world of people living in near perfect ecological harmony with nature. HR AOT. #

2468. Crace, Jim. *The gift of stones*. NY: Collier, 1990, c1988. 169 p. 0020311605
When a boy in a village of prehistoric stonecutters loses an arm, he becomes "useless" except for telling stories while the cutters work. When bronze tools make the stonecutters obsolete they turn to him to find a new direction and a new meaning for their lives. HR

2469. Grahn, Judy. *Mundane's world*. Freedom, CA: Crossing, 1988. 191 p. 0895943166
The book focuses on women in a prehistoric world where plants, animals, and humans all have a voice and what matters most is their interrelationships.

2470. Harrison, Sue. *Mother Earth, father sky*. NY: Doubleday, 1990. 313 p. 0385411596
Sequels: *My sister the moon* , *Brother wind* (published too late in 1994 for review). Chagak, an Alaskan woman of the last ice age, is forced to bear the child of an enemy who helped massacre her family. This is the story of her spiritual journey and quest for survival against the elements and enemies, an account of life in the ice age, and a love story.

2471. ___. *My sister the moon*. NY: Doubleday, 1992. 449 p. 0385420862
When young Kiin is torn between Chagak's sons, she endures banishment and cruelty and is forced to fight for the future of her child. Based on the Aleut sea otter incest legend, moon myths, and the Raven-trickster legends.

2472. Mackey, Mary. *The year the horses came*. SF: Harper, 1993. 377 p. 0062507354
The peaceful life of nature-loving, Goddess-worshiping neolithic Brittany is

threatened by an incursion of warlike "beastmen" from the steppes.

2473. Pierce, Meredith Ann. *Birth of the Firebringer.* NY: Four Winds, 1985. 234 p. 0027746100
Sequel: *Dark moon.*
Aljan, the headstrong son of the prince of the unicorns, becomes a warrior and discovers his destiny in his people's struggle against the hideous wyrms usurping their land. #

2474. ___. *Dark moon.* Boston: Joy Street, 1992. 237 p. 0316707449
Jan, the prince of the unicorns, pursues his destiny to save his kind from their enemies by seeking fire in a distant land of two-footed creatures. #

2475. ___. *The woman who loved reindeer.* NY: T. Doherty, 1989, c1985. 242 p. 0812503058
When her sister-in-law brings her a strange golden baby to care for, a young girl is unaware that this child will help her fulfill her destiny as leader of her people.

2476. Rosny, J. H. *Quest for fire : a novel of prehistoric times.* NY: Ballantine, 1982, c1920. 135 p. 034530067X
Translation of *La guerre du feu.*
Two rival warriors of a prehistoric tribe set off on a dangerous and distant search for fire, an element essential to tribal unity and leadership. They encounter other tribes, mastodons, and various adventures along the way.

2477. Sarabande, William. *The first Americans.* NY: Bantam, 1987-89. 3 v.
Contents: v.1. *Beyond the sea of ice* (0553268899) — v.2. *Corridor of storms* (0553271598) — v.3. *Forbidden land* (0553282069).
After much of their tribe is trampled by mammoths, Torka and Lonit search for a new home. v.2. They travel across the tundra to hunt mammoths and rely on a boy's natural psychic powers to defeat a magician.

2478. Shuler, Linda Lay. *She who remembers.* NY: Arbor House, 1988. 400 p. 0877958920
Sequel:*Voice of the eagle.*
Kwami is banished from her tribe, wanders the ancient Southwest, has adventures with the Toltec trader Kokopeli, and becomes the keeper and teller of ancient Earth secrets that only women can know. #

2479. ___. *Voice of the eagle.* NY: Morrow, 1992. 654 p. 0688095194
Kwami uses her understanding of nature and consequent spiritual powers to protect the Zuni people from marauding tribes. #

2480. Thomas, Elizabeth Marshall. *The animal wife*. Boston: Houghton Mifflin, 1990. 289 p. 0395524539
When a young hunter abducts a naked woman from a pond he learns more about women, life, and nature than he bargained for.

2481. ___. *Reindeer Moon*. Boston: Houghton Mifflin, 1987. 338 p. 0395421128
Sequel: *The animal wife*.
Prehistoric life seen through the eyes of a gifted rebel. Describes a world where animals and humans are physically and spiritually interdependent.

2482. Wolverton, Dave. *Path of the hero*. NY: Bantam, 1993. 420 p. 0553561294
In 2681 the Alliance of Nations send specimens of dinosaurs and other extinct species genetically engineered by paleobotanists to the planet Anee. The half-human, half-Neanderthal residents are oppressed by the slave masters. When Tull leads the slaves in rebellion he must face the "Creators" which are actually machines designed to destroy all living things.

21. Environmental Action

By the mid 1970s, veterans of the anti-war, feminist, and American Indian movements had begun to radicalize environmentalism. Simultaneously, ecofiction took a similarly radical turn with the publication of Nichols' *Milagro Beanfield War*, Abbey's *The Monkey Wrench Gang*, Callenbach's *Ecotopia*, Eastlake's *Dancers in the Scalp House*, and Hoban's *Turtle Diary*.

Publication and readership of such titles has proliferated, and radical environmental action has become a fairly common theme in General fiction, Ecofeminist novels, Mysteries, Native American and Australian fiction, the New West and even the once bucolic field of Country Living. See the introductions to those chapters for specifics.

Of the approximately 100 adult level novels about environmental action in this and other chapters, nearly half are about land development conflicts, with 21 about endangered species or animal rights (most being mysteries), and several about nuclear power or weapons, forest or rainforest conservation, dams, pollution, wilderness preservation, or oil spills. The rest are general or combine several subjects. Ecotage is featured in novels by Abbey, Eastlake, Nichols, Dodge, Kesey, Matthiessen, Metz, Rosemary, and Wallingford, and by King (2002) and McNickle (2015) in the Native American chapter. Also see "Ecotage" in the subject index. Books envisioning a new ecologically oriented society have been written by Callenbach, Froese, Le Guin, Robinson and many others.

Environmental protection is also a common theme in contemporary young adult fiction, but the distribution by subject is quite different. Animals or the animal rights movement are featured in nearly half, with others about pollution, land development, nuclear power or weapons, and forest preservation. The remainder are general or combine several subjects. Ecotage is discussed in books by Kelleher and Taylor. Dealing with family or school conflicts, accepting yourself and people different from you, and developing a personal identity, three longtime concerns of young adult literature, are incorporated in nearly half these books.

Books for adults:

2483. Abbey, Edward. *Black sun*. NY: Buccaneer, 1991, c1971. 159 p. 1568490828

This is both a bitterweet love story and a nature tale about a National Park ranger set primarily around the Grand Canyon.

2484. ___. *The brave cowboy : an old tale in a new time.* NY: Avon, 1992, c1956. 297 p. 0380714590
Sequel: *Good news.*
A stunning first novel about Jack Burns, a man whose values of independence, honesty, heroism, and love for the land place him squarely in conflict with land developers, middle class consumers, and the government. HR. #

2485. ___. *Fire on the mountain.* NY: Avon, 1992, c1962. 181 p. 0380714604
An adolescent boy describes his love for the New Mexico mountains and wildlife while relating the tale of his cowboy grandfather's struggle to keep the Air Force from taking over his land.

2486. ___. *The fool's progress : an honest novel.* NY: Avon, 1990, c1988. 513 p. 0380708566
A picaresque novel which chronicles an aging man's voyages from Arizona to West Virginia, Europe, and back again with his dying dog, Solstice. Depicts a series of adventures in nature, in the arms of women, and in the depths of his own soul. Recommended for mature adults.

2487. ___. *Good news.* NY: Plume, 1991, c1980. 242 p. 052548521X
Prequel: *The brave cowboy.*
The now elderly brave cowboy and his Indian partner search for Jack's son in a near-future dystopian Southwest which has been ravaged by overpopulation and the destruction of the natural environment. Recommended for adults with a sense of irony.

2488. ___. *Hayduke lives!* Boston: Little, Brown, 1990. 308 p. 0316004111 0316004138 (pbk)
Prequel: *monkey wrench gang.*
As Seldom Seen, Doc, and Bonnie end their time on probation, the presumably dead Hayduke re-emerges from hiding. They rejoin forces to battle uranium strip mining near the Grand Canyon and related industrial development of the wilderness. HR AMT. #

2489. ___. *The monkey wrench gang.* Ill. by R. Crumb. Salt Lake City: Dream Garden, 1990, c1975. 356 p. : ill., map. 094268818X
Sequel: *Hayduke lives!*
Four environmentalists frustrated with the failure of legal tactics to stop ecological disasters resort to covert guerilla action. HR AMT. #

2490. Allende, Isabel. *The house of the spirits.* NY: Bantam, 1986. 448 p.

0553258656
Translation of *La casa de los espiritus*.
Two Chilean families relationship as they struggle through the ups and downs of mariage from the turn of the century through the military coup responsible for killing Salvatore Allende.

2491. Amado, Jorge. *Tieta, the goat girl*. NY: Avon, 1988, c1979. 671p. 0380754770
Translation of *Tieta do Agreste, pastora de cabras*.
When a madam from Sao Paulo returns to her hometown she learns that Germans plan to build a titanium plant on a local magnificent beach. Tieta and her band of of locals fight this plan using sex, drunkenness, violence, and conservation.

2492. Arthur, Elizabeth. *Bad guys*. NY: Knopf, 1986. 261 p. 0394554426
Three self-styled "terrorists" invade a juvenile work camp on an Aleutian island. Arthur combines satiric comedy, exciting action sequences, and vivid descriptions of nature. AOT. #

2493. ___. *Binding spell*. NY: Doubleday, 1988. 372 p. 038524844X
Attempting to thwart the international banking community's efforts to destroy family farms and poison the land, Howell Bourne kidnaps two visiting Russian scientists. He organizes them, an aspiring pagan witch, an anti-nuclear psychologist, an herbalist, and others to oppose the proposed development.

2494. Bergon, Frank. *The temptations of St. Ed & Brother S*. Reno: Univ. of Nevada, 1993. 305 p. 0874172268
Two contemporary American monks struggle with their own inner demons while protesting the creation of a nuclear waste dump in Nevada. This deep but rollicking novel which blends comedy, philosophy, and environmental action is HR.

2495. Bodett, Tom. *The big garage on Clear Shot : growing up, growing old, and going fishing at the end of the road*. NY: Morrow, 1990. 299 p. 0688095259
AR: *End of the road, As far as you can go without a passport*.
This otherwise humorous picaresque of life in an Alaskan town (featuring a loser turned successful artist, anglers, a militant vegetarian, and a junkyard tycoon) takes a more serious tone as a result of the peoples' anger and frustration over the Exxon-Valdez oil spill.

2496. Buckley, Christopher. *Thank you for smoking*. NY: Random House, 1994. 272 p. 0679431748
This dark but hilarious satire follows a PR man for the tobacco industry as

he plans and executes public relations campaigns, appears on talk shows, and is eventually kidnapped by anti-smoking terrorists who plan to kill him by covering him with nicotine patches.

2497. Callenbach, Ernest. *Ecotopia emerging*. Berkeley: Banyan Tree, 1981, 1979. 326 p. 0960432035 (pbk) 0960432043
Sequel: *Ecotopia*.
Harnessing solar power creates cheap, plentiful, non-polluting energy. It is opposed by utility executives and their political allies who see that it will not only limit their wealth, but lead to a more democratic society. AOT#

2498. ___. *Ecotopia : the notebooks and reports of William Weston*. NY: Bantam, 1990. 181 p. 0553348477
In the early 21st century, Oregon, Washington, and Northern California secede to form an ecologically and socially progressive, decentralized society which relies on solar power and other appropriate technology. This novel is one of the cornerstones of modern eco-fiction and is HR AOT. #

2499. Cameron, Sara. *Natural enemies*. Atlanta: Turner, 1993. 251 p. 1878685376
Edward Abbey Award for Ecofiction and Turner Tomorrow Award, 1993.
A naturalist, a reporter, and a detective pursue the muderers of the head of the Kenyan Wildlife Service and fight the ivory trade. HR AMT. #

2500. Carpentier, Alejo. *The lost steps*. NY: Noonday, 1989, c1956. 278 p. 0374521999
Translation of *Los pasos perdidos*.
A composer from New York City and his mistress escape from their humdrum lives to the backwaters of the Orinoco River, presumably to collect native musical instruments. The further they go upriver the more completely they disappear from time itself.

2501. Choyce, Lesley. *Downwind*. St. John's, NF.: Creative, 1984. 140 p. 0920021085
An energy crisis created by an oil embargo is the genesis of a secret project to build nuclear plants in Canada to sell electricity to New England, bypassing safety and regulatory procedures.. Recommended for adults, particularly anti-nuclear activists.

2502. ___. *Transcendental anarchy : confessions of a metaphysical tourist*.
Kingston, Ont.: Quarry, 1993. 240 p. 1550820737
AR *An avalanche of ocean*, *Magnificent obsessions*, *December six: the Halifax solution*.
This collection of autobiographical essays from an expatriate American author living in Nova Scotia is infused with an inherent love of nature, con-

cern for the environment, and joy in all aspects of life. This book is a great home remedy for burnt out activists

2503. Crace, Jim. *Arcadia.* NY: Collier, 1993. 311 p. 0020192002
When an octagenagian billionaire tries to replace a rustic open air "Soap Market" with an antiseptic mall, he encounters surprising opposition from his assistant. He smashes the resistance to build the mall, but a new farmer's market "Soap Two, just like a film" emerges in its shadows. A tough but rewarding read. HR.

2504. Douglas, Marjory Stoneman. *Nine Florida stories.* Gainesville: Univ. of Florida, 1990. 240 p. 081300988X 0813009944 (pbk)
The famous Florida naturalist depicts threats to the Everglades from developers, organized crime, poachers, and politicians.

2505. Dodge, Jim. *Stone junction : an alchemical potboiler.* NY: Atlantic Monthly, 1990. 355 p. 0871133318
AR: *Not fade away.*
A young man becomes a member of the arcane Alliance of Magicians and Outlaws, a group of subversives opposed to the prevailing authoritarian, technologically-oriented, nature-defiling power structure. HR for adventurous adults and mature teens, although some parents may object to the drug use. #

2506. Drinkard, Michael. *Disobedience.* NY: Norton, 1993. 349 p. 039303478X
In this multigenerational novel Eliza Tibbets starts an orange grove in 1885. One hundred years later her rebel great granddaughter clashes with her upscale husband. How will their mentally gifted teenage son react to his father's attempts to sell the last orange groves to developers?

2507. Durrell, Gerald Malcolm. *The mockery bird.* West Seneca, NY: Ulverscroft, 1983, c1982. 373 p. 0708909019
AR teens: *The talking parcel.*
Durrell uses wit and satire to portray the struggle between wildlife preservationists and developers.

2508. Eastlake, William. *Dancers in the scalp house.* Flint, Mich.: Bamberger, 1989, c1975. 245 p. 0917453190
A group of traditional Navajo, a nuclear physicist, and a teacher band together to drive developers from the reservation and restore the land. When the massive Atlas Dam which will cover their homeland nears completion they concoct a nuclear "gift" to present at the opening ceremony. HR AMT. #

2509. Froese, Robert. *The hour of blue.* Unity, ME: North Country, 1990. 252 p. 0945980213 0945980221 (pbk)
When a forest inventory in Maine yields shocking results, a Forest Service programmer and a marine biologist find a loose conspiracy of trees, dolphins, and psychic children working to help the Earth protect itself. Gaia is striking back in self-defense. #

2510. Garland, Hamlin. *Cavanagh, forest ranger.* Temecula, CA: Reprint Services, 1988, c1910. 300 p. 0781212405
Presents the conflicts between rangers trying to conserve natural resources and ranchers who resented resources being "locked up." In two subplots Cavanagh romances the city–educated daughter of mountaineer parents, and they convince her parents that their simple ways are better than those of the ranchers who look down upon them.

2511. Grass, Gunter. *The rat.* San Diego: Harcourt Brace Jovanovich, 1987. 371 p. 0151759200
Translation of *Die Rattin.*
This fascinating story within a story within a story begins when Grass is given a rat for Christmas. Subplots include feminists on a research ship, the return of the aging tin drummer, a video of the re-greening of Germany by the Brothers Grimm and their characters, and arguments with the rat that lead him to visions of an apocalypse. HR.

2512. Harrison, Jim. *A good day to die.* NY: Delta, 1989, c1973. 176 p. 0440550211
An idealistic poet, a disturbed Vietnam vet "and the girl who only loved one of them-at first" set out from Florida to save the Grand Canyon from a dam with a case of dynamite. Macho and gritty.

2513. Hiaasen, Carl. *Tourist season.* NY: Putnam, 1986. 272 p. 0399131450
AR:*Trap line.*
A group of Florida activists concerned by the furious pace of land development, habitat destruction, species loss, and drug smuggling decides that tourists are partly to blame.

2514. Hoban, Russell. *Turtle diary.* NY: Pocket, 1986, c1975. 192 p. 0671618334
Two reticent people, a bookstore clerk and an author of children's books, conspire to return the turtles in a London zoo to the sea. AMT. #

2515. Hoffman, Arthur L. *Tail tigerswallow and the great tobacco war.* Albuquerque: Amador, 1988. 157 p. 0938513044
Concerned that tobacco companies are poisoning America, one person

resorts to guerilla warfare, including cutting down billboards.

2516. Johnson, Paul. *Operation remission*. Delhi, NY: Nefyn & Shaw, 1993. 330 p. 0963797433
A man dying from leukemia caused by atomic testing in the fifties becomes an unwitting accomplice in a plan to irradiate the entire United States Congress.

2517. Kesey, Ken. *Sailor song*. NY: Viking, 1992. 535 p. 0670835218
A Hollywood film company in Alaska to film an Inuit children's story promises to bring development and wealth to the local people, but strange and forboding events begin to take place. Can eco warrior Ike Salas and the Loyal Order of Underdogs avert environmental disaster and the impending destruction of Earth? Some sexual and drug activity, but okay for AMT. #

2518. Matthee, Dalene. *Circles in a forest*. NY: Ballantine, 1985. 336 p. 0345326520
The son of Dutch colonists in South Africa in the 1880s learns to love the forest providing sanctuary for him and wild animals. He rebels against and is banished from his forest-burning family. In intertwining narratives he wanders the forest as a boy and later establishes an almost mystical connection with a cunning elephant named Old Foot. AOT. #

2519. Matthiessen, Peter. *At play in the fields of the Lord*. NY: Vintage, 1991. 373 p. 0678737413
National Book Award nominee.
Two sleazy American mercenaries, one of them a half-breed Indian named Louis Moon, are hired to bomb villages in the Amazon rainforest. After ingesting psychedelic yage, Moon "goes native," and turns against the developers, government, and missionaries who are destroying the land and culture. HR AMT. #

2520. Metz, Don. *King of the mountain*. NY: Harper & Row, 1990. 289 p. 0060163771
When Walker Owen returns to his boyhood home in Vermont he learns that developers, a disabled Vietnam vet whose family had cherished the land, and the son of an Abenaki shaman have conspired to destroy the forest to build condos. Walker and a local farmer fail in legal efforts to stop the condos and resort to ecotage.

2521. Mortimer, John Clifford. *Titmuss regained*. NY: Penguin, 1991, c1990. 280 p. 014014921X 0140148809 (pbk)
Sequel: *Paradise postponed*.
A conservative Member of Parliament buys a country estate and then

learns that an adjacent "model community" has been proposed. Will he vote with his cronies or "go green" to save the estate?

2522. Mueller, Marnie. *Green fires : assault on Eden, a novel of the Ecuadorian rain forest*. Willimantic, CT: Curbstone, 1994. 320 p. 1880684160
A former Peace Corps volunteer and her German husband have an eventful honeymoon when they meet another German who claims to live deep in the rain forest as a protector of the native people, who are being napalmed and otherwise harrassed by an oil company.

2523. Nichols, John Treadwell. *The magic journey*. NY: Holt, 1978. 529 p. 0030153565 0030428661
During an economic downturn, a bus filled with dynamite blows up, creating a gushing hot spring in dry New Mexico. The little town of Chamisville becomes a tourist mecca. When the locals rise up against the developers, a series of strange alliances and crimes take place. AMT. #

2524. ___. *The Milagro beanfield war*. NY: Holt, 1994, c1974. 464 p. : ill. 0805028056
Sequels: *The magic journey*, *The Nirvana blues*.
When a New Mexico farmer illegally diverts some water for his parched fields, he inspires the longtime local farmers to organize to fight the developers and politicians who had legally stolen all the water from the farmers in the first place. This humorous but powerful eco-fiction classic is HR AMT. #

2525. ___. *The Nirvana blues*. NY: Ballantine, 1983, c1981. 596 p. 0345304659
By the late 1970s most of the Chicanos and hippies have been driven out of Chiamasville by developers and yuppies. Joe Miniver wants to buy land back from them to farm, but lacking cash or credit he becomes involved in a cocaine scam. AMT. #

2526. Platt, Randall Beth. *Out of a forest clearing : an environmental tale*. Santa Barbara: John Daniel, 1991. 220 p. 093678489X
Alarmed by nuclear weapons and nuclear power, the Council of Beasts gathers an assortment of animals to close a nuclear submarine base, but how do you get people to listen to animals?#

2527. Posey, Carl A. *Bushmaster fall*. NY: Fine, 1992. 311 p. 1556112459
A British scientist working in the Bolivian rain forest discovers an American radioactive experiment which threatens to poison the entire area. He must temporarily keep his discovery secret lest the Americans (abetted by a druglord and the Shining Path Maoists) kill him and destroy the rain forest to destroy the evidence.

2528. Poyer, David. *Winter in the heart*. NY: TOR, 1993. 350 p. 0312854218
When Thunder Oil begins secretly dumping toxic waste along Pennsylvania's backroads, a retired oil field worker and a troubled teen team up to stop them.

2529. Proffitt, Nicholas. *Edge of Eden*. NY: Bantam, 1991, c1990. 448 p. : ill. 055328715X
A Kenyan police inspector and former white hunter join forces to track down a viscious ivory poacher. Includes vivid descriptions of Kenyan wildlife and tribal lore.

2530. Roberts, Thomas A. *Shy moon*. Sarasota: Pineapple, 1989. 263 p. 0910923736
An American ecologist attempts to protect the desert environment from leaks in the Trans-Jordan Pipeline and becomes involved in the Israeli-Palestinian conflict.

2531. Rosemary, Kristine. *The war against gravity*. Seattle: Black Heron, 1993. 193 p. 0930773209 0930773241
This thriller set in the Pacific Northwest pits the powers of nature and environmental activists against the logging industry, white collar criminals, corrupt officials, and drug dealers.

2532. Salmon, M. H. *Home is the river*. San Lorenzo, NM: High-Lonesome, 1989. 250 p. 0944383033
A mountain man fights the destruction of a wilderness river in New Mexico against sunbelt boosters.

2533. Siddons, Anne Rivers. *King's oak*. NY: Harper & Row, 1990. 623 p. 0060162481
After a disasterous first marriage, Diane falls in love with a man as committed to protecting the wilderness as she is.

2534. Smith, Wilbur. *Elephant song*. NY: Random House, 1991. 498 p. 0679408991
A naturalist and an anthropologist struggle to stop the international ivory trade which is destroying wild elephant herds, the natural environment, and the native people who rely upon elephants for their survival.

2535. Vollmann, William T. *You bright and risen angels : a cartoon*. NY: Penguin, 1988, c1987. 635 p. : ill. 0140110879
Insects are pitted against electrical generation and environmental activists against their opponents in this allegory of nature vs. technology.

2536. Waters, Frank*. *People of the valley*. Athens, OH: Swallow, 1984, c1941. 201 p. 0804002428
An old Mexican woman who has "grown up wild" in an isolated valley is called upon by her people to help oppose construction of a dam. She initially resists, but realizes the futility of resistance and persuades them to move to a higher valley to prepare them for transistion to the modern world. HR.

2537. Willard, Nancy. *Sister water*. NY: Knopf, 1993. 255 p. 0679407022
When a developer plans to build a mall on the site of a museum of history and natural history in Ann Arbor, a mother and daughter become involved in efforts to save it. Can they possibly be behind the murders and appearances of strange water creatures, or is Mother Nature striking back?

2538. Yamashita, Karen Tei. *Brazil-Maru*. Minneapolis: Coffee House, 1992. 248 p. 1566890004
In 1925 a group of Japanese immigrants establishes a utopian farm community in Brazil. #

2539. ___. *Through the arc of the rain forest*. Minneapolis: Coffee House, 1990. 212 p. 091827382X
When an Amazon rain forest native discovers the curative power of feathers, a three armed American entrepeneur enlists the assistance of Japanese immigrant settlers to corner the feather market. The Indians, Japanese, and the rain forest itself then face extermination. Yamashita's cautionary message is couched in a clever satire. HR. #

Books for young adults:

2540. Ames, Mildred. *Who will speak for the lamb?* NY: Harper & Row, 1989. 216 p. 0060201126
Although at first antagonistic to one another, high school senior and former model Julie and college freshman Jeff find their mutual interest in an activist group fighting for animal rights gives both of them a chance to become individuals and escape the shadow of domineering mothers.

2541. Benchley, Nathaniel. *Kilroy and the gull*. NY: Harper & Row, 1977. 118 p. : ill. 0060205024 0060205032
AR teens: *Demo and the dolphin*. AR adults: *All over again, Lassiter's folly*.
Captured and trained to be a marineland sensation, a killer whale escapes to life on the open sea with his friend, a sea gull. GR 4-8.

2542. Bond, Nancy. *The voyage begun*. NY: Atheneum, 1984, c1981. 319 p.

0689502044
Living in the near future when the energy supply has been almost depleted, a teenage boy explores the deserted colonies near his father's Cape
Cod research station and begins to understand the long-term effects of air
pollution on the land and the people.

2543. Choyce, Lesley. *Clearcut danger.* Halifax, N.S.: Formac, 1992. 135 p.
0887802133
A pulp and paper company promises new jobs for an economically
depressed Nova Scotia town, but when two teenagers discover the possible
environmental damage, health problems, and corruption the project would
probably entail, they find the courage to oppose it. Contains anti-sexist,
anti-racist, and pro-Indian themes. HR.

2544. Collier, James Lincoln. *When the stars begin to fall.* NY: Dell, 1989,
c1986. 160 p. 0385295162
Angry and frustrated that his entire family is considered to be poor trash,
Harry defies his father and attempts to prove that a factory is polluting
their small Adirondack community. GR 6-9.

2545. Cooper, Clare. *Earthchange.* Minneapolis: Lerner, 1986, c1985. 96 p.
0822507307
After a catastrophe turns Earth into an inhospitable wilderness, young
Rose sets out to find help for her grandmother and a baby, warding off
wolves and fierce humans, and finally reaching a group of survivors with
scientific interests in restoring Earth to its original beauty. Notable for its
subtle anti-sexism message. HR.

2546. Dorame, Anthony. *Peril at Thunder Ridge.* Santa Fe: Red Crane, 1993.
128 p. : ill. 1878610260
While attending an environmental camp, a Native American teenager and
his friends discover the misuse of forest resources and become immersed
in a mission to expose the violators.

2547. Durell, Ann; George, Jean Craighead; and Paterson, Katherine. eds.
The big book for our planet. NY: Dutton, 1993. 136 p. : ill 0ill 0531059952
0531085953
Contains nearly 30 stories, poems, and non-fiction pieces by such notables
as Natalie Babbitt, Marilyn Sachs, and Jane Yolen. Depicts some of the
environmental problems now plaguing Earth, e.g. overpopulation, tamper-
ing with nature, litter, pollution, and waste disposal. They explore solutions and recommend various environmental actions. HR.

2548. Elish, Dan. *The great squirrel uprising.* NY: Orchard, 1992. 113 p.

0525451196
With the help of a sympathetic 10-year-old girl, Scruff the squirrel leads a consortium of squirrels and birds in a blockade of New York's Central Park to protest the litter there. GR 4-8.

2549. Garden, Nancy. *Peace, o river.* NY: Farrar Straus,Giroux, 1986. 245 p. 0374357633
Returning to her small Massachusetts town after four years, a 16-year-old reunites with a childhood friend and together they try to stop the dangerously escalating hostilities between their town and the less affluent community across the river. If the people fail to unite, how can they stop a planned nuclear waste dump from being built? HR.

2550. Herzig, Alison Cragin. *Shadows on the pond.* Boston: Little, Brown, 1985. 244 p. 0316358959
Dangerous intruders threaten to destroy the secluded beaver pond which is Jill's only refuge from her many problems.

2551. Hurmence, Belinda. *The nightwalker.* NY: Clarion, 1988. 140 p. 0899197329
A 12-year-old wonders if her sleepwalking brother is setting the mysterious fires that are leveling the fishermen's shacks on their island home off North Carolina. This becomes part of a larger controversy of whether to protect the land as a nature preserve, or the livelihoods of the part-Indian anglers. GR 4-9.

2552. Kelleher, Victor. *Rescue!* NY: Dial, 1992. 217 p. 0803709005
After freeing two ailing baboons from a research station in Africa, David and Jess are unexpectedly caught up in a struggle for survival in the bush. Couched in a survival adventure is a consideration of appropriate and counterproductive means of activism. GR 6-10.

2553. Keller, Beverly. *Fowl play, Desdemona.* NY: Harper, 1991, c1989. 183 p. 0064403939
While speculating on the merits of her father's new girlfriend, Dez teams up with a vegetarian animal rights activist to design posters for the school play. This humorous, nondidactic treatment of animal rights and vegetarianism is HR GR 4-8.

2554. Keller, Janet. *Necessary risks.* Atlanta: Turner, 1993. 163 p. 1878685384
A veterinarian is forced to overcome her own resentment over her parents' activism and death, and the resentment of people who may lose their jobs when she and her neighbors confront an agrochemical company which threatens their health and lives. AMT. #

2555. Killingsworth, Monte. *Eli's songs*. NY: McElderry, 1991. 137 p. 0689505272
Shipped off to relatives in Oregon while his father is touring with a rock band, a 12-year-old comes to love the magnificent trees of a nearby forest and tries to prevent their imminent destruction. GR 4-9.

2556. Klass, David. *California Blue*. NY: Scholastic, 1994. 200 p. 0590466887
When a 17-year-old discovers a new sub-species of butterfly which may necessitate closing the mill where his dying father works, they find themselves on opposite sides of the environmental conflict. HR.

2557. Langton, Jane. *The fragile flag*. NY: Harper, 1989, c1984. 275 p. 0064403114
A nine-year-old girl leads a march of children from Massachusetts to Washington, in protest against the President's new missile which is capable of destroying the Earth. HR GR 4-9.

2558. Lasky, Kathryn. *Home free*. NY: Dell, 1988, c1985. 245 p. 0440200385
A 15-year-old's fight to save a wilderness area for endangered eagles helps an autistic girl return to reality and reveals her strange hidden power. HR.

2559. ___. *Shadows in the water*. San Diego: Harcourt Brace Jovanovich, 1992. 211 p. 015273533X 0152735348 (pbk)
The two sets of Starbuck twins use their telepathic powers and the aid of some endangered dolphins to help their father catch a gang dumping toxic waste in the Florida Keys. HR.

2560. Levin, Betty. *The trouble with Gramary*. NY: Greenwillow, 1988. 198 p. 0688073727
Large print and braille eds. also available from Greenwillow.
Merkka's longing for a conventional existence is threatened by the art projects of her stubborn sculptor grandmother whose scrap metal collection offends the other citizens in their small Maine village. #

2561. Lipsyte, Robert. *The chemo kid*. NY: HarperCollins, 1992. 167 p. 0060202858
When the drugs that he takes as part of his chemotherapy suddenly transform him from wimp into superhero, a 16-year-old and his friends plot to rid the town of its most lethal environmental hazard: toxic waste in the water supply.

2562. Lisle, Janet Taylor. *Forest*. NY: Orchard, 1993. 150 p. 053106803X
Twelve-year-old Amber's invasion of an organized forest community of squirrels starts a war between humans and beasts, despite the protests of

an unconventional and imaginative squirrel named Woodbine.

2563. ___. *The great Dimpole oak*. NY: Orchard, 1987. 135 p. : ill. 053105716X
The citizens of Dimpole rally together to save an historic oak tree from being cut down. GR 4-8.

2564. Mikaelsen, Ben. *Rescue Josh McGuire*. NY: Hyperion, 1991. 266 p.
1562820990 1562821008
When 13-year-old Josh runs away to the mountains of Montana with an orphaned bear cub destined for laboratory testing, they both must fight for their lives in a sudden snowstorm. GR 6-9.

2565. Moeri, Louise. *Downwind*. NY: Dell, 1987, c1984. 121 p. 0440921325
After fleeing their California home to escape a possible radiation leak from a nuclear power plant, a 12-year-old and his family find themselves caught up in circumstances perhaps even more threatening to them. GR 4-9.

2566. Nimmo, Jenny. *Ultramarine*. NY: Dutton, 1992. 199 p. 0525448691
A brother and sister learn about their past when a mysterious man appears to walk out of the sea to help them rescue birds endangered by an oil spill. GR 6-9.

2567. Pace, Sue. *The last oasis*. NY: Delacorte, 1993. 230 p. : maps.
0385308817
In the near future, environmental decay and resource shortages have led to individual suffering and government repression. Two young people attempt to escape to the wilderness in Idaho for a better and freer life.

2568. Paulsen, Gary. *Sentries*. NY: Puffin, 1987, c1986. 165 p. 0590440950
The common theme of nuclear disaster and human vulnerability inter-weaves the lives of four young people (an Ojibway Indian, an illegal Mexi-can migrant worker, a rock musician, and a sheep rancher's daughter) with the lives of three veterans of past wars.

2569. Powell, Pamela. *The turtle watchers*. NY: Viking, 1992. 115 p.
067084294X
Three sisters living on a Caribbean island band together to protect a nest of leatherback turtle eggs from poachers and natural enemies. GR 6-9.

2570. Ryden, Hope. *Backyard rescue*. NY: Tambourine, 1994. 128 p. : ill.
0688128807
Ten-year-olds Greta and Lindsay are best friends who share an interest in rescuing and rehabilitating wildlife. When Greta's unemployed father gets a job in another town, Lindsay is afraid she'll lose Greta's friendship. GR 4-8.

2571. Sharpe, Susan. *Trouble at Marsh Harbor.* NY: Puffin, 1991. 170 p. : ill. 0140347887
Original title: *Waterman's boy.*
Ten-year-old Ben wants to be just like his father, a waterman on Chesapeake Bay, but crabs, rock-fish and oysters are not as plentiful as before. Ben decides to help a scientist who is looking for the source of an oil leak in the bay. Although written for GR 4-7 reading level, it retains its interest for older teens, too.

2572. Spinelli, Jerry. *Night of the whale.* NY: Dell, 1988, c1985. 147 p. 0440200717
Six rowdy high school seniors staying at a beach house for the summer are determined to devote themselves to partying until the night they encounter a group of beached whales. GR 9-12.

2573. St. Antoine, Sara. *The Green Musketeers and the fabulous frogs.* NY: Bantam, 1994. 131 p. 0553159992
Sequel: *The Green Musketeers and the incredible energy escapade.*
A group of fifth graders try to save the unique Big Bog tree frog from extinction by getting it declared the town mascot. Somewhat simplistic and overidealized. GR 4-8.

2574. Sutherland, Robert D. *Sticklewort and feverfew.* Normal, IL: Pikestaff, 1980. 355 p. : ill. 0936044004 0936044012 (pbk)
The human and animal inhabitants of a small town band together to save their homes from the pollution created by the Sudge-Buddle factory. This book also has an antisexism message.

2575. Taylor, John Robert. *Hairline cracks.* NY: Lodestar, 1990. 137 p. 0525673040
When Sam and a friend begin to investigate his mother's disappearance, they put themselves in danger as they learn about hairline cracks in the silos of a nuclear plant and the ruthlessness of men who want this kept secret. GR 6-9.

2576. Taylor, Theodore. *The hostage.* NY: Dell, 1991, c1987. 158 p. 0440209234
A 14-year-old has second thoughts about harboring a killer whale that he and his father captured and plan to sell to a sea amusement park. HR. #

2577. ___. *Sniper.* San Diego: Harcourt Brace Jovanovich, 1989. 227 p. 0152764208
ALA Best Books for Young Adults, California Young Reader Medal.
A 15-year-old must rely on his own resources and knowledge of nature

when a mysterious sniper begins shooting the big cats in his family's private zoological preserve. HR. #

2578. ___. *The weirdo*. San Diego: Harcourt Brace Jovanovich, 1991. 289 p. 0152949526
ALA Best Book for Young Adults, Edgar Allen Poe Award for Best Young Adult Mystery.
A 17-year-old fights to save the black bears in the Powhaten National Wildlife Refuge. HR. #

2579. Webster, Elizabeth. *Dolphin sunrise*. NY: St. Martin's, 1993. 368 p. 0312092768
Young Englishman Matt Ferguson forms friendships with a marine biologist, a sea captain, and a dolphin named Flite. He becomes a save the dolphins activist and travels to California where he and a young woman protesting military uses of dolphins must summon all their courage to overcome threats to their work. #

2580. West, Tracey. *Fire in the valley*. NY: Silver Moon, 1993. 80 p. 1881889327
Twelve-year-old Sarah's feelings about life on her family's farm change when she writes a letter to President Teddy Roosevelt, to protest diverting water for use by the growing city of Los Angeles.

22. Literary Studies

This chapter covers biographical and critical discussion by or about individual authors or about environmental themes in literature. These themes include ecofeminism effects, forests, environmental reform, the influence of place and landscape on writing, and the effect of nature on creativity. These are just a few of many similar studies, and were selected because of their current or potential widespread interest both within and beyond the academy. *Wilderness Tapestry* combines preservation literature, fiction, and literary criticism. O'Grady's *Pilgrims to the Wild*, Ronald's *The New West of Edward Abbey*, Rusho's *Everett Reuss, a Vagabond for Beauty*, Stegner's *Where the Bluebird Sings to the Lemonade Springs*, and Turner's *Spirit of Place* are particularly lively, and are recommended for all adults. Rusho's book is a real-life adventure which should also appeal to adventurous older teens and adults. Also see *Seeking Awareness in American Nature Writing* (1230)

Books for adults:

2581. Craven, Margaret. *Again calls the owl*. New York: Dell, 1984, c1980. 120 p 0440300746
Craven describes how her eyesight miraculously improved while reading about the Kwakiutl Indians and how she was driven to write a novel about their vanishing culture. Her development of courage and a reverence for nature is reflected in the companion novel I *heard the owl call my name*.

2582. Devine, Maureen. *Woman and nature : literary reconceptualizations*. Metuchen, NJ: Scarecrow, 1992. 250 p. 0810826127
Studies the relationships between woman and nature as portrayed in Atwood's *Surfacing*, Laurence's *The diviners*, Piercy's *Woman on the edge of time*, Gearhart's *The wanderground*, Robinson's *Housekeeping*, Walker's *The color purple*, and Naylor's *The women of Brewster Place*.

2583. Dobie, J. Frank. *Guide to life and literature of the Southwest, Rev. and enl. in both knowledge and wisdom*. Dallas: SMU, 1981, c1942. 232 p. 080740369
AR: *Coronado's children, The longhorns, Cow people, Tales of old-time Texas, Some part of myself, Out of the old rock, Prefaces*.
Analyzes southwestern literature and culture from a perspective which

combines history, natural history, folklore and social criticism. Consists primarily of a detailed, annotated bibliography. HR.

2584. Fairbanks, Carol. *Prairie women : images in American and Canadian fiction*. New Haven, CT: Yale, 1986. 300 p. : ill. 0300033745
Attempts to see the lives of prairie women in Canada and the United States in order to understand historic literature through a feminist perspective.

2585. Fowles, John. *The tree*. St. Albans, GB: Sumach, 1992. 94 p. : col. photos. 0712652329
In this photoessay a celebrated British author discusses childhood and adult recollections of trees and forests, the disparity between language and nature, "seeing nature whole," perceiving nature through art and science, and his own fiction.

2586. Haines, John Meade. *Living off the country : essays on poetry and place*. Ann Arbor: Univ. of Mich., 1981. 188 p. 0472063332
Contains essays by a renowned nature poet on "Alaska and the Wilderness," "Poets and Poetry," and "Autobiographical Sketches" along with an interview by David Stark and Robert Hedin.

2587. Halper, Jon. ed. *Gary Snyder : dimensions of a life*. SF: Sierra Club, 1991. 451 p. 0871566362 0871566168 (pbk.)
Contains over 60 brief selections by Jim Dodge, Ursula Le Guin, Allen Ginsburg, Michael McClure, Anne Waldman, Wendell Barry, Carol Koda, Charlene Spretnak, Dave Foreman, Peter Berg, Wes Jackson, and others. Covers Snyder's life, poetry, philosophy, and influence on both literature and such concerns as conservation, living close to the land, and bioregionalism.

2588. Harrison, Robert Pogue. *Forests : the shadow of civilization*. Chicago: Univ. of Chicago, 1992. 301 p. 0226318060
Investigates the role that forests have played in the human imagination. Analyzes such classics works and authors as *Gilgamesh*, Greek tragedy, Dante, Rousseau, Wordsworth, Nietzsche, and Thoreau.

2589. Haslam, Gerald W. ed. *The other California : the Great Central Valley in literature*. exp. ed. Reno: Univ. of Nevada, 1994, c1990. 221 p. 087417225X
Contains 19 essays about authors and common literary themes of the San Joaquin River Valley. Authors discussed include Saroyan, Didion, Everson, Kingston, Soto, and Rodriguez.

2590. Hepworth, James, and McNamee, Gregory. eds. *Resist much, obey little : some notes on Edward Abbey*. Tucson: Harbinger House, 1989. 152 p. 0943173450

Wendell Berry, Gary Snyder, William Eastlake, Barry Lopez, and eleven other writers consider Abbey's writing, life, and influence. Includes four interviews with Abbey. HR.

2591. O'Grady, John P. *Pilgrims to the wild : Everett Ruess, Henry David Thoreau, John Muir, Clarence King, Mary Austin.* Salt Lake City: Univ. of Utah, 1993. 169 p 0874804124
A series of meditations focused upon literary excursions into the wild as secular pilgrimages. Considers the erotic link between humans and wilderness and the wildness within humans.

2592. Peck, H. Daniel. ed. *The Green American tradition : essays and poems for Sherman Paul.* Baton Rouge: LSU, 1989. 357 p. : ill. 0867115134
A large and diverse collection of essays and poems honoring a leading critic and promoter of nature writing and American literature

2593. Powell, Lawrence Clark. *Southwest classics, the creative literature of the arid lands : essays on the books and their writers.* Tucson: Univ. of Arizona, 1982, c1974. 370 p. : ill. 0816507953
AR: *California classics, Books West Southwest, Bookman's progress.*
Provides a personalized analysis of novels by 26 SW authors. The relationship between author, landscape, and nature is a persistant theme. Authors covered include Harvey Fergusson, Mabel Dodge Luhan, D.H. Lawrence, Haniel Long, Willa Cather, Oliver LaFarge, John C. Van Dyke, Joseph Wood Krutch, J. Frank Dobie, and others. HR.

2594. Ronald, Ann. *The new West of Edward Abbey.* Reno: Univ. of Nevada, 1988, c1982. 255 p. 0874171318
Ronald considers Abbey's early fiction, his non-fiction, and his later fiction up to *Good News.* Analyzes Abbey's influence on the environmental movement and western literature. HR.

2595. Rusho, W. L., and Ruess, Everett. *Everett Ruess, a vagabond for beauty.* Salt Lake City: P. Smith, 1985, c1983. 228 p. : ill., map. 0879051434
In 1934, a young poet/amateur naturalist wandered into the Slickrock Country never to be officially seen again. Did he die or just escape civilization? Rusho combines Ruess's journals and letters with exhaustive personal research to try to untangle the mystery. For AOT interested in real-life mystery stories. HR. #

2596. Schofield, Edmund A., and Baron, Robert C. eds. *Thoreau's world and ours : a natural legacy.* Golden, CO: North American, 1993. 405 p. 1555919030
Papers from the Thoreau Society Jubilee covering various aspects of the world of Thoreau and his contemporaries, their relationship with Native

Americans, their influence on subsequent nature writiers, and the ecological history of Walden Pond. HR.

2597. Stegner, Wallace Earle. *Where the bluebird sings to the lemonade springs : living and writing in the West.* New York: Penguin, 1993, c1992. 250 p. 0140174028
Essays and articles by Stegner divided into three sections: autobiographical; "Habitat," covering the environment of the west and the conservation movement; "Witnesses," discussing western and nature writers such as Steinbeck, George R. Stewart, Walter Van Tilburg Clark, Norman MacLean, and Wendell Berry. HR.

2598. Turner, Frederick. *Spirit of place : the making of an American literary landscape.* Washington: Island, 1992, c1989. 370 p. 1559631805
Demonstrates the many roles played by the landscape in the works of Thoreau, Twain, George Washington Cable, Cather, Sandoz, Faulkner, Steinbeck, William Carlos Williams, and Silko. Essential for the study of Western American literature. HR.

2599. Western Literature Assn. *A literary history of the American West.* Fort Worth: TCU, 1987. 1406 p. : ill. 087565021X
Contains nearly 100 bio-critical-bibliographic essays by various authors on "Encountering the West" (the oral tradition, genre writing, and literary historiography); "Settled in: many Wests" (authors by region); and "Rediscovering the West" (ethnic literatures and present trends). Essential for Western history or literature scholars. HR.

2600. White, Jonathan. *Talking on the water : conversations about nature and creativity.* SF: Sierra Club, 1993. 288 p. : photos. 0871565153
White discussses the human relation to nature, nature in literature, ecological lessons from the land, American Indian ecological concepts, and related topics with Gretel Ehrlich, Kenneth Brower, Ursula Le Guin, Gary Snyder, Peter Matthiessen and others. HR.

2601. Zeveloff, Samuel I.; Vause, L. Mikel; and McVaugh, William H. eds. *Wilderness tapestry : an eclectic approach to preservation.* Reno: Univ. of Nevada, 1992. 306 p. 0874172004
Contains 19 essays on the philosophy of preservation, analyzing wilderness literature, historical and societal aspects of wilderness, innovative management, and possible future directions. Authors include Terrell Dixon, Ann Ronald, William Kittredge, Terry Tempest Williams, and David Mech. HR.

Author Index

Author Index

Subject Index

Title Index

Entries 1-1711 are nonfiction, 1712-2601 fiction. An "n" after a title indicates that the title is noted in another record. Most of these are out of print, but some are late 1994 or 1995 imprints. Please note that subtitles have only been retained to distinguish between otherwise identical titles.

Earth Works

DATE